Principles of Tribology

Principles of Tribology

Edited by

J. Halling

Professor of Engineering Tribology
University of Salford

First edition 1975
Paperback edition 1978
Reprinted 1979, 1983

Published by
THE MACMILLAN PRESS LTD
London and Basingstoke
Companies and representatives
throughout the world

ISBN 0 333 15496 7 (hard cover)
     0 333 24686 1 (paper cover)

Printed in Hong Kong

This book is dedicated
to the Lima Arms whose
efficacious personal
lubrication service
has often eased the
path to its completion

Contents

Foreword

by H. Peter Jost

Chairman: Committee on Tribology, Department of Industry, 1966–74
President: International Tribology Council

Without tribology, in other words, without 'interacting surfaces in relative motion', that is, surfaces rolling on each other, surfaces sliding over each other and surfaces rubbing on each other, life would be impossible.

This truism applies equally to heavy machinery and to a precision mechanism; to a brake and to a rocket; to a mechanical and to a human joint.

Friction and wear, the principal constituents of tribology, have been with us since time immemorial, so have been our efforts to control the former and minimise the latter. However, it is only less than nine years ago, that the modern interdisciplinary concept of tribology was recognised in the United Kingdom, since when its progress has spread like wildfire throughout the industrial world.

The main reason behind this development, which is also the principal reason for the importance of this book, is the recognition of the close interrelationship between tribological design principles and practices on the one hand, and their economic effect on the other. The days of single disciplinary designs and of design by trial and error are gone forever. Modern products must, during their design stages, have incorporated all the factors that lead to a satisfactory control of friction and prevention of wear.

For this to be accomplished, Professor Halling and his co-authors have provided, in a single volume, nearly all the basic theory needed for a thorough grounding in the subject. It will therefore be a book most useful for every engineering student; in addition, I believe that *Principles of Tribology* will be found very useful by practising engineers in industrial design and research departments. For here, the concentration of valuable information in one volume, will eliminate the consultation of several sources of books and scientific papers, a saving of valuable time that will be appreciated by those employed in industry and others working against the clock.

Similarly, this book should be of considerable value to workers in many other fields, whose occupation brings them face to face—often unpleasantly so—with the realities of tribological problems and who may find, in this volume, fundamental answers to at least some of their problems.

The original Report, that bears my name, estimated that by the better application of tribological principles and practices, industry in the United Kingdom could save around £515 million per annum (at 1965 values). During the years since its publication, it has become apparent that this estimate of savings has been too conservative. Indeed, in a recent Report, commissioned by the Congress of the United States of America, it was stated that there was considerable scope for savings of losses through tribological causes (friction and wear), which were estimated to cost the U.S. economy around $100 billion per annum, of which $20 billion were in materials.

The savings through the correct application of tribological principles, as outlined in this book, can be considerable. A recent report, commissioned by the S.S.R.C., concluded that 'for individual enterprises—even efficient ones—the rate of return in improvements in the tribological characteristics of their capital equipment could be very high indeed—far higher than is customary on ordinary industrial investments'.

Modern machines, mechanisms and equipment must be reliable. I firmly believe that the application of the ground rules, contained in *Principles of Tribology*, will materially contribute towards greater reliability of industrial products and therefore to the economy of this country.

I congratulate Professor Halling and his co-authors on their work, which I can warmly recommend not only to those desiring to become engineers, but also to those professional engineers and others whose work is connected with the control of friction and the prevention of wear, in other words, in work where the minimisation of breakdowns, replacements and outages through tribological causes are of importance.

London   H. PETER JOST

Preface

Tribology is a new word, not yet in common usage, but it deals with problems which man has encountered throughout the whole of his history. The word was introduced to focus attention on the problem of carrying load across solid interfaces in relative motion. The ingredients of the subject are therefore well-established batches of knowledge which occur in a wide range of texts on such topics as lubrication, friction, wear, contact mechanics, surface physics, and chemistry. The subject is truly interdisciplinary since the basic knowledge from physics, chemistry, mathematics, materials science and engineering is used to study problems in all branches of engineering, in medicine, and in almost all aspects of our daily life from the cleaning of our teeth to the slicing of our golf drive.

The awareness of the social and economic importance of this subject has resulted in its introduction into several courses at colleges and universities. This book is an attempt to bring together in a single volume those topics which are currently scattered throughout the scientific literature. In particular this volume concentrates on the basic principles of the subject, using practical examples only to demonstrate the physical manifestations of these principles. This book should, therefore, prove a useful supplement to other volumes dealing with such specific practical applications as the design of bearings. The book is basically aimed at the final-year level of undergraduate courses in engineering, but the authors hope that it will prove of interest to a wider readership. Research workers, new to the subject, and designers and development engineers seeking background knowledge to the existing literature should all find the subject matter relevant.

The authors have deliberately excluded illustrations of standard equipment and techniques since these should already be familiar to final-year under-graduates and practising engineers. The notation may also appear to lack complete consistency but it has been considered desirable to use that notation which is already well established in the literature concerned with the various topics considered. This should facilitate the use of standard references, many of which are included with each chapter. The symbols used are clearly identified in the text of each chapter.

It may be that the relatively large number of authors has resulted in some variation in literary style, but since each chapter is more or less self-contained this should offer no problem to the reader. Since all the authors belong to the same department we believe that our collaborative efforts have resulted in a coherent philosophy for the book. In each topic the authors have concentrated on the important physical principles and have not thought it desirable to include rigorous development of the mathematical treatment, although the most significant mathematical formulations are included. Since the authors are mainly concerned with the engineering applications of tribology, it will be noted that this book does not include much information on the chemical aspects of tribology. This does not imply any denigration of the importance of such material but rather that such topics tend to be of a rather specialist interest.

The final chapter has been included as an indication of the relevance of the remainder of the book. Indeed this chapter could be taken out of sequence if the reader seeks to appreciate the value of the topics discussed in the various chapters.

A number of problems together with outline solutions have been included in an appendix to this book, since the final proof of assimilation of concepts must be our ability to handle them in given situations.

Finally the authors would like to acknowledge their deep gratitude for the enthusiastic and competent way in which Mrs L. M. Chadderton has transcribed their often untidy drafts into a coherent manuscript.

Salford   J. HALLING

1

Introduction

1.1 TRIBOLOGY

What does the word mean? It is derived from the Greek word TRIBOS meaning rubbing, so that a literal translation would be 'the science of rubbing'. The word is so new as to appear in only the latest editions of dictionaries where it is there defined as 'the science and technology of interacting surfaces in relative motion and of related subjects and practices'. This latter definition, although embracing the literal translation, is of even wider significance and was created to bring together the interest in friction and wear of chemists, engineers, metallurgists, physicists and the like. This wide-ranging concern with tribology immediately illustrates the interdisciplinary nature of the subject. In a sense it is the name alone which is new because man's interest in the constituent parts of tribology is older than recorded history. Clearly, the invention of the wheel illustrates man's concern with reducing friction in translationary motion, and this invention certainly predates recorded history. That man should have been so concerned with the tribological problems of friction and wear is not surprising because our involvement with such phenomena affects almost every aspect of our lives.

These problems are not confined to the machines which we use, they also have profound influences on many other aspects of life. The action of animals' joints is clearly a tribological situation and cures for such diseases as arthritis already owe much to the tribologists' expertise. We also rely on the control of friction in our leisure pursuits, whether they be rock climbing or any form of ball game—such as the spinning of a cricket or tennis ball, the slicing of our drive in golf or our proficiency on skates or skis. Holding, cutting and brushing are other manifestations of the impact of tribology on our daily life, while the cleaning of our teeth is clearly a controlled wear-process where we wish to avoid wear of the enamel while wearing away unwanted films, etc. Even our ability to walk is dependent on the existence of appropriate friction, so that tribological effects have clearly had a major effect on the whole evolutionary process.

We may examine the effect of friction on the evolutionary process by considering the way in which developments in translation over the earth's surface have evolved against the timescale of history[1]. In figure 1.1 the resistance to translation is represented by a resistance/weight ratio which might

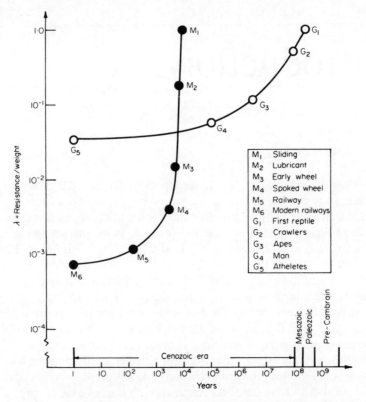

M_1 Sliding
M_2 Lubricant
M_3 Early wheel
M_4 Spoked wheel
M_5 Railway
M_6 Modern railways
G_1 First reptile
G_2 Crawlers
G_3 Apes
G_4 Man
G_5 Atheletes

Figure 1.1 Improvements in land locomotion throughout history

be considered as an equivalent coefficient of friction λ. This clearly represents the ease of translation and its reduction with timescale is seen as evolution has developed from primeval sliding to the movements of a modern athlete, that is along line G. In this figure it should be noted that the scales are logarithmic rather than linear so that the shape of the curve is somewhat deceptive. Modern man starting some 10 000 years ago has used his inventiveness to achieve a much better performance than is obtained from the physio-logical developments in animals. Thus the use of lubricants and improvements in the design of wheels have clearly proved advantageous.

It is also interesting to plot the results in figure 1.1 against the speed achieved by each method of translation, shown in figure 1.2. Again we note

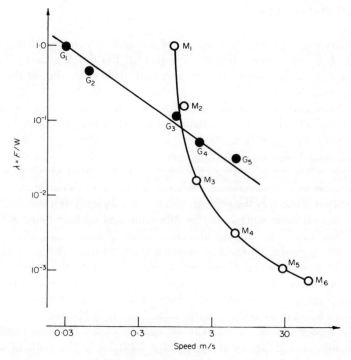

Figure 1.2   The results from figure 1.1 indicating speeds achieved

the two curves, one representing the evolutionary process and the other due to man's inventiveness. What is most interesting in this logarithmic plot is that the evolutionary line is approximately straight and has a negative slope of 45°. This means that

$$\log \lambda = -1 \times \log S + \log C \qquad (1.1)$$

where S is the speed of translation and C is some constant. This equation may thus be expressed

$$\lambda S = C \qquad (1.2)$$

But λS is the resisting force multiplied by speed and divided by the weight, that is, power divided by the weight carried, thus this straight line suggests that the evolutionary line is governed by a substantially *constant* power/weight ratio. Clearly this is reasonable, since the power/weight value is

3

defined by the same physiological processes in all animals, including man. The curve due to man's inventiveness does not suffer from such a restriction, since our inventions have benefited from a more successful application of scientific principles and the use of other than purely physiological materials.

1.2 HISTORICAL

Man's invention of the wheel as one of the earliest tribological devices has already been mentioned, but friction affected the development of civilised man in many other ways[2]. It is known that drills made during the palaeo-lithic period for drilling holes or producing fire were fitted with bearings made from antlers or bones, and potters' wheels or stones for grinding cereals, etc., clearly had a requirement for some form of bearing[3]. Records show the use of wheels from 3500 B.C., and yet it is interesting to note that the very advanced Inca civilisation of more recent date never did discover the principle of the wheel. Lubricants were also used from about this period, and a tomb in Egypt provided evidence of this fact. A chariot in this tomb still contained some of the original animal-fat lubricant in its wheel bearings and it is also interesting that this lubricant was contaminated with road dirt in the form of quartz sand and compounds of aluminium, iron and lime.

In their monumental tasks of building the Egyptians also showed a clear appreciation of tribological principles[4]. Surviving illustrations in the form of bas-reliefs show the use of rollers and sledges to transport their heavy weights. Figure 1.3 illustrates one example of such transportation; here 172 slaves are being used to drag a large statue weighing about 6×10^5 newtons along a wooden track. Closer examination shows one man (standing on the sledge supporting the statue) pouring a liquid into the path of motion; perhaps one of the earliest lubrication engineers. If we assume that each slave pulls with about 800 newtons we can see that this picture would suggest a coefficient of friction of

$$\mu = \frac{172 \times 800}{6 \times 10^5} \simeq 0.23$$

This is just about the value we would expect for a lubricated wooden slide, thus we can infer that this picture is a true record of what actually occurred. The somewhat artificial arrangement of the slaves probably arises from the artist's inability to draw perspective.

In 1928 fragments of what must have been a ball thrust-bearing were found in Lake Nimi near Rome, and probably date from about A.D. 40. The bearing is shown in figure 1.4. It was probably used to support a statue in a sculptor's workshop thus facilitating rotation during the sculpting process. There is little further evidence of tribological development until the time of Leonardo da Vinci (1452–1519), who first postulated a scientific approach to friction. He recognised that the friction force is proportional to

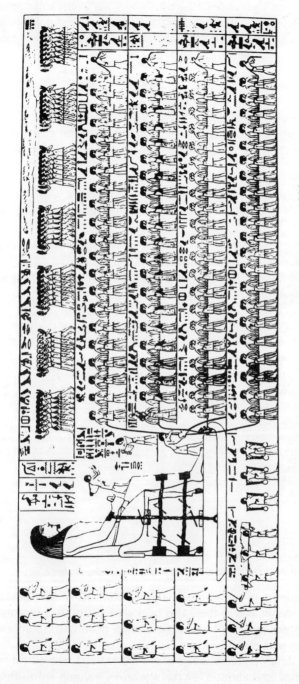

Figure 1.3   Transporting an Egyptian colossus, c. 1900 B.C.

5

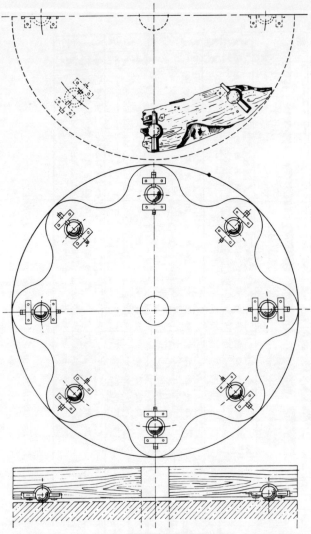

Figure 1.4 Detail of fragment and theoretical reconstruction of Lake Nemi bearing, estimated diameter nearly 1 metre (reproduced from Ucelli, Le Navi di Nemi, *Libreria Dello Stato, Roma)*

load and independent of the nominal area of contact. It was almost two hundred years however before these two laws were enunciated by Amonton in 1699, who, independently of Leonardo da Vinci, postulated these laws and is usually credited with their discovery. The eighteenth century saw considerable tribological development because of the increasing involvement of man with new machines, and about 1780 the work of Coulomb led to the third law of friction which suggested that friction was independent of velocity.

These three laws are still used today as reasonably true and are to be found in the teaching of friction in elementary texts in physics and engineering. Many other developments occurred during this century, particularly in the use of improved bearing-materials. As early as 1684 Robert Hooke suggested the combination of steel shafts and bell-metal bushes as preferable to wood shod with iron for wheel bearings. Further developments were undoubtedly associated with the growth of industrialisation in the latter part of the eighteenth century.

With lubricated bearings, although the essential laws of viscous flow had earlier been postulated by Newton, any scientific understanding of their operation did not occur until the end of the nineteenth century. Indeed our understanding of the principles of hydrodynamic lubrication can be seen to date from the experimental studies of Beauchamp Tower (1883)[6] and the perspicacious theoretical interpretation by Osborne Reynolds (1886)[7]. Other work by Stokes and Petroff[8] about this time was also very pertinent to these developments. Subsequent development in hydrodynamic-bearing theory and practice was extremely rapid in an attempt to meet the increasing demand for reliable bearings in the new machinery then being developed, some of which is still operational to this day.

Since the beginning of this century, and spurred on by the industrial demand for better tribology, our knowledge in all areas of tribology has expanded enormously. In this context ball-bearings, which were first introduced for industrial applications about 1700, have now reached an unusual peak of efficiency. They are available in a wide range of sizes and today offer a very cheap and flexible solution to many tribological design problems. Other developments in the quality and service characteristics of lubricants have also added to our toolkit of tribological solutions, but the pace of modern industrial society is such that increasing demands for higher speeds and loads, often in hostile environments such as nuclear reactors and space vehicles, still necessitate further development of the subject. Today we take as the norm that a car engine should last about 100 000 miles, whereas less than twenty-five years ago the life expectancy was only a third of this value. It is also interesting to note that a modern car contains upward of 2000 tribological contacts, so that it is not surprising that this subject is of increasing importance to engineers engaged in a wide variety of engineering disciplines.

1.3 TRIBOLOGY IN INDUSTRY

Before discussing the prevalence and types of tribological problems in industry we must first take a somewhat global view of the phenomena. In essence we are dealing with the interaction of *two* solid surfaces within a given environment which results in two outward manifestations.

(a) There is an energy dissipation which is the resistance to motion and is indicated by the coefficient of friction. This energy dissipation results in a heat release at the contact and a small, but sometimes significant, amount of noise. It should be emphasised that since two solid surfaces are always involved, parameters such as the coefficient of friction must relate to the pair of interacting materials. To talk about the coefficient of friction of steel without reference to the mating solid is scientifically incorrect and misleading. It might also be remarked that the idea of frictionless surfaces is scientifically impossible, and the often-stated implication that low friction is associated with the smoothness of the surface is also basically incorrect.

(b) During the sliding process all surfaces are to a greater or lesser extent changed in their basic characteristics. They may become smoother or rougher, have physical properties such as their hardness altered, and some material may be lost in the so-called wear processes. Such surface changes may be either beneficial, as happens when surfaces 'run in' to produce near ideal operation, or disastrous, as happens when surface failure occurs, necessitating the replacement of components.

From the foregoing it may be thought that both friction and wear are always disadvantageous, but this is not the case. In many engineering applications we are able to employ friction to fulfil required functions. Thus brakes, clutches, driving wheels on trains, cars, etc. all operate because of the existence of friction, while the commonplace nut and bolt only work because of the friction between the two. In the same way the wear of machinery is sometimes advantageous. The initial wear resulting in better mating of components (running-in), is evidently desirable, while the fact that components wear-out provides a strong motivation to replace obsolescent machinery. In its most extreme form this leads to the concept of 'planned obsolescence' where designers endeavour to use the wear phenomena to provide machinery with a specified life-span.

A fairly general misconception exists that friction and wear, which must be related in some way since they both arise from the interaction of surfaces, are simply related such that high friction means high wear. That this is not the case is clearly shown in table 1.1 where it is seen that the lowest friction is not associated with the lowest wear. These results are also interesting in that they show the very large differences in wear rates for materials whose friction coefficients vary in a fairly modest manner. The complexity of the relationship between friction and wear is also demonstrated by the peculiar results which sometimes occur when, with the same materials, friction may decrease after a given running time and this reduction in friction is associated with an increase in the wear rate.

In almost all industrial situations the wear effects are more important than the frictional losses because they tend to have the greater economic consequences. Higher friction can often be tolerated, with its slightly higher running costs, provided there are consequent savings by increased wear life of the machine. Some of the industrial situations where friction and wear are important are indicated by the following categories.

TABLE 1.1 FRICTION AND WEAR FROM PIN ON RING
TESTS. RINGS ARE HARDENED TOOL STEEL EXCEPT IN
TESTS 1 AND 7

	Materials	Coefficient of friction	Wear rate $cm^3/cm \times 10^{-12}$
1	Mild steel on mild steel	0.62	157,000
2	60/40 leaded brass	0.24	24,000
3	PTFE	0.18	2,000
4	Stellite	0.60	320
5	Ferritic stainless steel	0.53	270
6	Polyethylene	0.65	30
7	Tungsten carbide on itself	0.35	2

Load 400 g   Speed 180 cm/s

1.3.1 Energy Losses

While friction may not always be a prime consideration, the tribologist has
continuously to be aware of the absolute magnitude of the energy losses
involved and any likely consequences of the ensuing heat release. Thus in a
500 MW generating-set a frictional dissipation of only one tenth of one per
cent represents 0·5 MW or the heat equivalent of 500 single-bar electric
fires. This heat is confined to a fairly localised region of the system and
therefore requires special consideration from the designer. Indeed in many
machines the tribological heat release can result in undesirable thermal
distortions.

1.3.2 Wear

In industrialised societies about one-half of the gross national product is
used to replace the results of wear and similar effects, so that these phenomena
are central to the development of such societies. Savings by improvements
in the life of machinery also have the economic advantage that they are
deflationary, a particularly attractive feature in these days of inflation. It
is of course true that wear may also be desirable as a spur to the replacement
of obsolescent machinery, that is, the concept of 'planned obsolescence'.
Such concepts do not however exclude the need to understand wear processes,
since planned wear or wear control of any kind implies an appreciation of the
basic rules of the wear processes.

1.3.3 Control

There is a requirement in sophisticated machinery for the precise control of
parts in relative motion. This clearly involves one in proper tribological
design and is a factor which is increasingly important for systems using

automatic control methods for the positioning of machine elements. Where parts are moving at very slow speeds such problems are accentuated due to instabilities in the frictional behaviour.

1.3.4 Environment

It is not possible to solve all engineering tribological problems by a liberal supply of oil. In many applications the use of such a lubricant is either excluded or technically undesirable. Typical examples of such situations arise either in whole or in part with space vehicles, nuclear reactors, chemical plants, textile plants, food processing machines and many other forms of machinery. Indeed the tribologists' problems arise increasingly from the designers' requirement for higher load-capacity, higher speeds and the operation in difficult and sometimes 'hostile' environments.

1.3.5 Friction Devices

As already mentioned, in many situations friction is a prior requirement for the functioning of the device. In such cases the heat release and the rate of wear become the primary design considerations, as in the design of automobile brakes and clutches.

1.4 ECONOMIC CONSIDERATIONS

The economic significance of tribology is so obvious as to require little detailed discussion, but it is still true that the savings in individual cases are generally small, and we all find difficulty in applying the maxim 'save the pennies and the pounds will look after themselves'. It is because of the enormous prevalence of tribological contacts in machinery that the small savings on each contact can add up to significant sums for the nation as a whole. Thus it is not surprising to find that about one-half of the world's energy production is used to overcome friction in some form. In a very real sense we therefore see that savings by better tribological design could have considerable significance in the conservation debate which concerns the whole future of mankind. For these reasons governments of industrial societies have placed increasing emphasis on the economic aspects of tribology.

In Britain the recent acceleration in tribological activities owes much to the 1966 D.E.S. report, *Lubrication (Tribology), Education and Research.*[9] This report (the Jost Report) suggested that this country could save no less than £515 million per annum by better tribological practices. This enormous sum arises from the constituent savings shown in figure 1.5, and it must always be remembered that such costs include the loss of production etc which is consequent on tribological failures in modern industry. It is also

10

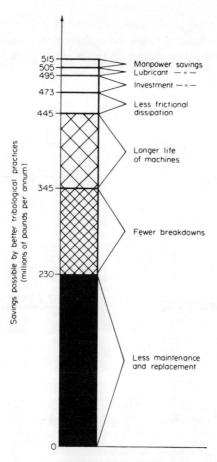

Figure 1.5 Savings indicated by the Jost Report[9]

apposite to note that these savings do not require new research but only the application of current knowledge. As research in this area expands, even greater savings could be anticipated. Experience since 1966 suggests that these figures were not unduly optimistic and the introduction of the word tribology and the attendant publicity have already had considerable effect on entrenched attitudes and in the best interest of the nation's economy.

1.5 TRIBOLOGICAL SOLUTIONS

Perhaps the most significant effect of the introduction of the word tribology has been to introduce a problem-orientated view of any system. We thus tend to think: 'What is the best solution to the problem of carrying load across the interface with acceptable friction and wear?' rather than heretofore

11

where the lubrication engineer naturally tended to have a predisposition to adopt a lubrication solution. We now therefore categorise the available solutions to tribological problems in the manner shown in figure 1.6 and discussed in the following text.

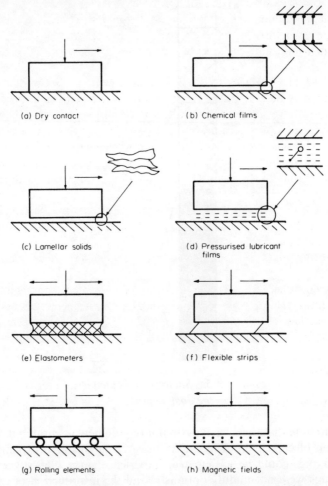

(a) Dry contact

(b) Chemical films

(c) Lamellar solids

(d) Pressurised lubricant films

(e) Elastometers

(f) Flexible strips

(g) Rolling elements

(h) Magnetic fields

Figure 1.6   Methods of solution of tribological problems

Figure 1.6a We may choose the contacting materials because they have intrinsically low friction and/or wear characteristics, although this may mean accepting lower load-carrying capacities as for instance when plastic materials are employed. In many cases it is possible to use materials as surface layers supported on substrates which fulfil the basic structural requirements of the particular component. This method is employed in the well-known bearing shells used in automotive engines.

Figure 1.6b We may apply chemical films which protect the surfaces and in part reduce the intimate contact of the base materials. In such systems the thermal stability of these films is important due to the high local temperatures which are created at the points of intimate contact during sliding.

Figure 1.6c Solid surface coatings may be used since they have low resistance to transverse shear, for example soft metal layers or lamellar solids such as graphite and molybdenum disulphide. These latter materials have a layer structure rather like a pack of playing cards with strength to carry normal load and weakness along planes at right-angles to facilitate sliding.

Figure 1.6d The surfaces may be separated with a continuous film of fluid, which may be either a liquid, a vapour or a gas. In such systems the fluid film must have a built-in pressure to withstand the effects of the applied normal load. Such pressures may be provided by two distinct mechanisms. The most obvious is to supply the fluid at a pressure generated by an external pumping system as used in the externally pressurised bearing, often referred to as the hydrostatic or aerostatic bearing. Alternatively the pressure may be generated by the motion of the surfaces themselves as they tend to drag the fluid into a converging gap. This action of the hydrodynamic bearing is more dependent on the viscosity properties of the fluid than in the hydrostatic case. In both these fluid-film applications a wide range of fluids such as water, oil, air and even liquid metals in nuclear reactors, have been successfully employed.

Figure 1.6e Where the degree of transverse displacement is of fairly small amplitude the surfaces may be separated by elastometers bonded to the two surfaces. This clearly offers an excellent tribological solution and alternative designs might incorporate flexible elastic strips as shown in figure 1.6f.

Figure 1.6g A widely employed tribological solution is to interpose rolling elements such as balls, cylinders and the like between the two surfaces. The wide range of rolling contact bearings available is evidence of the value of this particular solution.

Figure 1.6h The carrying of load without mechanical contact is clearly possible by using magnetic and similar force fields. Such a bearing is to be found in the domestic electricity-supply meter.

Having this range of solutions in mind the designer may now consider such factors as the load to be carried, the speed, the nature of the environment and any limitations on friction and wear, in arriving at the most appropriate answer to his design problem. It sometimes happens, however, that by a complete change of design philosophy particular tribology problems may be either eliminated or at least modified while still achieving the desired effect from the resulting machine. This concept is illustrated in figure 1.7 where a power unit is required to drive an aeroplane. Here the tribological implications of three possible engines are considered, namely, the internal combustion reciprocating engine, the jet engine and the ram jet. The shaded areas illustrate where tribological contacts have to be considered between the

13

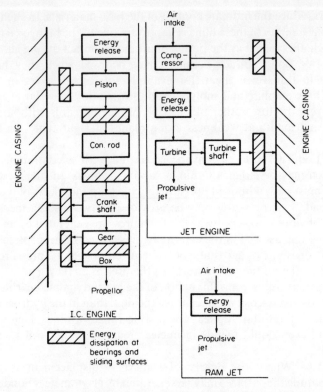

Figure 1.7   Energy dissipation in devices for propulsion

various engine components and it is seen how these are reduced in number, although sometimes becoming more complex, as we move from one design to the next.

In the foregoing it has been assumed that surfaces are simply smooth planes defining the boundaries between solids and their environment. Unfortunately all engineering surfaces are rather more complex than this, having geometries with hills and valleys and equally complex physical and chemical properties which are seldom uniform throughout the depth of the material. It is these features of the surfaces which contribute to the complexity of the subject of tribology. The remainder of this book is therefore devoted to considering the nature of such surfaces and to the underlying scientific principles involved in the various topics discussed above.

REFERENCES

1. J. Halling, The roll (role) of the wheel. *Wear*, **24**, (1973), 53.
2. D. Dowson, Tribology—Inaugural Lecture. University of Leeds Press, (1969).
3. C. St. C. Davison, Bearings since the Stone Age. *Engineering*, (Jan. 1957), 2.

4. C. St. C. Davison, Transporting sixty-ton statues in early Assyria and Egypt. *Technology and Culture*, **2**, (1961), No. 1.
5. J. H. Harris, *The Lubrication of Rolling Bearings*. Shell-Mex and B.P., London, (1967).
6. B. Tower, Report on Friction Experiments. *Proc. Instn mech. Engrs*, (1884), 632.
7. O. O. Reynolds, On the theory of lubrication and its application to Mr. Beauchamp Tower's experiments. *Phil. Trans. R. Soc.*, London, (1886), p. 117.
8. N. P. Petroff, Friction in machines and the effects of the lubricant. *Engng J.*, (1883), St. Petersburg.
9. *Lubrication (Tribology) Education and Research*, D.E.S. Report, (1966), H.M.S.O.

2

Surface Properties and Measurement

2.1 THE NATURE OF METAL SURFACES

In the field of tribology it is usually necessary to widen the simple interpretation of a surface as being a geometric plane separating two media. A surface must be recognised as a layer that grows organically out of the solid and has physical properties of considerable functional significance. The surface layer of metals is known to consist of several zones having physico-chemical characteristics peculiar to the bulk material itself.

At the base of the surface layer, (see figure 2.1) there is a zone of work-hardened material on top of which is a region of amorphous or micro-crystalline structure. This so-called Bielby layer is produced by the melting and surface flow during machining of molecular layers which are subsequently hardened by quenching as they are deposited on the cool underlying material. This basic structure is usually contaminated by the products of chemical reaction with the atmosphere and is covered by dust particles and molecular films deposited from the environment. Finally, the surface contains atoms of gas, which live with the surface and have properties somewhat dissimilar to those of the gaseous environment. In addition, the whole texture of the surface layer has a geometric property characterised by a series of irregularities having different amplitudes and frequency of occurrence. This particular property, the surface texture, is of fundamental importance in the study of

16

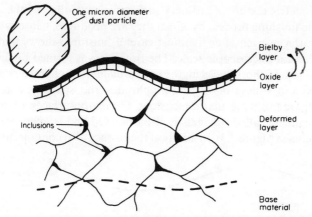

Figure 2.1   Typical surface layers

friction, wear and lubrication and thus a knowledge of, and a definition of, the topographic features of surfaces is vital.

Some feel for the scale of the various surface properties is given in figure 2.2, and it should be noted that the vertical scale is a log scale in ascending orders of magnitude. Furthermore such terms as oxide films must be carefully defined since they are complex layers which, on steel for example, tend to have an increasing oxygen/metal ratio as the atmosphere is approached.

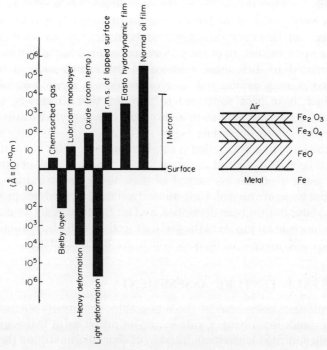

Figure 2.2   Order of magnitude of surface features

17

The geometric texture of ordinary surfaces is controlled by the characteristics of the finishing process by which they are produced. Close examination of these surfaces, even after the most careful finishing, shows that they are still rough on a microscopic scale. The roughness is formed by fluctuations in the surface, of short wavelengths (*microroughness*), characterised by hills (asperities) and valleys of varying amplitudes and spacings and these are large compared to molecular dimensions. On many surfaces a longer wavelength roughness called waviness is also observed and is often referred to as *macroroughness* (figure 2.3). In addition the surface also contains undulations

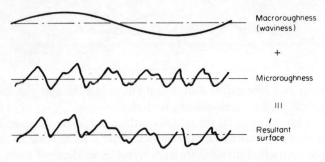

Figure 2.3   Components of surface geometry

of very long wavelengths caused by the vibrations of the workpiece or tool during the preparation of the surface. The distribution of the asperities over the surface can be either directional or homogeneous in all directions, depending upon the nature of the processing method. Surfaces which have been submitted to directional methods of processing, such as turning, milling and planing, exhibit a definite orientation of asperity distribution. Those which have been submitted to non-directional methods, such as electropolishing and lapping, show an isotropic or equiprobable distribution in all directions along the surface.

From the foregoing description it can be appreciated that the surfaces of metals can be as complicated as the surface of the earth and, obviously, in order to make a precise assessment of their topographic features, three-dimensional maps are needed. Unfortunately, simple methods for producing such maps have not yet been developed, and for the time being we must rely on a two-dimensional profile of the surface together with observations using the various microscopic techniques which are discussed later.

2.2   SURFACE TEXTURE ASSESSMENT

The quantitative assessment of the topographic features of surfaces is of vital importance for solving a wide variety of problems in tribology. Tribological phenomena such as friction and wear depend primarily on the nature of the real area of contact between the surfaces which is, in turn, dependent

18

upon the distributions, sizes and shapes of the asperities. Measurements of these features will, therefore, prove indispensable in the study of all kinds of surface contact phenomena. Indeed, they offer a valuable analysis of the conditions for elastic or plastic contact of metal surfaces and information on the size of the interstices between them. Furthermore, the application of surface measurement to such problems has produced correlation, both qualitative and quantitative, between theoretical arguments and experimental evidence.

Many methods are available for the measurement of the micro- or macro-geometrical features of surfaces. They include optical methods using electron, interference or reflection microscopy and mechanical methods such as oblique sectioning and profilometry. The optical methods show certain advantages in that they can provide a three-dimensional appreciation of the surface. However, the obtaining of quantitative assessment of surfaces by these methods is still relatively tedious.

Because of the difficulties in representing every irregularity within the whole plane of the surface, it is necessary to accept measurements which are based on a small sample of the surface, always provided that its size is large enough to be representative, that is, there must be a high degree of probability that the surface lying beyond the sample is similar to that which lies within it. Most of the existing methods of surface measurement give either a high-resolution picture of a small, and often unrepresentative, sample of the surface or a representative sample with correspondingly lower resolution, (table 2.1). This difficulty is overcome with profilometers which provide a representative length of the surface and a high resolution in a plane normal to it[1]. The higher vertical magnification results in a distortion of the recorded shape of the profile and thereby sometimes leads to a physical misconception as to the general character of the actual surface profile. Most surfaces have asperities with gentle slopes rather than the jagged character-istics observed on profilometer traces, (figure 2.4). The major disadvantage of profilometers however, is that they are restricted to a single line sample which may not be representative of the whole surface if the texture has character-istics dependent upon the orientation of the record. Nevertheless, for iso-tropic surfaces they provide useful measurements of the various parameters which are necessary for their characterisation.

It is also appropriate to record a difference between a peak on a profilo-meter trace and an actual summit on the real surface. Since the stylus will register a peak even when it only traverses a shoulder of an actual summit, it will be clearly recognised that the number of true summits is considerably less than the number of peaks recorded. This distinction may easily be shown by taking a series of line profiles in the same direction but with each traverse slightly displaced from its neighbour. In this way, the information in each of the line tracings may be consolidated to produce a true three-dimensional contour map of the surface and thereby clearly identifies the actual summits, (figure 2.5)[2].

TABLE 2.1 SURFACE MEASUREMENT METHODS

Method	Resolution (μm)		Comments
	Lateral	Vertical	
Optical microscope	0.25–0.35	0.18–0.35	Results depend on the quality of the optical system and hence depth of field of photomicrographs
Light profile	0.25	0.25	Using optimum microscopic conditions
Oblique section	0.25	0.025	Sectioning angle $\simeq \tan^{-1} 0.1$ and optimum microscopic conditions
Interference microscope	0.25	0.025	Requires high specular reflectivity of surface
Multiple beam interference	5	0.005	Requires high specular reflectivity of surface and no angular deviations greater than 5°
Reflection electron microscope	0.03–0.04	0.03	Vertical resolution from profiles.
		0.02–0.008	Vertical resolution from shadows
Electron microscope	0.005	0.005	Vertical resolution using stereo device.
		0.0025	Vertical resolution using shadowed replica angle $\tan^{-1} 0.3$
Profilometers	1.3–2.5	0.005–0.25	Finite size of the stylus is ultimate limit on resolution

The best-known profilometry method for obtaining surface profiles is the stylus method, in which a fine diamond stylus traverses the surface and its vertical movements are recorded, usually by electrical systems[3]. The *Talysurf* (Rank, Taylor, Hobson) is one of the most popular instruments of this type. In such systems the vertical measurements must be recorded with respect to some appropriate datum. The two most common methods for the establishment of such datums are

(a) The use of datum-generating attachments which ensure accurate horizontal motion of the stylus-support system.
(b) The use of large-radius skids or flat shoes which rest on and traverse the surface being measured, thereby generating the general level of the surface texture.

20

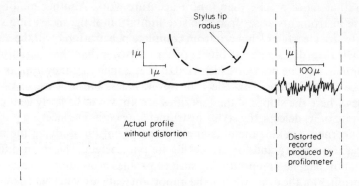

Figure 2.4

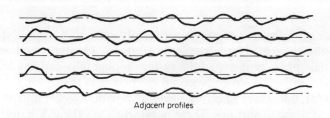

Adjacent profiles

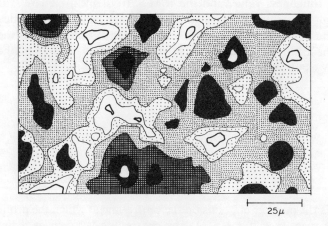

*Figure 2.5   Typical surface map (lighter areas corre-
spond to higher surface regions)*

21

Clearly the use of such devices creates some experimental errors, although (a) is rather more precise than (b) for accurate work. Another major error in these instruments arises from the size limitation of the measuring stylus. Its finite size clearly precludes it from complete penetration of all the valleys, thus their recorded shape will often appear narrower than their actual shape. In a similar way the peaks are distorted in shape such that they appear wider than they actually are. This effect, however, is less harmful with smoother surfaces where the slopes of the asperities are known to be fairly gentle.

In spite of its defects, the stylus instrument, as far as engineering purposes are concerned, remains the outstanding instrument for studying the nature of surface geometry and for evaluating its parameters. The main difficulty does not, in fact, lie in reproducing surface profiles accurately, it lies in using these profiles for the evaluation of the important features which will determine the functional behaviour of real surfaces.

2.3 SURFACE PARAMETERS

The choice of surface parameters is necessarily influenced to a certain extent by the method chosen to reveal the surface features. The stylus method reveals only a single plane property of the surface topography and, consequently, its characterisation has to be based on the nature of the ensuing single line profile. In production engineering one of two parameters is usually used to define the texture of surfaces. These parameters are the C.L.A. (centre-line average) roughness value, and the R.M.S. (root mean square) value. The C.L.A. value is defined as the arithmetic average value of the vertical deviation of the profile from the centre line, and the R.M.S. value as the square root of the arithmetic mean of the square of this deviation. In mathematical form they can be written as

$$\text{C.L.A.} = \frac{1}{n} \sum_{i=1}^{n} |z_i|$$

(2.1)

$$\text{R.M.S.} = \left[\frac{1}{n} \sum_{i=1}^{n} (z_i)^2 \right]^{1/2}$$

where n is the number of points on the centre-line at which the profile deviation z_i is measured, (figure 2.6). The centre-line is taken as a line which divides the profile in such a way that the sums of the enclosed areas above and below it are equal.

These parameters are seen to be primarily concerned with the relative departure of the profile in the vertical direction only; they do not provide any information about the slopes, shapes and sizes of the asperities or about

22

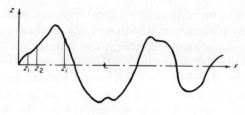

Figure 2.6

the frequency and regularity of their occurrence. It is possible, therefore, for surfaces of widely differing profiles to give the same C.L.A. or R.M.S. values, (figure 2.7). These single numerical parameters are mainly useful for classifying surfaces of the same type which are produced by the same method.

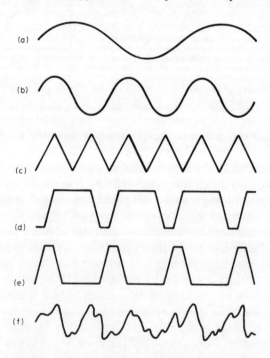

Figure 2.7 Various surfaces having the same C.L.A. value

Lapped surfaces, for example, differing only in the grades of the lapping compounds to which they have been submitted, will have the same pattern of roughness and a single parameter will be adequate to characterise them. For a more rigorous definition of surface profiles, however, the C.L.A. and R.M.S. parameters are inadequate and more information will be required. It is important to realise that the choice of parameters for specifying the

23

surface texture will be controlled by the functional requirements of the surface, that is, by measurements of those features of the surface which are known to be significant in any given practical situation.

Abbott and Firestone[1], the founders of profilometry, were concerned with the wear behaviour of surfaces, and they chose their parameters accordingly. Using the profile of a surface they constructed its bearing-area curve by measuring the fraction of the sample length which lies inside the profile at various positions above the lowest point on the profile, (figure 2.8). By

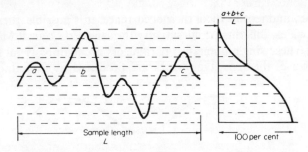

Figure 2.8   Method of deriving the bearing area curve

dividing the surface into three height-zones, containing the highest 25 per cent, middle 50 per cent and lowest 25 per cent of the bearing area, they identified three parameters—the peak, medial and valley occurrences respectively. It is interesting to find that they were well aware, some thirty-nine years ago, of the inadequacy of single numerical parameters to define surface textures. Meyers[4] was concerned with friction between sliding surfaces and suggested that the slope and curvature of the profile should affect the frictional behaviour of surfaces and found strong correlation between slopes and friction.

Recent work on the dry contact of surfaces has shown that this phenomenon may be largely explained by the shapes of the peaks and their distribution through the upper decile of the texture. Indeed, in such problems the properties of the valleys are almost insignificant. For such a situation, therefore, the required surface parameters must incorporate knowledge of the number of peaks of any given height level in the upper part of the texture, a shape factor for the peaks and a statement of the frequency of occurrence of such peaks in the plane of the surface. This information would be particularly important in considering a large range of engineering problems such as tribological situations, electrical contacts, thermal contacts and the stiffness of joints created by mating surfaces. Conversely, in the context of stress concentrations in surfaces under load, it would be anticipated that the shape and occurrence of the valleys would be the most significant parameters. For surfaces separated by lubricated films, or those which are to be covered by layers such as paint, it is apparent that we shall be concerned with the whole of the texture characteristics.

It should be pointed out that the profiles of surfaces can be considered, in statistical terms, as stationary processes of the random type and therefore in order to formulate the various parameters discussed earlier, it is necessary to apply statistical methods in describing the properties of surfaces.

2.4 THE STATISTICAL PROPERTIES OF SURFACES

It was mentioned earlier that surface profiles often reveal both a periodic and a random component in their geometric variation. Thus the periodicity is most marked in fine shaped surfaces and in diamond turning, whereas a considerable degree of randomness is apparent on abraded surfaces such as those produced by grinding. It has become common practice to break up the periodic variations into frequency bands such as waviness and roughness. In a sense it would be desirable not to make such distinctions, but this would necessitate the evaluation of a sample of the surface which includes all the frequency variations, and this inevitably leads to a need for the measurement of the whole of the surface. This latter requirement is unacceptable and the division into frequency bands is therefore a practical necessity. In what follows, we shall restrict ourselves to consideration of the higher frequency spectra, that is, the roughness of the surface.

Before any quantitative statements can be made about surface profiles a datum line must first be established. For this purpose the centre-line of the texture offers the most usual choice, although any line parallel to this line would be satisfactory. Let us first of all consider the properties of surface profiles in the vertical plane.

2.4.1 The Height Distribution of Surface Textures

Surface textures can be adequately described in terms of the distribution function of the heights of their profiles. Indeed, Abbott and Firestone's bearing-area curve is, in statistical terms, the cumulative distribution of the all-ordinate distribution curve. This can be written as

$$F(z) = \int_{-\infty}^{\infty} \psi(z)\, \mathrm{d}z$$

where z refers to the heights of the profile measured from the centre-line and $\psi(z)$ is the probability density function of the distribution of these heights.

The practical derivation of such distribution curves involves taking measurements of z_1, z_2, etc. at some discrete interval l and summing the number of ordinates at any given height level, (figure 2.9). In effect this may be interpreted as converting the continuous analogue signal of the profile into a discrete digital record taken at intervals l. Self-evidently the

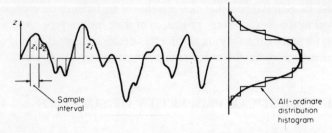

Figure 2.9   *Method of deriving the all-ordinate distribution*

distribution curve is then seen to be the best smooth curve drawn through the histogram produced by such a sampling procedure.

Many surfaces tend to exhibit a normal, gaussian, distribution of texture heights. Figure 2.10 shows how a gaussian distribution is a reasonable fit

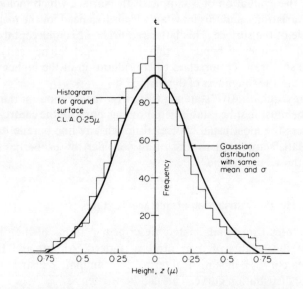

Figure 2.10   *Typical distribution curve for a ground surface*

for the histogram obtained for the all-ordinate distribution of a ground surface. The curve of the gaussian distribution or its density function is given by

$$\psi(z) = \psi_0(z)e^{-z^2/2\sigma^2}$$

where σ is the standard deviation of the distribution, which is defined in the same way as the R.M.S. value given by equation 2.1, and σ^2 is the variance. $\psi_0(z)$ can be calculated from the fact that the area of the curve must be equal

26

to the total number of data summed over the chosen scale. The area of the gaussian curve is

$$\int_{-\infty}^{\infty} \psi_0(z)e^{-x^2/2\sigma^2} = \psi_0(z)\sigma(2\pi)^{1/2}$$

Therefore the curve

$$\psi(z) = \frac{1}{\sigma(2\pi)^{1/2}} \, e^{-z^2/2\sigma^2} \tag{2.2}$$

has unit area and the gaussian distribution curve is usually written in this standard form. Since this curve encloses unit area, the foregoing procedure has simply adjusted the vertical scale of the distribution curve to produce a probability density function, that is the size of any ordinate on this curve indicates the probability of occurrence of that event in the total population. It is worth recording that the origin of the above curve is located at the centre-line or the mean of the distribution. If we wish to write the curve with reference to some other point as origin, then

$$\psi(z) = \frac{1}{\sigma(2\pi)^{1/2}} \, e^{-(z-m)^2/2\sigma^2}$$

where m is the distance of the mean from the value chosen as origin.

The values of the ordinates $\psi(z)$ of the gaussian distribution curve and those of the corresponding areas are found in most books on statistics. The form of the gaussian distribution necessitates a spread for the events from $-\infty$ to $+\infty$ which cannot happen with practical surfaces; in practice the distribution curve is truncated to finite limits of about $\pm 3\sigma$. Fortunately about 99.9 per cent of all events lie within this region and consequently the truncation would lead to negligible error while providing useful simplification.

It must be emphasised at this stage that although several common surface preparations produce near-gaussian distributions, many do not. There are surfaces whose textures show certain departures from the gaussian distribution and it is necessary to define some statistical parameters which are used for measuring such departures, as well as to consider some other simple distributions[5].

Moments, Skewness and Kurtosis

(a) Moments

The nth moment of the distribution curve $\psi(z)$ about the mean is defined as

$$M_n = \int_{-\infty}^{\infty} z^n \psi(z) \, dz \tag{2.3}$$

27

so that

$$\text{R.M.S.} \equiv \sigma \equiv \left[\int\limits_{-\infty}^{\infty} z^2 \psi(z) \, dz \right]^{1/2}$$

$$= [2\text{nd moment of } \psi(z)]^{1/2}$$

and

$$\text{C.L.A.} \equiv 2 \int\limits_{0}^{\infty} z \psi(z) \, dz$$

$$= \text{twice the 1st moment of half } \psi(z)$$

Obviously, the 1st moment of the whole of $\psi(z)$ about the mean is zero, and is in fact the method by which the centre-line of the distribution can be located.

(i) *Gaussian Distribution* For a gaussian distribution curve the nth moment is given by

$$M_n = \frac{1}{\sigma(2\pi)^{1/2}} \int\limits_{-\infty}^{\infty} z^n e^{-z^2/2\sigma} \, dz$$

If n is odd this vanishes, as it must for any symmetrical curve. If n is even then

$$M_n = \frac{n!}{2^{n/2}(n/2)!} \sigma^n \qquad (2.4)$$

Thus the 2nd moment becomes simply σ^2, the variance.

(ii) *Rectangular (Uniform) Distribution* The rectangular or uniform distribution implies that at any level there is an equal number of events, (figure 2.11). We can write

$$\psi(z) = c = \text{constant}$$

so that

$$\int\limits_{-\infty}^{\infty} \psi(z) \, dz = 1$$

For practical purposes let us suppose that this distribution is cut off at the finite limits $\pm h$, and then work out these limits in terms of the standard deviation σ. From the definition of the second moment we have

$$\sigma^2 = \int\limits_{-h}^{h} z^2 \psi(z) \, dz$$

$$= \int\limits_{-h}^{h} z^2 c \, dz = \frac{2ch^3}{3}$$

28

that is

$$c = \frac{3\sigma^2}{2h^3}$$

Now, making the area under the curve equal to unity gives

$$2h \frac{3\sigma^2}{2h^3} = 1$$

or

$$h = (3)^{1/2}\sigma$$

and the distribution curve can be written in the standard form

$$\psi(z) = \frac{1}{2(3)^{1/2}\sigma}$$

(iii) *The Triangular (Linear) Distribution* If the number of events varies linearly with height then the distribution of the events is said to be triangular or linear, (figure 2.11).

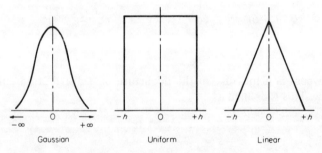

Gaussian   Uniform   Linear

Figure 2.11 Forms of distribution

Again let us consider the finite limits $\pm h$ of such a distribution. For this we can write

$$\psi(z) = c - \frac{c}{h}z \qquad 0 < z < h$$

$$= c + \frac{c}{h}z \qquad -h < z < 0$$

such that

$$\int_{-h}^{h} \psi(z)\,\mathrm{d}z = 1$$

29

Now

$$\sigma^2 = \int_{-h}^{h} z^2 \psi(z)\, \mathrm{d}z$$

$$= \int_{0}^{h} z^2\left(c - \frac{c}{h}z\right)\mathrm{d}z + \int_{-h}^{0} z^2\left(c + \frac{c}{h}z\right)\mathrm{d}z = \frac{ch^3}{6}$$

that is

$$c = \frac{6\sigma^2}{h^3}$$

Making the area under the distribution curve equal to unity gives

$$h \times c = 1 \quad \text{or} \quad h = (6)^{1/2}\sigma$$

The triangular distribution can therefore be written as

$$\psi(z) = \frac{1}{(6)^{1/2}\sigma} - \frac{z}{6\sigma^2} \qquad 0 < z < (6)^{1/2}\sigma$$

$$= \frac{1}{(6)^{1/2}\sigma} + \frac{z}{6\sigma^2} \qquad -(6)^{1/2}\sigma < z < 0$$

(b) Skewness

The skewness is a measure of the departure of a distribution curve from symmetry. This is defined as

$$s = \frac{\displaystyle\int_{-\infty}^{\infty} z^3 \psi(z)\, \mathrm{d}z}{\sigma^3}$$

$$= \frac{\text{3rd moment of } \psi(z)}{\sigma^3} \tag{2.5}$$

For a large class of moderately skewed distribution curves the skewness can be calculated from an empirical relationship given by

$$s = \frac{3\,(\text{mean} - \text{median})}{\sigma} \tag{2.6}$$

where the median represents that value of the variable whose ordinate divides the area under the distribution curve into two equal parts.

Clearly, symmetrical including gaussian distribution curves have zero skewness; unsymmetrical curves can have either negative or positive skewness, as shown in figure 2.12.

30

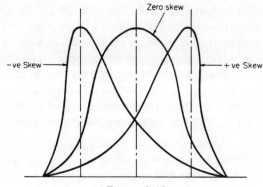

Figure 2.12

(c) Kurtosis

The kurtosis is a measure of the hump on a distribution curve. This is defined
as

$$k = \frac{\int\limits_{-\infty}^{\infty} z^4 \psi(z)\, dz}{\sigma^4}$$

$$= \frac{\text{4th moment of } \psi(z)}{\sigma^4}$$

(2.7)

For a gaussian distribution, using equation 2.4, we find that

$$k = \frac{1}{\sigma^4} \times \frac{4 \times 3 \times 2}{2 \times 2 \times 2} \sigma^4 = 3$$

that is a gaussian distribution curve has a kurtosis of 3 which is usually
taken as a standard value for the kurtosis. Curves with values of k less than 3
are called platykurtic and those with k greater than 3 are called lepto-
kurtic, (figure 2.13).

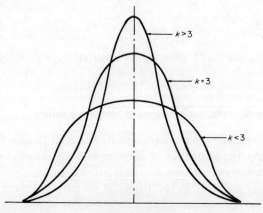

Figure 2.13

31

In plotting distribution curves of events on actual surfaces it has been shown that this involves drawing smooth curves through the histograms obtained by summing the frequency of such events at discrete intervals. This is obviously a fairly tedious procedure and a more useful presentation is to plot the distributions on probability paper. This has the advantage that the scales are adjusted such that a truly gaussian distribution degenerates into a straight line. Figure 2.14 shows the distribution of events on probability paper for gaussian, rectangular and triangular distribution of events.

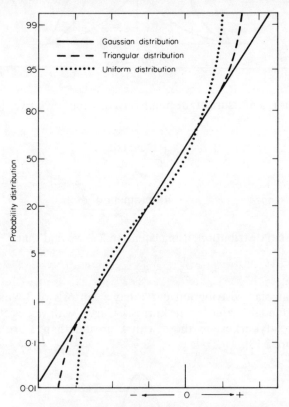

Figure 2.14   Distributions plotted on probability paper

2.4.2   The Autocorrelation Function of Surface Profiles

The effect of increasing the sampling interval l, although to some extent affecting the ensuing frequency histogram and distribution curves, can be most clearly seen by considering the correlation between adjacent ordinates[7]. This may best be achieved by plotting the variation of the autocorrelation function, $R(l)$ against the sampling interval l. The autocorrelation function for a single profile is obtained by delaying the profile relative to itself by some

32

fixed interval, then multiplying the original profile by the delayed one and averaging the product values over a representative length of the profile. Thus

$$R(l) = E[z(x)z(x + l)] \qquad (2.8a)$$

where E indicates the expected (average) value and $z(x)$ is the height of the profile at a given coordinate x along the mean line $z(x) = 0$ and $z(x + l)$ is the height at an adjacent coordinate $(x + l)$ taken at an interval l from the previous one. If the value of the ordinates at discrete intervals l is known this may be interpreted as

$$R(l) = \frac{1}{N - l} \sum_{x=1}^{N-l} z(x) \times z(x + l) \qquad (2.8b)$$

where N is the total number of ordinates in a sample length L. For a continuous function, equation 2.8b may be more usefully written

$$R(l) = \lim_{L \to \infty} \frac{1}{L} \int_{-L/2}^{L/2} z(x) \times z(x + l) \, dx \qquad (2.8c)$$

It can easily be seen that when $l = 0$, $R(l)$ reduces to the variance σ^2 or mean square value of the profile. The autocorrelation function is therefore usually plotted in its standardised form, $r(l)$, where

$$r(l) = \frac{R(l)}{R(l_0)} = \frac{R(l)}{\sigma^2}$$

A typical plot of the autocorrelation function for two different profiles is shown in figure 2.15. The shape of this function is most useful in revealing

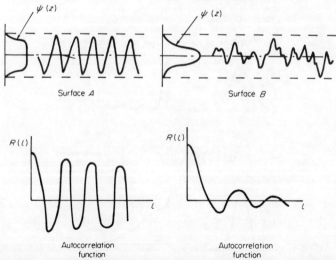

Figure 2.15   *Typical surfaces and the resulting autocorrelation functions*

33

some of the characteristics of the profile. The general decay of the function indicates a decrease of correlation as l increases and is an indication of the random component of the surface profile, while the oscillatory component of the function indicates any inherent periodicity of the profile. Figure 2.16 shows some typical results for actual machined surfaces where the auto-correlation functions can be seen to be demonstrating the general features of the profiles[8]. For this reason it is possible to describe the features of any surface profile by two characteristics: the height distribution function $\psi(z)$ and the autocorrelation function $r(l)$.

The main difficulty which may prevent the use of the autocorrelation function in practical applications lies in the large volume of digital informa-

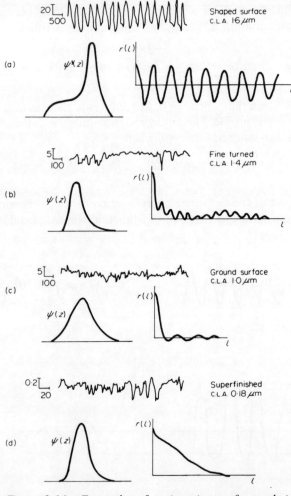

Figure 2.16 Examples of engineering surfaces, their distributions and autocorrelation functions

34

tion which must be obtained from the surface profile for its evaluation. However, the autocorrelation function may be broken up into two terms such that the random component of the surface profile may be expressed by an exponential decay term and the periodic component by a trigonometric term. For many practical applications, for example the dry hertzian contact of surfaces, the random component of the surface may be of primary importance, and consequently the problem can be simplified by neglecting the periodic component of the profile. The surface will then be statistically defined by an exponential autocorrelation function.

In certain situations the profile may be required to be presented in terms of its frequency domains. To this end a plot of the power spectra C is useful. The power spectrum is obtainable from the autocorrelation function using the expression

$$C(\omega) = \frac{2}{\pi} \int_0^\infty R(l) \cos \omega l \, \mathrm{d}l$$

where $\omega = 2\pi f$, f being the frequency of occurrence. The power spectrum is in fact the Laplace transform of the autocorrelation function. For the purposes of this book the use of the power spectra will not be pursued further, although it clearly has considerable possibilities for the defining of surface contours.

2.4.3 The Distribution of the Peaks, Valleys, Curvatures and Slopes

Surface profiles might be considered as comprising a certain number of peaks of varying heights and an equal number of valleys of varying depths. These features may therefore be assessed and represented by their appropriate distribution curves which can be described by the same sort of characteristics as were used previously for the all-ordinate distributions, for example, σ_p and σ_v are the standard deviations of the peak and valley distributions. These distribution curves will obviously involve summing all the peaks or valleys at any given height level, (figure 2.17).

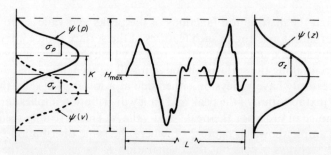

Figure 2.17   Peak and valley distributions of a surface

35

As with the all-ordinate distributions of surface textures, the peak and valley distributions often follow the gaussian curve. Figure 2.18 shows how close to gaussian distributions are the distributions for the peaks and all-ordinates for a ground surface.

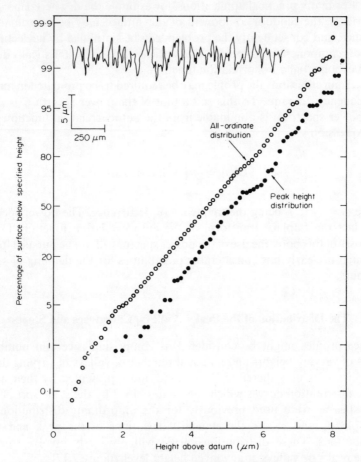

Figure 2.18   Distributions of an engineering surface

Although the distribution functions of the peaks and valleys provide the basic information relating to such features of the surface profile, they do not include any specific statement on the order in which the particular peaks and valleys occur along the surface; although it will be recognised that any peak must necessarily have valleys on either side at a lower level than the peak. Thus the juxtapcsition of the peak and valley distribution implies some limit on the choice of order for the peaks and valleys. This is readily seen by the simple example shown in figure 2.19, where the same distribution of peaks and valleys always occurs but their separation K varies. Where there is no overlap of these distributions, figure 2.19a, any peak may be associated with

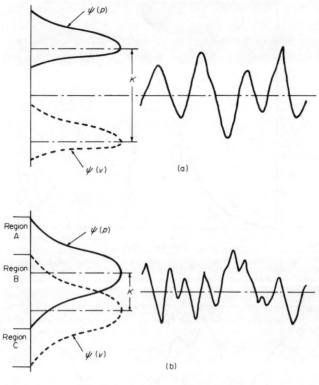

Figure 2.19   The effect of the interaction of the peak and
the valley distributions

any two valleys so that the order of occurrence of the peaks along the surface
is in no way defined. Where the distributions overlap, figure 2.19b, it is seen
that in regions A and C the foregoing still applies. However in region B
one cannot associate a valley in the upper part of this region with a peak in
the lower part of the same region, since peaks cannot be lower than their
adjacent valleys. Thus it is seen that the greater the degree of overlap of the
peak and valley distributions the more the order in which the peak and
valley occur along the surface is implied.

Similar distribution curves can be obtained for the curvatures of the peaks
and valleys. This may be done by simple curve-fitting—such as parabaloids
or circular arcs—to the peaks or valleys and assessing the way in which the
curvature varies along the profile. A typical distribution curve for the curvature of the peaks of real surfaces is shown in figure 2.20, together with the
gaussian distribution of the same mean and standard deviation.

The way in which the slope varies along the profile can also be described
in terms of its distribution curve. The derivation of this curve will involve
measuring the slope at each point along the surface profile. The distribution
curve may then be characterised by its standard deviation $\acute{\sigma}$ (figure 2.21).

37

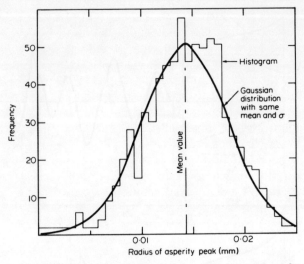

Figure 2.20   Distribution of the asperity peak radii

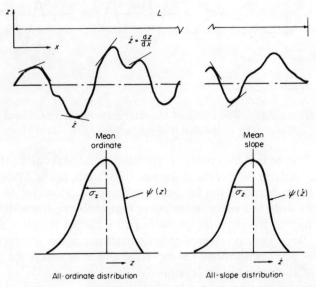

Figure 2.21   Derivation of the slope distribution of surfaces

2.5 MEASUREMENT OF SURFACE PARAMETERS

Abbott and Firestone measured their bearing-area curve manually using an optical comparator. For each height considered a line was drawn along the profile and the fraction of the sample length embraced by the profile at that height was measured. However, the practical derivation of such curves, in

38

the form of all-ordinate distributions of surface profiles, and indeed of the distribution curves of all surface parameters, peaks, valleys, curvatures, etc., involves the measurement of a sequence of ordinates along the profile. Such a process can be very tedious since a considerable amount of digital information is required, but with the aid of modern electronics, the stylus instrument can be modified to make the operation fairly easy. The usual practice is to feed the analogue electrical signal from the *Talysurf* instrument through a digital voltmeter to a high-speed paper-tape punch to get a permanent numerical record of a sequence of profile ordinates at some discrete sampling interval. The data on the paper tape can then be analysed by a digital computer for the required information. The height distributions for the peaks and valleys and their curvatures can be derived by using the technique of three-point analysis. A peak may be defined when the middle of three successive ordinates is higher than those on either side. Similarly, the peak curvatures can be obtained by fitting a curve to the profile using the three-point analysis.

1. E. J. Abbot and F. A. Firestone. Specifying surface quality. *Mech. Engng*, **55**, (1933), 569.
2. J. B. P. Williamson. The microtopography of surfaces. *Proc. Instn mech. Engrs*, **182** (3K), (1967–8), 21.
3. R. H. Reason. The trend of surface measurement. *J. Instn. Prod. Engrs*, (May, 1954).
4. N. O. Meyers. Characterization of surface roughness. *Wear*, **5**, (1962), 182.
5. J. Halling. The specification of surface quality—quo vadis?. *Prod. Engr*, **51**, No. 5, (1972), p. 171.
6. J. Halling and M. El-Refaie. A statistical model for engineering surfaces. *Tribology Convention, Paper No. C60/71*, Instn. mech. Engrs., (May, 1971).
7. D. J. Whitehouse and J. F. Archard. The properties of random surfaces of significance in their contact. *Proc. R. Soc.*, **316A**, (1970), 97.
8. J. Peklenik. New developments in surface characterization and measurements by means of random process analysis. *Proc. Instn mech. Engrs*, **182** (3K), (1967–8), 108.

3

Contact of Surfaces

3.1 INTRODUCTION

It is clear that any study of tribology must incorporate a detailed understanding of the mechanics of contact of solid bodies. This involves an understanding of the nature of the associated deformations and the stresses induced by any applied loading to bodies of a wide variety of geometric shapes. In particular we are concerned not only with the deformation and stresses at the surfaces of solids but also throughout the depth of the surface layers. Any load inducing a deformation of solids may readily be resolved into a normal and a tangential component, and it is generally convenient to consider these two influences separately with respect to the stresses and deformation which they induce, and then by superimposing the two obtain the total effect. In such cases the principle of superposition is acceptable since the systems are essentially statically determinate.

Solid materials subjected to loads deform in either an elastic or plastic manner; the former deformation being characterised by simple linear relations between stress and strain and being basically reversible. With plastic deformation the stress–strain relations are more complex and some deformation persists even after removal of the load. In most contact situations we find a mixture of both elastic and plastic deformations. Thus the loads applied to solids in contact may induce a general elastic behaviour in the bulk of the solid bodies, but since the actual contact must occur at the tips of the surface asperities these may be subjected to localised plastic deformation at their tips. The amount of plastic/elastic deformation must obviously depend on the value of the applied load and the degree of plastic deformation increases as

40

the load is increased. Thus, in metal-working processes when the nominal contact pressures are exceedingly large, the amount of plastic deformation of the surface is increased.

In much of the following we shall be considering the deformation patterns induced by loads applied to cylinders and spheres. This study is valuable for two reasons

(a) Many engineering contacts are concerned with the contact of bodies defined by circular arcs such as wheels on tracks, rolling element type bearings, gear-teeth contacts, many variable-speed drives and belt and rope drives.
(b) All solid bodies have surface asperities which may be considered as very small spherically shaped protuberances. Thus, the contact of essentially flat bodies reduces to the study of an array of roughly spherical contacts where we shall concentrate on the deformation of the tips of such spherical asperities.

Finally, in sliding contacts we are all aware that the work done in overcoming any friction is finally manifested as a heat release; try sliding down a rope and this will become painfully obvious. We shall therefore be very interested in the nature of this heat release, the temperatures which are induced and the distribution of these temperatures throughout the bulk of the contacting solids. In sliding down a rope the surface temperature of our skin is clearly much in excess of the temperature of the bulk of our hands.

From the foregoing it is clear that a detailed study of surface contacts will necessitate a relatively detailed understanding of such topics as elastic and plastic deformations and the nature of heat conduction due to moving heat sources. Clearly it is not possible to develop all these arguments in depth in a book of this type, so that recourse will often be made to physical arguments where the essential conclusions will be stated. In each case the reader will be able to find extensive verification of such results in standard works on elasticity, plasticity and heat conduction. It is not the purpose of this book to develop such arguments, but only to demonstrate the value of such knowledge in the particular problems with which we are concerned.

3.2 STRESS DISTRIBUTION DUE TO LOADING

In almost all our studies we shall be concerned with effects within the outermost layers of the surface (typically within the outermost millimetre or so of the surface). The effects at several centimetres below the surface are of only secondary importance so that we may often treat the surfaces, from a physical standpoint, as though they represent the surfaces of bodies of essentially infinite depths, that is, they may be considered as semi-infinite bodies. This device enables us to concentrate on the details of the surface contact of solids rather than considering their overall geometric shape and thereby leads to considerable mathematical simplification.

Consider a single normal line load P per unit length in the plane xz and applied at a point O' defined by the coordinates (ε, O) on the surface $(z = 0)$ of a semi-infinite solid and having the same value for all values of y, see figure 3.1a. The elastic stress field in the plane xz is readily obtained. Considering a unit length in the y direction, the radial stress σ_r will be given by[1]

$$\sigma_r = -\frac{2P}{\pi r} \cos \theta, \qquad (3.1)$$

the tangential stress σ_θ and the shearing stress $\tau_{r\theta}$ being equal to zero.

This represents a state of simple radial compressive stress, the stress increasing with decreasing radius r and decreasing angle θ. The use of the two-dimensional Mohr's circle of stress, figure 3.1b, for these stresses gives

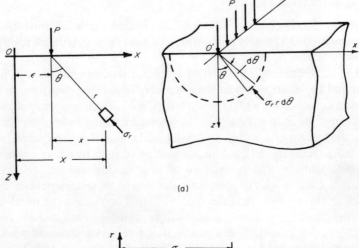

(a)

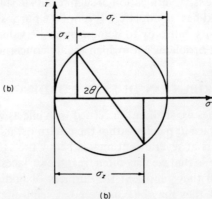

(b)

(b)

Figure 3.1   Stress distribution due to a line load acting on a semi-infinite body

42

us the resulting cartesian stresses with respect to O'

$$\sigma_x = \frac{\sigma_r}{2}(1 - \cos 2\theta) = \sigma_r \sin^2 \theta$$

$$= -\frac{2P}{\pi}\left(\frac{\sin^2 \theta \cos \theta}{r}\right) = \frac{2P}{\pi}\left[\frac{zx^2}{(x^2 + z^2)^2}\right]$$

$$\sigma_z = \frac{\sigma_r}{2}(1 + \cos 2\theta) = \sigma_r \cos^2 \theta$$

$$= -\frac{2P}{\pi}\left(\frac{\cos^3 \theta}{r}\right) = -\frac{2P}{\pi}\left[\frac{z^3}{(x^2 + z^2)^2}\right]$$

$$\tau_{xz} = \frac{\sigma_r}{2}\sin(2\theta) = \sigma_r \sin\theta \cos\theta = -\frac{2P}{\pi}\left(\frac{\sin\theta \cos^2\theta}{r}\right)$$

$$= -\frac{2P}{\pi}\left[\frac{z^2 x}{(x^2 + z^2)^2}\right] \tag{3.2}$$

or with respect to the original origin O as

$$\sigma_X = -\frac{2P}{\pi}\left[\frac{Z(X - \varepsilon)^2}{[(X - \varepsilon)^2 + Z^2]^2}\right]$$

$$\sigma_Z = -\frac{2P}{\pi}\left[\frac{Z^3}{[(X - \varepsilon)^2 + Z^2]^2}\right] \tag{3.3}$$

$$\tau_{XZ} = -\frac{2P}{\pi}\left[\frac{Z^2(X - \varepsilon)}{[(X - \varepsilon)^2 + Z^2]}\right]$$

A similar approach may be applied to obtain the stresses due to a single tangential line load T acting at O', (figure 3.2) where

$$\sigma_r = -\frac{2T}{\pi r}\cos\theta'$$

$$\sigma_{\theta'} = \tau_{r\theta'} = 0 \tag{3.4}$$

and

$$\sigma_X = -\frac{2T}{\pi}\left[\frac{Z^2(X - \varepsilon)}{[(X - \varepsilon)^2 + Z^2]^2}\right]$$

$$\sigma_Z = -\frac{2T}{\pi}\left[\frac{(X - \varepsilon)^3}{[(X - \varepsilon)^2 + Z^2]^2}\right] \tag{3.5}$$

$$\tau_{XZ} = -\frac{2T}{\pi}\left[\frac{Z(X - \varepsilon)^2}{[(X - \varepsilon)^2 + Z^2]^2}\right]$$

If we make $T = \mu P$, where μ is the appropriate coefficient of friction and add the stress components due to P and T at any point (x, y) we will clearly

43

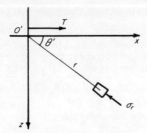

*Figure 3.2 Stress due
to a tangential line load
acting on a semi-infinite
body*

have the stress distribution arising in a simple frictional contact. The solution, however, suffers from one serious drawback. If we examine equations 3.1 and 3.4 we find that at O' ($r = 0$) the stresses are infinite and such a situation is obviously inadmissible. This arises from our basic assumption that the load acts at a single point, that is, over zero contact area. In real cases we must therefore always have some finite area of contact and this clearly changes our initial problem. Fortunately, we are still able to use our original solution for the new situation.

Consider a uniformly distributed load giving rise to a contact pressure p over a region O to a on the surface ($z = 0$) of a semi-infinite solid, figure 3.3. Taking a length along the y direction equal to unity, it will be recognised that the total load P is given by

$$P = \int_0^a p \, \mathrm{d}x = pa$$

If we consider a vanishingly small load $p \, \mathrm{d}\varepsilon$ at some point defined by the coordinates (ε, O) we may obtain the stress at any point (X, Z) due to this

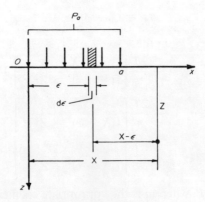

Figure 3.3

44

load using equations 3.3. In this case P will be replaced by $p\,d\varepsilon$. The total stress at a point X, Z due to the distributed load P is then clearly obtained by the summation of the effects of all the $p\,d\varepsilon$ loads acting at different values of ε from O to a, or mathematically speaking

$$\sigma_X = -\frac{2p}{\pi} \int_0^a \left[\frac{Z(X-\varepsilon)^2}{[(X-\varepsilon)^2 + Z^2]^2} \right] d\varepsilon$$

$$\sigma_Z = -\frac{2p}{\pi} \int_0^a \left[\frac{Z^3}{[(X-\varepsilon)^2 + Z^2]^2} \right] d\varepsilon \qquad (3.6)$$

$$\tau_{XZ} = -\frac{2p}{\pi} \int_0^a \left[\frac{Z^2(X-\varepsilon)}{[(X-\varepsilon)^2 + Z^2]^2} \right] d\varepsilon$$

In a similar way if we consider a tangential load $T = \mu P$ distributed over the region O to a, then at every point it follows that $t\,dx = \mu p\,dx$ (figure 3.4) and

$$T = \int_0^a t\,dx = \int_0^a \mu p\,dx = \mu P$$

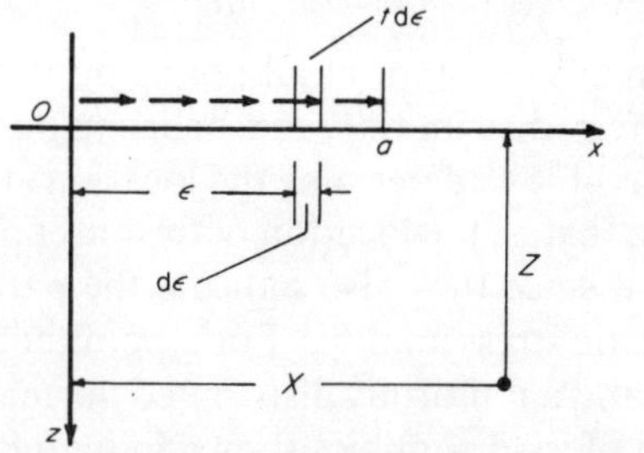

Figure 3.4

By using equations 3.5 for each elemental tangential load $t\,d\varepsilon$ acting on element $d\varepsilon\,(0, \varepsilon)$ we can obtain the stresses at any point (X, Z) due to the total distributed load T, thus

$$\sigma_X = -\frac{2t}{\pi} \int_0^a \left[\frac{Z^2(X-\varepsilon)}{[(X-\varepsilon)^2 + Z^2]^2} \right] d\varepsilon$$

$$\sigma_Z = -\frac{2t}{\pi} \int_0^a \left[\frac{(X-\varepsilon)^3}{[(X-\varepsilon)^2 + Z^2]^2} \right] d\varepsilon \qquad (3.7)$$

$$\tau_{XZ} = -\frac{2t}{\pi} \int_0^a \left[\frac{Z(X-\varepsilon)^2}{[(X-\varepsilon)^2 + Z^2]^2} \right] d\varepsilon$$

For a sliding contact subjected to a normal load P uniformly distributed over the contact O to a, the total stresses are the sum of the stresses given by equations 3.6 and 3.7 above.

It is clear that the use of the basic normal and tangential point load solutions may be used to obtain the resultant stress distribution for any type of load distribution over the contact region. All of the preceding solutions have assumed elastic behaviour of the bodies, but we have already anticipated that we are also interested in the possible plastic effects in such situations. The simplest criterion for the onset of plastic deformation assumes that this occurs when the maximum shear stress reaches the critical shear stress for the material k, where $k = Y/2$, Y being the tensile yield stress. For the cases considered above, where plane strain conditions apply, the maximum shear stresses always occur in the xz plane. The maximum shear stress in this plane is simply the radius of the Mohr's circle of stress, (figure 3.1b), that is

$$\tau_{max} = \frac{\sigma_r}{2} = -\frac{P}{\pi r} \cos \theta$$

If we consider a circle of diameter b drawn in the manner shown in figure 3.5a, we find that $r = b \cos \theta$ and

$$\tau_{max} = -\frac{P}{\pi b}$$

that is, the stress remains constant at all points on the circle. It is therefore useful to plot the stress distribution as isochromatics or lines of constant τ_{max}, and it is then possible to determine the location at which τ_{max} will reach its limiting value of k, that is, the location of the onset of plastic deformation. These plots are useful since they also indicate the pattern of isochromatics obtained in photoelastic stress analysis. For a point normal load, not strictly achieved in practice, and a uniformly distributed normal load, calculations of τ_{max} yield the pattern of isochromatics shown in figures 3.5b and c. It is seen that in both cases the material will first reach a yield condition at the surface where increasing load gives $\tau_{max} = k$, the yield shear stress of the material.

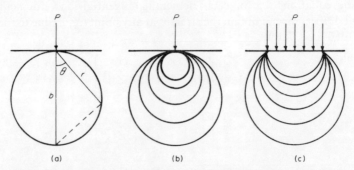

(a)   (b)   (c)

Figure 3.5   Patterns of lines of constant maximum shear stress (isochromatics)

46

3.3 DISPLACEMENTS DUE TO LOADING

Having obtained the stress distribution we can now obtain the displacements in a solid using the usual equations which relate the strain e and the corresponding displacements. Thus for a single normal load P acting at O', figure 3.1a, the horizontal and vertical displacements u and w are given by

$$\frac{\partial u}{\partial r} = e_r = \frac{1}{E}(\sigma_r - v\sigma_\theta) = -\frac{2P}{\pi r E}\cos\theta$$

$$\frac{u}{r} + \frac{\partial w}{r\partial\theta} = e_\theta = \frac{1}{E}(\sigma_\theta - v\sigma_r) = v\frac{2P}{\pi r E}\cos\theta$$

$$r\frac{\partial u}{\partial\theta} + \frac{\partial w}{\partial r} - \frac{w}{r} = \gamma_{r\theta} = \frac{1}{G}\tau_{r\theta} = 0$$

To solve these equations we require a knowledge of the boundary conditions. For this we assume that points on the z-axis, that is, at $\theta = 0$, have no lateral displacements and that at a point on the z-axis at a distance b from the origin there is no vertical displacement. Clearly we are interested in the displacements occurring at the boundary, $z = 0$, of the solid, thus by putting $\theta = \pm\pi/2$ in the solution of the above equations it can be shown that the horizontal displacement is given by

$$(u)_{z=0} = -\frac{(1-v)P}{2E} \tag{3.8}$$

This indicates that at all points on the boundary of the solid there is a constant displacement directed toward the origin. We can also find that the vertical displacement of a point on the boundary $z = 0$ at a distance x from the origin is given by

$$(w)_{z=0} = \frac{2P}{\pi E}\log\frac{b}{x} - \frac{(1+v)P}{\pi E} \tag{3.9}$$

Here we also find that at the point of load application ($x = 0$) the vertical displacement becomes infinitely large. As mentioned earlier, this is due to our assumption of a point load, but in practice the load is usually distributed over a finite area. In this case if the load is distributed over the region O to a, figure 3.3, giving rise to a contact pressure p, the vertical displacement at any point (X, O) produced by an element of load $p\, d\varepsilon$ at a distance ε from point O is known from equation 3.9 by substituting $p\, d\varepsilon$ for P and $(X - \varepsilon)$ for x so that the total displacement at point (X, O) is given by

$$(w)_{z=0} = \frac{2}{\pi E}\int_O^a p\log\frac{b}{X-\varepsilon}\,d\varepsilon - \frac{(1+v)}{\pi E}\int_O^a p\, d\varepsilon \tag{3.10}$$

In the foregoing we have concerned ourselves with problems of a two-dimensional nature where the deforming solid is considered to be subjected

47

to plane strain conditions. It is clear that the three-dimensional analogues of these problems necessitate a more complex treatment, however, the results are of great importance and are often met in tribological situations. For the purposes of this book it will suffice to state these results, detailed treatments can be found in standard books on elasticity. Thus, for a point normal load P acting on a semi-infinite solid, the horizontal and vertical displacements along the boundary $z = 0$ at a distance x from the point of load application are found to be given by

$$(u)_{z=0} = -\frac{(1 - 2v)(1 + v)P}{2\pi Ex} \tag{3.11}$$

$$(w)_{z=0} = \frac{(1 - v^2)P}{\pi Ex} \tag{3.12}$$

For a load distributed over a region of the boundary, giving rise to a pressure p acting on an element of area dA taken at some distance x from a point, the vertical displacement at that point is given by

$$(w)_{z=0} = \frac{(1 - v^2)}{\pi E} \int \frac{p \, dA}{x} \tag{3.13}$$

The integral in the above equation is, in fact, an elliptic integral.

3.4 HERTZIAN CONTACTS

As mentioned earlier we are particularly interested in the problems of contact between bodies whose geometry is defined by circular arcs. This problem was first solved by Hertz for elastic contact and is generally referred to as hertzian contact.

Consider the contact of two identical cylinders under conditions of plane strain. From arguments of symmetry we can see that the zone of contact will be created by compression of the cylinders to generate a straight line, that is, produce a plane contact zone, figure 3.6a. Although this is no longer strictly true for a cylinder in contact with a plane, the error is such as to be negligible and for this situation a plane contact zone may be assumed. One feature of such contacts becomes immediately apparent, in that as we increase the load the width of the contact zone will increase and, by noting that the deformation at the centre of the zone is larger than that at the extremities, we would expect a contact pressure which is no longer constant. This problem is therefore more complicated than the preceding cases and, before considering the stress distribution, we need to define both the contact pressure distribution and the actual size of the contact zone for any given applied load. Detailed treatment of this problem is found in books on elasticity, but for our purposes we shall use a simpler physical argument.

For two identical elastic cylinders in contact under a normal load P

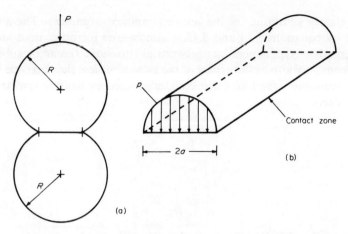

Figure 3.6   Pressure distribution for the contact of two
cylinders

per unit axial length, let the resulting plane contact zone have a width of $2a$, figure 3.6b. Since the normal deformation at the centre of the contact zone is greater than at the extremities, it is not surprising that the actual contact pressure distribution p has the form[1]

$$p = \frac{2P}{\pi a}\left(1 - \frac{x^2}{a^2}\right)^{1/2} \qquad (3.14)$$

Without this information it is physically apparent that the stresses in such a system would be such that

$$\text{stress} \propto \left(\frac{P}{a}\right)$$

Considering the deformation, we note that increasing the load would increase a and thereby increase the strain so that we would expect the non-dimensional strain to be related by

$$\text{strain} \propto \left(\frac{a}{R}\right)$$

where R is the radius of the cylinder. From these relations for stress and strain we have

$$\frac{P}{a} \propto E\left(\frac{a}{R}\right) \qquad \text{or} \qquad a^2 \propto \frac{PR}{E}$$

The actual solution for this case is in fact[1]

$$a^2 = \frac{4PR(1 - v^2)}{\pi E} \qquad (3.15)$$

49

which clearly substantiates the above simplified argument. The solution defined by equations 3.14 and 3.15 is almost true for other than identical cylinders, that is, plane contact geometries. Provided the angle subtended by the contact width at the centre of the cylinder is less than 30°, the results may be reasonably used for other contact geometries, figure 3.7, by using E' and R' where

$$\frac{1}{E'} = \frac{1 - v_1^2}{E_1} + \frac{1 - v_2^2}{E_2}$$

and

$$\frac{1}{R'} = \frac{1}{R_1} + \frac{1}{R_2}$$

whence

$$a^2 = \frac{4PR'}{\pi E'}$$

For the case of a cylinder on a plane, the radius of the plane is taken as infinity, thereby R' becomes the radius of the cylinder only and for concave curvatures the radius is taken as negative. It is also worthwhile noting that when $E \to \infty$, the solids become rigid leading to a single point contact where $a \to 0$.

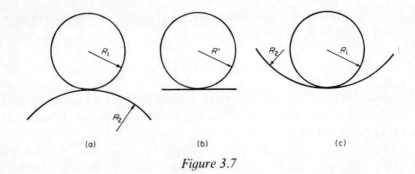

Figure 3.7

3.4.1 Stress Distribution in Hertzian Contacts

Earlier in this chapter we have seen that the onset of plastic deformation may be associated with the maximum shear stress reaching a critical value k. We therefore proceed to study the distribution of the maximum shear stress for a body loaded by a pressure distribution given by equation 3.14, acting over the contact zone $-a$ to $+a$. Using equations 3.6 for elemental loads $p\,d\varepsilon$ and integrating for the actual distribution of P will give the cartesian stress distribution within the body. Thus

$$\sigma_x = -\frac{4P}{\pi^2 a}\int_{-a}^{0}\left[1-\frac{\varepsilon^2}{a^2}\right]^{1/2}\left[\frac{Z(X+\varepsilon)^2}{[(X+\varepsilon)^2+Z^2]^2}\right]d\varepsilon$$

$$-\frac{4P}{\pi^2 a}\int_{0}^{a}\left[1-\frac{\varepsilon^2}{a^2}\right]^{1/2}\left[\frac{Z(X-\varepsilon)^2}{[(X-\varepsilon)^2+Z^2]^2}\right]d\varepsilon$$

$$\sigma_z = -\frac{4P}{\pi^2 a}\int_{-a}^{0}\left[1-\frac{\varepsilon^2}{a^2}\right]^{1/2}\left[\frac{Z^3}{[(X+\varepsilon)^2+Z^2]^2}\right]d\varepsilon$$

$$-\frac{4P}{\pi^2 a}\int_{0}^{a}\left[1-\frac{\varepsilon^2}{a^2}\right]^{1/2}\left[\frac{Z^3}{[(X-\varepsilon)^2+Z^2]^2}\right]d\varepsilon$$

$$\tau_{xz} = -\frac{4P}{\pi^2 a}\int_{-a}^{0}\left[1-\frac{\varepsilon^2}{a^2}\right]^{1/2}\left[\frac{Z^2(X+\varepsilon)}{[(X+\varepsilon)^2+Z^2]^2}\right]d\varepsilon$$

$$-\frac{4P}{\pi^2 a}\int_{0}^{a}\left[1-\frac{\varepsilon^2}{a^2}\right]^{1/2}\left[\frac{Z^2(X-\varepsilon)}{[(X-\varepsilon)^2+Z^2]^2}\right]d\varepsilon \qquad (3.16)$$

Again, the maximum shear stress for plane strain conditions is given by the radius of the Mohr stress circle, see figure 3.1b, that is

$$\tau_{max} = \left[\left(\frac{\sigma_x - \sigma_z}{2}\right)^2 + \tau_{xz}^2\right]^{1/2} \qquad (3.17)$$

where σ_x, σ_z and τ_{xz} are defined by equations 3.16. Equation 3.17 therefore defines the values of τ_{max} at all points. The evaluation of this equation will enable us to draw the isochromatics and obtain the somewhat surprising result shown in figure 3.8. We note immediately that the greatest value of τ_{max} occurs below the surface at a distance of $0.67a$. Furthermore, we find that as the load is increased τ_{max} at this point will reach the value k when the maximum pressure at the centre of the contact zone p_o is $3.1k$. This is again a surprising result since if we had simple compression conditions we would expect the surface material to yield when p_o reaches a value of $2k$. This does not occur because the surface elements are subjected to compressive stresses in all three orthogonal directions, allowing p_o to be greater than $2k$ without producing yield. This is equivalent to saying that the hydrostatic component of stress at any point cannot contribute to the plastic deformation of the material at that point. This is a very important result since it means that contact pressures in excess of the yield value for the material do not result in plastic deformation, so that higher loads than might have been expected can be carried elastically with hertzian type contacts. Moreover, it will be recognised that even when yielding has taken place below the surface, very little plastic deformation can occur because the plastic zone is constrained by elastic material on all sides.

51

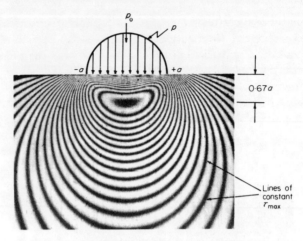

Figure 3.8   Actual isochromatics obtained for the contact of a cylinder and a plane due to normal load alone

As the load is increased further, the plastic zone will increase in size and ultimately spread to the surface of the body. Plastic flow may then occur fairly readily and the cylinder will indent the surface of the body. This will happen when the mean contact pressure p_m is about $6k$, that is, more than twice the contact pressure at which initial yield occurred[2]. The mean pressure under these conditions is essentially the indentation hardness value of the material, H, and this is why for metals we find that

$$H \simeq 6k \simeq 3Y$$

where Y is the uniaxial tensile yield strength of the material; see figure 3.9.

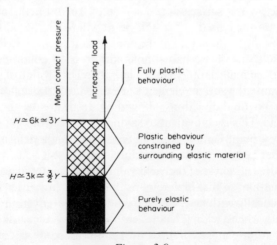

Figure 3.9

52

We have as yet only considered normal loads applied to hertzian contacts and it is apposite to ask what happens in the presence of both normal and tangential loads. The stress field due to a tangential load μP may readily be obtained by the method discussed earlier, since at all points it is clear that the tangential traction $t = \mu p$. Combining the stress distribution due to the normal and tangential loads and calculating the values of τ_{max} leads to the isochromatic pattern shown in figure 3.10. It is seen that the location of the

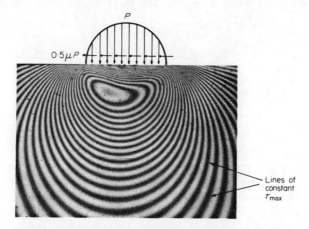

Figure 3.10 Actual isochromatics obtained for the contact of a cylinder and a plane due to combined normal and tangential loads where $T = 0.5\mu P$

greatest value of the maximum shear stress is now much nearer to the surface, thus the plastic deformation can take place more readily than in the previous case. In other words, macroscopic plastic deformation is facilitated by the presence of such friction tractions.

3.4.2 Conditions When $T \leqslant \mu P$

In many practical situations contacting bodies are subjected to tangential loads less than μP so that macroscopic sliding does not occur. This happens in situations where friction is used as the mechanism for preventing slip between mating components, for example, nuts and bolts, interference fits and friction drives such as clutches. Most textbooks explain this situation by allowing the coefficient of friction to increase from zero to a limiting value at which slip occurs. This is clearly inadmissible since it makes what should be a physical constant into a variable. In what follows we show that this assumption is not necessary if we recognise that the contacting bodies are deformable rather than rigid. The mechanism will be explained by considering a cylinder pressed against a plane and subjected to a tangential load less than μP.

We shall start by assuming the answer to our problem and subsequently justify its validity. What happens is that within the contact zone there exists a central area in which no slip occurs, while at the two extremities a small degree of slip takes place, see figure 3.11. The coexistence of a zone of sticking and zones of microslip is possible because of the deformable nature of the materials in contact and the deformation pattern being such as to allow slip at the extremities of the contact zone. As the value of T increases, the areas of slip increase until, when $T = \mu P$, they meet at the centre of the contact

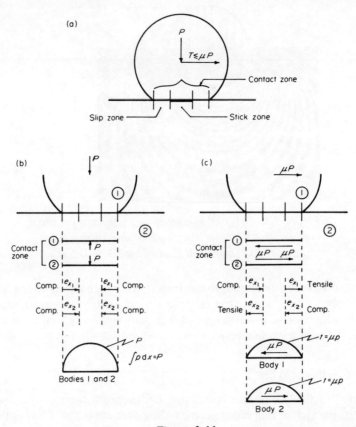

Figure 3.11

and macroslip takes place throughout the contact zone. With this model it is possible for μ to have a constant value wherever slip occurs, that is, within the slip regions $t = \mu p$ while within the stick region $t < \mu p$. Since T is the integral of t over the contact zone this can be seen from figure 3.11 to satisfy the requirements of the problem with μ always having a constant value.

For the case when $T = \mu P$, the distribution of normal pressure p and tangential traction $t = \mu p$ are as shown in figure 3.11b and 3.11c. Increasing the normal load induces equal compression strains e_x in both bodies so that no slip occurs due to this effect. With the tangential load, however, we see that since these must be opposite in direction on the two bodies they will cause the patterns of strain shown in figure 3.11c, so that slip must occur everywhere throughout the contact zone. For $T < \mu P$ we therefore see that the central stick areas must have zero strain in both bodies and this consequently defines the type of distribution for the tangential tractions.

The distribution of tangential traction which results in zero strain over the central stick area is illustrated by the argument shown in figure 3.12. Consider a tangential load $T = \mu P$ which results in the traction distribution

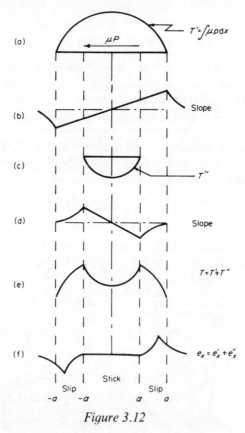

Figure 3.12

shown in figure 3.12a. The strain e'_x resulting from such a distribution is shown in Figure 3.12b. It will be noted that this strain follows a linear law within the width of the contact zone and that the slope is, from $-a$ to $+a$, $e'_x = \mu x/2R$. Now consider a similar shape of tangential traction distribution, T'', in the opposite sense applied over the region $-\alpha$ to $+\alpha$. The strains, e''_x, resulting from such a distribution are as shown, and it will be seen that the slope of the strain distribution is exactly opposite in sign, but of the same magnitude, as that due to T'. Thus by adding the strains e'_x and e''_x one obtains a zero strain over the region $-\alpha$ to $+\alpha$ and a tangential traction in this region where $t < \mu p$. In the remainder of the contact zone one obtains $t = \mu p$ and strains which are non-zero and consequently area of slip, figures 3.12e and 3.12f.

The above arguments show that even when no macroscopic motion occurs some degree of microslip exists when $T < \mu P$ and this gives rise to a mechanism known as fretting. For more complicated contact geometries these arguments are still qualitatively correct and microslip will occur at the extremities of the contact zone.

3.4.3 General Three-Dimensional Contact

For simplicity we have so far concerned ourselves with two-dimensional hertzian contacts. In many practical situations we must be able to deal with the more complicated three-dimensional problems and the following outlines the main results of such analysis. In general, the patterns of behaviour are analogous, but some of our preceding formulae must be modified.

If we subject two identical spheres to a normal load N, the area of contact will clearly be a plane circle of radius a and the pressure distribution is hemispherical in form, figure 3.13, and is given by[1]

$$p = \frac{3N}{2\pi a^2} \left(1 - \frac{x^2}{a^2} - \frac{z^2}{a^2} \right)^{1/2} \tag{3.18}$$

The value of a is given by

$$a = \left(\frac{3NR}{8E'} \right)^{1/3} \tag{3.19}$$

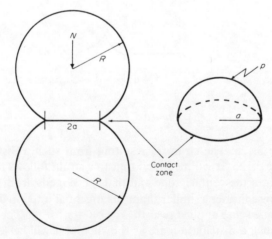

Figure 3.13   The pressure distribution due to
the contact of spheres

We can recognise that this is correct by applying the simple analysis used previously.

Although the contact of two dissimilar spheres does not result in a plane contact area, the results of equations 3.18 and 3.19 still hold with substantial accuracy, but in this case we have

$$a = \left(\frac{3NR'}{4E'}\right)^{1/3}$$

(3.20)

where R' is related to the radii of the spheres R_1 and R_2 by

$$\frac{1}{R'} = \frac{1}{R_1} + \frac{1}{R_2}$$

For the contact of a sphere and a plane, R' is merely the radius of the sphere.

A more general approach to the problem is to consider the contact of two bodies 1 and 2 whose contact geometry is defined by the principal radii of curvature of each body in two orthogonal planes, figure 3.14. The contact

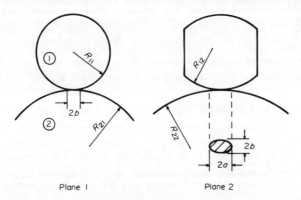

Plane 1   Plane 2

(R_{12} radius of body 1 in plane 2 etc.)

Figure 3.14

geometry is now clearly elliptical in shape and the contact pressure distribution is given by

$$p = \frac{3N}{2\pi ab}\left(1 - \frac{x^2}{a^2} - \frac{z^2}{b^2}\right)^{1/2}$$

(3.21)

The size of the contact ellipse is defined by the semi-major and semi-minor axes a and b which are given by

$$a = k_a\left[\frac{3N}{4E'(A + B)}\right]^{1/3}$$

$$b = k_b\left[\frac{3N}{4E'(A + B)}\right]^{1/3}$$

(3.22)

57

where k_a and k_b are constants which depend on the values of the principal curvatures of the contacting bodies and on the angle ϕ between the normal planes which contain these curvatures. If we denote the principal radii of curvature of body 1 by R_{11} and R_{12} and those of body 2 by R_{21} and R_{22}, then the constants A and B may be defined by

$$B - A = \frac{1}{2}\left[\left(\frac{1}{R_{11}} - \frac{1}{R_{12}}\right)^2 + \left(\frac{1}{R_{21}} - \frac{1}{R_{22}}\right)^2\right.$$

$$\left. + 2\left(\frac{1}{R_{11}} - \frac{1}{R_{12}}\right)\left(\frac{1}{R_{21}} - \frac{1}{R_{22}}\right)\cos 2\phi\right]^{1/2} \quad (3.23)$$

$$A + B = \frac{1}{2}\left(\frac{1}{R_{11}} + \frac{1}{R_{12}} + \frac{1}{R_{21}} + \frac{1}{R_{22}}\right)$$

In these equations any concave curvature is taken as negative. The coefficients k_a and k_b in equations 3.22 are numbers which depend on the ratio $(B - A)/(A + B)$ and these coefficients can be obtained by introducing an auxiliary angle γ defined by

$$\cos \gamma = \frac{B - A}{A + B}$$

Thus, using equations 3.23, the value of γ can be easily determined. To evaluate the values of k_a and k_b corresponding to a certain value of γ we require complicated numerical calculations involving elliptic integrals. The results of such calculations are given in figure 3.15.

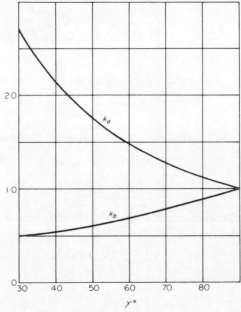

Figure 3.15

58

When dealing with such complicated geometries it is clear that our assumption of a plane area of contact will no longer be true. As mentioned earlier, while the pressure distribution and the size of the contact as determined from hertzian theory are still substantially correct, we sometimes need to know the actual shapes of such contacts. For materials having the same elastic properties it is sufficient to assume that the deformed surface, which has some common radius R_c, is about midway between the two original surfaces, figure 3.16. Thus the value of the common radius of curvature is given by[1]

$$R_c = \frac{2R_1 R_2}{R_1 - R_2} \qquad (3.24)$$

It is clear that for two identical spheres the above equation gives the expected result of a plane contact area. Where concave curvatures occur, the radius is taken as negative.

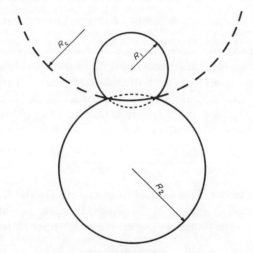

Figure 3.16

Subsequently, we shall find that it is necessary to define the *normal approach* of a sphere due to the application of normal load and the consequent deformation. Consider the contact of a sphere and a plane, figure 3.17. It is easily seen that the separation u of the surfaces at a distance r from the centre of the contact zone is given by

$$u = R - (R^2 - r^2)^{1/2}$$

$$= R - R\left(1 - \frac{r^2}{R^2}\right)^{1/2} = R - R + \frac{r^2}{2R} - \cdots$$

If r is small compared to R then

$$u = \frac{r^2}{2R} \qquad (3.25)$$

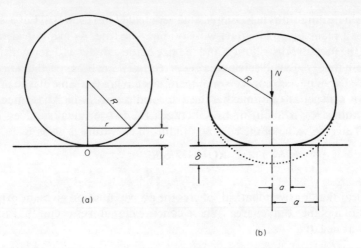

Figure 3.17   The elastic contact between a sphere and a plane

The normal approach is defined as the distance which points on the two bodies remote from the deformation zone move together on application of a normal load. This arises from the flattening and general displacement of the surface within the deformation region. If a is the radius of the contact zone and w is the displacement of the sphere at the boundary of this zone then the normal approach δ will be given by

$$\delta = u + w = \frac{a^2}{2R} + w \qquad (3.26)$$

Obviously, at the centre of the contact zone, δ is given by the degree of deformation and it is therefore reasonable to assume that the normal approach will be proportional to the flattening of the sphere. In other words

$$\delta \propto \frac{a^2}{R}$$

From the results given by equation 3.20 we have

$$a \propto \left(\frac{NR}{E'}\right)^{1/3}$$

so that

$$\delta \propto \left(\frac{N^2}{E'^2 R}\right)^{1/3}$$

The exact results show that

$$\delta = \left(\frac{9N^2}{16E'^2 R}\right)^{1/3}$$

or

$$N = \tfrac{4}{3}E'R^{1/2}\delta^{3/2} \qquad (3.27)$$

60

Combining equations 3.20 and 3.27, the area of contact, A, will be given by

$$A = \pi a^2 = \pi R \delta \qquad (3.28)$$

It will be noted that the relation given by equation 3.28 indicates that the surface outside the contact region is displaced in such a way that the actual area of contact is only one half of the geometrical area, which is clearly equal to $\pi \alpha^2 = 2\pi R \delta$.

3.5 THE CONTACT OF ROUGH SURFACES

Although in general all surfaces have roughness, we shall find some simplification if we consider the contact of a single rough surface with a perfectly smooth surface. The results from such an argument are then reasonably indicative of the effects to be expected from real surfaces. Moreover, the problem will be simplified further by introducing a theoretical model for the rough surface in which the asperities are considered as spherical segments so that their elastic deformation characteristics may be defined by the hertzian theory. We shall also assume that there is no interaction between separate asperities, that is, the displacement due to a load on one asperity does not affect the heights of the neighbouring asperities.

Consider a surface of unit nominal area to consist of an array of identical spherical asperities all of the same height z with respect to some reference plane XX', figure 3.18. As the smooth surface approaches due to the application of load we see that the normal approach will be given by $(z - d)$, where

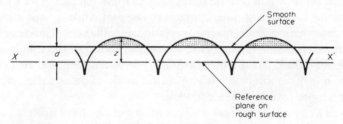

Figure 3.18   Contact between a smooth plane and a idealised rough surface

d is the current separation between the smooth surface and the reference plane. Clearly, each asperity is deformed equally and carries the same load N_i so that for η asperities per unit area the total load N will be·equal to ηN_i. For each asperity, the load N_i and the area of contact A_i are known from the hertzian theory, see equations 3.27 and 3.28. Thus if β is the asperity radius we have

$$N_i = \tfrac{4}{3} E' \beta^{1/2} (z - d)^{3/2}$$

and

$$A_i = \pi \beta (z - d)$$

61

and the total load will be given by

$$N = \tfrac{4}{3}\eta E'\beta^{1/2}\left(\frac{A_i}{\pi\beta}\right)^{3/2}$$

that is, the load is related to the total real area of contact $A = \eta A_i$ by

$$N = \frac{4E'}{3\pi^{3/2}\eta^{1/2}\beta}\,A^{3/2} \tag{3.29}$$

This result indicates that the real area of contact is related to the two-thirds power of the load, when the deformation is elastic.

If the loads are such that the asperities are deformed plastically under a constant flow pressure H, which is closely related to the hardness, we assume that the displaced material moves vertically down and does not spread horizontally so that the area of contact A' will be equal to the geometrical area $2\pi\beta\delta$. The individual load N'_i will then be given by

$$N'_i = HA'_i = 2H\pi\beta(z - d)$$

Thus

$$N' = \eta N'_i = \eta HA'_i = HA' = 2HA \tag{3.30}$$

that is, the real area of contact is linearly related to the load.

It must be pointed out at this stage that the contact of rough surfaces should be expected to give a linear relationship between real area of contact and load, a result which is basic to the laws of friction (chapter 4). From our simple model of rough surface contact we see that while a plastic mode of asperity deformation gives this linear relationship, the elastic mode does not. This is due to our simple and hence unrealistic model of the rough surface. Later we shall find that the proportionality between load and real contact area can in fact be obtained with an elastic mode of deformation when we consider a more realistic surface model.

It is well known that on real surfaces the asperities have different heights indicated by a probability distribution of their peak heights. We must therefore modify our previous surface model accordingly and the analysis of its contact must now include a probability statement as to the number of asperities in contact[3]. If the separation between the smooth surface and reference plane is d, then there will be contact at any asperity whose height was originally greater than d, figure 3.19. If $\phi(z)$ is the probability density of the asperity peak height distribution then the probability that a particular asperity has a height between z and $z + \mathrm{d}z$ above the reference plane will be $\phi(z)\,\mathrm{d}z$. Thus, the probability of contact for any asperity of height z is

$$\mathrm{prob}(z > d) = \int_d^\infty \phi(z)\,\mathrm{d}z$$

62

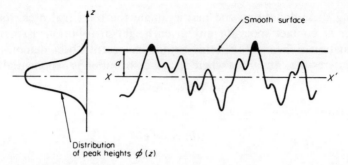

Figure 3.19   Contact between a smooth plane and a rough surface

If we consider a unit nominal area of the surfaces containing η asperities, the number of contacts n will be given by

$$n = \eta \int_d^\infty \phi(z) \, dz \qquad (3.31)$$

Since the normal approach is $(z - d)$ for any asperity and N_i and A_i are known from equations 3.27 and 3.28, the total area of contact and the expected load will be given by

$$A = \pi \eta \beta \int_d^\infty (z - d)\phi(z) \, dz \qquad (3.32)$$

and

$$N = \tfrac{4}{3}\eta \beta^{1/2} E' \int_d^\infty (z - d)^{3/2}\phi(z) \, dz \qquad (3.33)$$

It is convenient and usual to express these equations in terms of standardised variables by putting $h = d/\sigma$ and $s = z/\sigma$, σ being the standard deviation of the peak height distribution of the surface. Thus

$$n = \eta F_0(h)$$
$$A = \pi \eta \beta \sigma F_1(h)$$
$$N = \tfrac{4}{3}\eta \beta^{1/2} \sigma^{3/2} E' F_{3/2}(h)$$

where

$$F_m(h) = \int_h^\infty (s - h)^m \phi^*(s) \, ds$$

$\phi^*(s)$ being the probability density standardised by scaling it to give a unit standard deviation.

63

Using these equations one may evaluate the total real area, load and number of contact spots for any given height distribution. Experimental confirmation of the validity of this method has recently been demonstrated[4].

An interesting case arises where such a distribution is exponential, that is

$$\phi^*(s) = e^{-s}$$

In this case we have

$$F_m(h) = m!\, e^{-h}$$

so that

$$n = \eta e^{-h}$$
$$A = \pi\eta\beta\sigma e^{-h}$$
$$N = \pi^{1/2}\eta\beta^{1/2}\sigma^{3/2}E' e^{-h}$$

These equations give

$$N = C_1 A \quad \text{and} \quad N = C_2 n$$

where C_1 and C_2 are constants of the system. We find, therefore, that even though the asperities are deforming elastically, there is exact linearity between the load and the real area of contact.

For other distributions of asperity heights such a simple relationship will not apply, but for distributions approaching an exponential shape it will be substantially true. For many practical surfaces the distribution of asperity peak heights is near to a gaussian shape (see chapter 2) and this itself is near enough to exponential at the outermost decile of the distribution for the above result to be valid.

Where the asperities obey a plastic deformation law, equations 3.32 and 3.33 are modified to become

$$A' = 2\pi\eta\beta \int_d^\infty (z - d)\phi(z)\,\mathrm{d}z \tag{3.34}$$

$$N' = 2\pi\eta\beta H \int_d^\infty (z - d)\phi(z)\,\mathrm{d}z \tag{3.35}$$

We see immediately that the load is linearly related to the real area of contact by $N' = HA'$ and this result is totally independent of the height distribution $\phi(z)$, see equation 3.30.

We have so far based our analysis of surface contact on a theoretical model of the rough surface. An alternative approach to the problem is to apply the concept of profilometry using the surface bearing-area curve discussed in chapter 2. In the absence of asperity interaction, the bearing-area curve provides a direct method for determining the area of contact at any given normal approach. Thus, if the bearing-area curve or the all-ordinate distribution curve is denoted by $\psi(z)$ and the current separation between the smooth surface and the reference plane is d, then for a unit nominal surface area the real area of contact will be given by

$$A = \int_d^\infty \psi(z)\,dz \qquad (3.36)$$

so that for an ideal plastic deformation of the surface, the total load will be given by

$$N = H \int_d^\infty \psi(z)\,dz \qquad (3.37)$$

We may summarise the foregoing by saying that the relationship between the real area of contact and the load will be dependent on both the mode of deformation and the distribution of the surface profile. When the asperities deform plastically, the load is linearly related to the real area of contact for any distribution of asperity heights. When the asperities deform elastically, the linearity between load and real area of contact occurs only where the distribution approaches an exponential form and this is very often true for many practical surfaces. These results will be found to be of considerable significance when considering such effects as friction and wear.

3.6 CRITERION OF DEFORMATION MODE

It will of course be recognised that in most practical situations the higher asperities could be plastically deformed while the lower contacting asperities are still elastic. Thus we obtain a mixed plastic–elastic system in most real contacts, where the greater the load and hence the normal approach, the greater the number of plastic contacts. We therefore see that the normal approach will in a sense be an indicator of the degree of plasticity which exists. Referring to equations 3.27 and 3.28 we find that for an elastic asperity contact the mean pressure p_m is given by

$$p_m = \frac{4E'\delta^{1/2}}{3\pi\beta^{1/2}}$$

or

$$\delta^{1/2} = \frac{3\pi\beta^{1/2}p_m}{4E'} \qquad (3.38)$$

For a spherical contact we know that the transmission from purely elastic to completely plastic behaviour occurs over a range of loading. Plasticity is initially sub-surface when the maximum contact pressure is $3.1k$ or a mean pressure of about Y, and becomes macroscopic when the mean pressure is

65

about $3Y$, that is, the hardness value of the material (see section 3.6). Thus in equation 3.38 we see that the transition from elastic to fully plastic behaviour occurs in a range of values of $\delta^{1/2}$, and the initial deviation from elastic behaviour occurs when $p_m = H/3$ where

$$\delta^{1/2} = 0.78\left(\frac{\beta^{1/2}H}{E'}\right)$$

Recognising that the elastic–fully plastic transition is not instantaneous we shall assume a transition point where

$$\delta^{1/2} \simeq \frac{H}{E'}(\beta)^{1/2}$$

It is convenient to standardise this by dividing both sides by $\sigma^{1/2}$ so that

$$\delta*^{1/2} = \left(\frac{\delta}{\sigma}\right)^{1/2} = \frac{H}{E'}\left(\frac{\beta}{\sigma}\right)^{1/2}$$

This parameter decreases as the surface roughness as specified by σ increases, and it is usual to define a function Ω, the *plasticity index*, as the inverse of $\delta*^{1/2}$ so that

$$\Omega = \frac{E'}{H}\left(\frac{\sigma}{\beta}\right)^{1/2} \qquad (3.39)$$

This index is clearly indicative of the onset of significant plastic deformation, being large when the contact is basically plastic and small, say less than unity, when the contact is essentially elastic. The plasticity index is clearly a useful parameter since it is a non-dimensional grouping relating both the *current* physical and geometrical properties of the surface. In a wear-process study there are obviously advantages in making measurements of Ω at discrete time intervals, since the trend of such a parameter will then indicate whether the surfaces are approaching an elastic state, that is, a run-in condition, see figure 3.20.

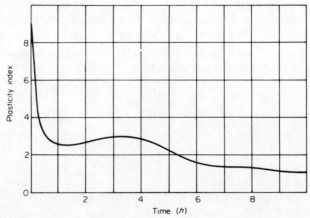

Figure 3.20   The variation of the plasticity index during a wear process

66

3.7 THERMAL EFFECTS

In a sliding situation most of the work done against friction will be manifested as a heat release with a consequent rise in temperature. Clearly, since the heat release is basically a continuous process, temperature gradients will develop in the contacting bodies with the highest temperature occurring at the point of heat release (heat source), that is, the contact surface. The points at which heat is released will obviously depend on the overall geometry of the contacting bodies. If we consider, for instance, our previous surface contact model of a smooth plane and a rough flat surface then heat will be released at each contact spot whose size is determined either elastically or plastically. Thus each contact spot may be treated as an independent heat source and the individual temperature analysis may then be applied to obtain the temperature distribution for the general contact region. The same argument is obviously true for the contact of curved surfaces, however, in this case the contact spots will be so close together that at high normal loads the whole contact zone may be regarded as a single heat source. This argument is justified by the fact that in an analogous electrical example the results show that the collective resistance of a large number of closely grouped constrictions is nearly the same as the resistance of one equivalent large constriction. It will also be recognised that the temperature effects will depend on whether a body is stationary or moving with respect to the heat source. Thus if we consider the simple example of two discs in contact we shall obtain a physical feel for the problem. In such a consideration we must realise that the thermal effects happen to discrete particles of material and therefore by contemplating their contact history we can anticipate their overall behaviour which is clearly related, at any given time, to the integrated effect of their history.

Consider the situation shown in figure 3.21a where two contacting discs are assumed to be rolling with very small relative slip. It is clear that all

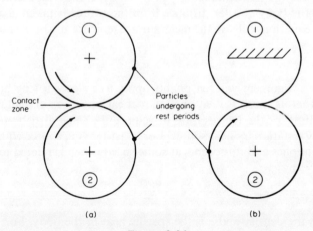

Figure 3.21

67

particles on the surfaces of both discs pass through the contact zone where heat is generated and afterwards undergo considerable periods of rest. Their temperature rise will therefore be relatively modest, partly because the frictional work is small, owing to the small relative sliding, and partly because the generated heat will be readily dissipated by heat losses during the rest periods. If one disc is fixed as in figure 3.21b, we will clearly have a state of pure sliding. In this case, surface particles on disc 2 will be subjected to relatively high temperatures, sometimes called flash temperatures, when passing through the contact zone and thereafter have a considerable rest period where cooling takes place. Surface particles on disc 1, on the other hand, never escape from the contact zone and are therefore subjected to a continuous build-up of temperature towards a steady state determined by the thermal properties of the whole system. In this example the contact zone, which is clearly fixed in space, can be regarded as a stationary heat source with respect to disc 1, and a moving heat source with respect to disc 2, and we shall treat the problem simply by considering these two distinct effects on the two bodies.

3.7.1 Stationary Heat Source

Consider a semi-infinite body subjected to a stationary heat source acting over a small circular surface area of radius a. Clearly, in this case heat will be supplied to the body through a fixed area and steady state conditions will exist. This problem is analogous to its electrical counterpart and the heat flow is considered as a flow of thermal current through a thermal resistance. If Q is the rate of heat supply then it can be easily shown that the mean temperature rise, $\Delta\theta$, at the surface is given by

$$\Delta\theta = \frac{Q}{4a\alpha}$$

where α is the thermal conductivity of the body. If the temperature of other points on the body at a far distance from the source is taken as zero, then the above equation will give the mean surface temperature of the body, that is

$$\theta_s = \frac{Q}{4a\alpha} \tag{3.40}$$

The same argument will reasonably apply to a slow-moving heat source provided that its speed V is so small that at each position of contact there is sufficient time for the temperature to acquire the same distribution as that induced by a stationary heat source. For a fast-moving source equation 3.40 no longer applies and this occurs at some speed which is related to a certain parameter ξ by[5]

$$\xi = \frac{a\rho c}{2\alpha} V \tag{3.41}$$

where ρ is the density and c is the specific heat of the body. For $\xi > 5$ the speed is considered high and we must apply moving heat source analysis.

3.7.2 Moving Heat Source

For a moving heat source traversing the surface of a semi-infinite body at a relatively high speed V, we can neglect the effects of the transverse flow of heat and the problem can be regarded as one of a linear heat flow. In this case, if the heat is supplied at a constant rate of q per unit area, then the mean temperature rise of a point on the surface of the body is given by[5]

$$\Delta\theta = \frac{2qt^{1/2}}{(\pi\alpha\rho c)^{1/2}}$$

where t is the time during which heat has been supplied. If heat is supplied through a circular area of radius a, then by putting $q = Q/\pi a^2$ and by considering the effective value of t for all points within this area we can obtain the mean surface temperature. The traverse time for any point (x, y) is given by

$$t = \frac{2x}{V} = \frac{2}{V}(a^2 - y^2)^{1/2}$$

and therefore the mean effective time is

$$t = \frac{1}{2a}\int_0^a \frac{2}{V}(a^2 - y^2)^{1/2}\,\mathrm{d}y = \frac{a\pi}{4V}$$

Therefore

$$\theta_\mathrm{m} = \frac{2Qa^{1/2}\pi^{1/2}}{2\pi a^2\pi^{1/2}(\alpha\rho c V)^{1/2}} = \frac{0.318Q}{a(a\alpha\rho c V)^{1/2}} \tag{3.42}$$

We may summarise the foregoing results by introducing a standardised surface temperature, θ^*, defined by

$$\theta^* = \frac{\rho c V}{\pi q}\theta$$

so that for a stationary heat source, using equation 3.40, we have

$$\theta_\mathrm{s}^* = \frac{a\rho c V}{4\alpha} = 0.5\xi \tag{3.43}$$

and for a moving heat source, using equation 3.42, we have

$$\theta_\mathrm{m}^* = \frac{0.318 \times (2a\rho c V)^{1/2}}{(2\alpha)^{1/2}} = 0.438\xi^{1/2} \tag{3.44}$$

where ξ is as defined by equation 3.41. Examining equations 3.43 and 3.44 we find that for small values of ξ or low speeds both stationary and moving heat sources produce similar temperatures on the surface of a semi-infinite body, while for larger values of ξ a stationary heat source produces considerably higher temperatures.

3.7.3 Application to Sliding Bodies

Let us consider the contact model shown in figure 3.22 where a single spherical asperity of body 1 makes contact, under a normal load P, with the surface of body 2 which is assumed to be sliding with a constant speed V. The amount of heat generated over the contact area A will be given by

$$Q = \frac{\mu P V}{J} \tag{3.45}$$

where μ is the coefficient of friction and J is the mechanical equivalent of heat. During sliding, the contact surface of body 1 will be continuously subjected to part of Q, say λQ, while the remainder, $(1 - \lambda)Q$, will be flowing

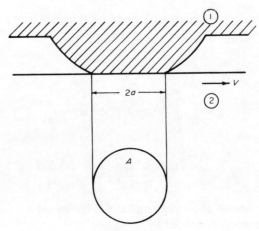

Figure 3.22

into body 2. For such a contact both bodies can be reasonably considered as semi-infinite so that their surface temperatures may be obtained by applying the results derived for stationary and moving heat sources. Thus, using equations 3.40 and 3.42 the surface temperature of bodies 1 and 2 will be given by

$$\theta_1 = \frac{\lambda \mu P V}{4a\alpha J} \tag{3.46}$$

and

$$\theta_2 = \frac{0.318(1 - \lambda)}{aJ} \left(\frac{V}{a\alpha\rho c} \right)^{1/2} \tag{3.47}$$

Now, it remains to define the value of λ. Clearly, this will depend on the heat transfer characteristics of the contacting bodies, but for simplicity we can

70

assume that λ is defined by the ratio of the thermal diffusivities of the two bodies, that is

$$\frac{\lambda}{1-\lambda} = \frac{\alpha_1/\rho_1 c_1}{\alpha_2/\rho_2 c_2}$$

Therefore, we may easily use equations 3.46 and 3.47 to determine the surface temperatures for the two bodies in contact.

REFERENCES

1. S. Timoshenko and J. N. Goodier. *Theory of Elasticity*. McGraw-Hill, New York, (1951).
2. D. Tabor. *The Hardness of Metals*. Oxford University Press, (1951).
3. J. A. Greenwood and J. B. P. Williamson. Contact of nominally flat surfaces. *Proc. R. Soc.*, **295**, A, (1966), 300.
4. K. A. Nuri and J. Halling. The normal approach between rough flat surfaces in contact. Conference on Mechanics of Contact Effects, Kiev, (June, 1973).
5. J. F. Archard. The temperature of rubbing surfaces. *Wear*, **2**, (1958–9), 438.

4

Friction Theories

4.1 INTRODUCTION

Friction is the resistance to motion which is experienced whenever one solid body slides over another. The resistive force, which is parallel to the direction of motion is called the 'friction force'. If the solid bodies are loaded together and a tangential force is applied, then the value of the tangential force which is required to initiate sliding is the 'static friction force'. The tangential force required to maintain sliding is the 'kinetic (or dynamic) friction force'. Kinetic friction is generally lower than static friction.

4.1.1 The Laws of Friction

It has been found experimentally that there are two basic 'laws' of friction, which are obeyed over a wide range of conditions. There are, however, a number of notable exceptions, examples of which will be included later. It should be stressed at this point that the two laws of friction are empirical in nature and, of course, no basic physical principles are violated in those cases where the laws of friction are not obeyed.

The first law states that the friction is independent of the apparent area of contact between the contacting bodies, and the second, that the friction force is proportional to the normal load between the bodies. Thus a brick can be slid as easily on its side as on its end and if the load between two sliding bodies is doubled then the friction force is doubled. These laws are often referred to as 'Amontons laws' after the French engineer Amontons who presented them in 1699[1]. Coulomb (1785) introduced a third law, that the kinetic friction is nearly independent of the speed of sliding, but this law has a smaller range of applicability than the first two.

4.1.2 Coefficient of Friction

The second law of friction enables us to define a coefficient of friction. The law states that the friction force F is proportional to the normal load W. That is

$$F \propto W$$

therefore

$$F = \mu W \tag{4.1}$$

where μ is a constant known as the 'coefficient of friction'. It must be stressed that μ is a constant only for a given pair of sliding materials under a given set of ambient conditions and varies for different materials and conditions. For example, a hard steel surface rubbing against a similar surface under normal atmospheric conditions would typically have a value of μ equal to about 0.6. The same combination rubbing under very high vacuum conditions would have a much higher value of μ. A graphite-on-graphite combination in normal atmosphere has a value of μ equal to about 0.1 but this rises to over 0.5 if the atmosphere is very dry.

It is the object of this chapter to develop theories of friction which can explain the variations in friction coefficient between different materials and under different conditions. Now let us examine the first law a little more closely and attempt to explain why the friction is independent of the apparent area of contact.

In chapter 2 it has been shown that nearly all surfaces are rough on a micro-scopic scale and true contact is obtained over a small fraction of the apparent contact area. Furthermore the real area of contact is independent of the apparent area of contact. Thus the first law of friction is explained since friction is related to the real area of contact.

4.1.3 Surface Roughness and Real Area of Contact

When two bodies are rubbed together, some form of interaction takes place at the contacting surfaces resulting in a resistance to relative motion. Most friction theories assume that the resistive force per unit area of contact is a constant. Thus

$$F = As$$

where F is the friction force
$\qquad A$ is the real area of contact
$\qquad s$ is the constant force per unit area, resisting relative motion, that is, the specific friction force

73

If the assumption that s is a constant can be justified, then we can see the importance of A. Let us now summarise some of the more important findings of chapter 3 with respect to the real area of contact.

(a) For a single spherical contact or an array of similar spheres all at the same height under loading conditions which produce elastic deformation only

$$A \propto W^{2/3}$$

(b) When the contact is wholly plastic

$$A \propto W$$

(c) For a surface whose asperity height distribution can be represented by an exponential function

$$A \propto W$$

whatever the mode of deformation, that is, either elastic or plastic.
(d) For a surface whose asperity height distribution is gaussian, we can again write with sufficient accuracy

$$A \propto W \text{ for all modes of deformation}$$

(e) Many practical surfaces have an asperity height distribution that is close to gaussian.

Results (a) and (b) above, combined with experimental findings confirming the second law of friction, led in the past to the conclusion that asperity contacts must be plastic. However, various physical arguments and experimental results indicated that under certain conditions, for example, well run-in surfaces, where it was found $F \propto W$, elastic conditions must exist. This apparent anomaly is now explained by (d) and (e) above. The analysis of the deformation of model and real surfaces has now made possible the formulation of reasonably realistic friction theories, and it is certain that any future improvements on current theories must be based on such analysis.

4.2 FRICTION MEASUREMENT

It is not the object of this chapter to present a comprehensive survey of friction measurement but a brief account of some of the available methods will now be given.

Any apparatus for measuring friction must be capable of supplying relative motion between two specimens, of applying a measurable normal load and of measuring the tangential resistance to motion. There are a large number of methods available and the final choice will depend largely on the exact conditions of rubbing contact under investigation. For example, probably the simplest arrangement is the tilting plane where a specimen is placed on a flat surface which is gradually tilted until sliding starts, figure 4.1. The coefficient of friction is then tan θ. This method is obviously unsuitable in those cases where a study of the variation of friction with continued rubbing is required but its simplicity makes it attractive in many cases.

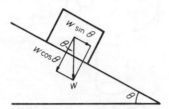

Figure 4.1 Measurement of friction using tilting plane, $\mu = \tan \theta$

Where continuous friction measurement is required over a period of time then an alternative approach must be used. Here one specimen, usually a disc or a cylinder, is driven continuously, while a second specimen, nominally stationary, is loaded against it. Commonly used combinations are crossed cylinders, pin-on-cylinder or -disc, and disc-on-disc. The loading of the stationary specimen can be by simple deadweight or, if the experimental conditions demand it, by some more complicated method such as hydrostatic or magnetic loading. The measurement of the friction force is usually accomplished by mounting the nominally stationary specimen so that a very small tangential movement, proportional to the frictional force, occurs. This small movement is measured and recorded. Two typical arrangements, a crossed cylinder and a pin-on-disc are illustrated in figures 4.2 and 4.3. In each case the specimen is mounted on leaf springs which allow a small movement in the direction of the friction force. The movement can be

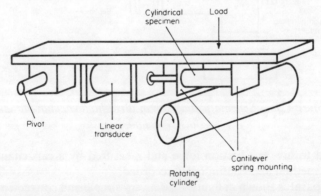

Figure 4.2 Simple crossed cylinder arrangement for the measurement of friction and wear

75

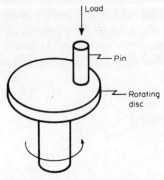

Figure 4.3a   Arrangement
of pin-and-disc machine

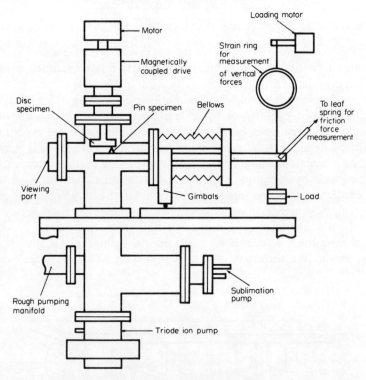

Figure 4.3b   Schematic diagram of ultra-high vacuum pin-and-
disc machine

calibrated to give the friction force and measured by a capacitance or in-
ductance method and continuously recorded.

The apparatus shown in figure 4.2 is a very simple and convenient method
of measuring friction. That shown in figure 4.3b has been used for ultra high
vacuum and controlled atmosphere friction tests. The motion is provided

by a magnetic drive through the walls of the chamber and the load applied and friction are measured outside the chamber wall, each force being transmitted through a bellows.

Just as is the case in wear testing, great care must be taken in ensuring cleanliness during these tests, since small amounts of contamination can significantly affect the measured friction, and this is why so much emphasis is placed on controlled atmosphere tests.

4.3 POSSIBLE CAUSES OF FRICTION

In section 4.1.3, we stated that friction must be due to some interaction between the opposing surfaces and that this results in resistance to relative motion. As the surfaces move relative to one another, work is done by the forces causing the motion, that is, there is an energy loss at the contacting surfaces. In considering the possible causes of friction it is convenient to consider separately the surface interaction and the mechanism of the energy loss.

4.3.1 Surface Interactions

When two surfaces are loaded together they can adhere over some part of the contact and this adhesion is therefore one form of surface interaction causing friction.

If no adhesion takes place then the only alternative interaction which results in a resistance to motion is one in which material must be deformed and displaced to accommodate the relative motion. We need consider only two interactions of this type. The first is asperity interlocking. Considering the situation illustrated in figure 4.4, it is obvious that relative motion cannot

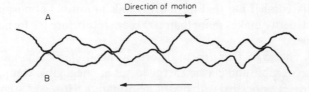

Figure 4.4   Asperity interlocking—motion cannot take place without deformation of the asperities

take place between surfaces A and B without displacement of the material of the asperities.

A second example of the displacement type of interaction is illustrated in figure 4.5. Here a hard sphere A is loaded against a relatively soft flat surface, B. In order for relative motion to take place some of the material B must be displaced. Although the surfaces of both sphere A and flat B will be rough on

Figure 4.5   Macro-displacement—a hard sphere A, loaded against a softer surface B, causes displacement of material B during motion

a microscopic scale the material displacement at the individual asperities will, in this case, be small compared to the 'macro-displacement'. Thus we have only two types of interaction, *Adhesion* and *Material Displacement*, although we will find it convenient to think of the material displacement as either, *Asperity Interlocking* or *Macro-Displacement*.

4.3.2   Types of Energy Loss

There are only three mechanisms which can cause appreciable loss of energy at the interacting surfaces. As relative motion takes place, material must be deformed. The deformation can be either elastic or plastic; additionally the material may be fractured. Plastic deformation will always be accompanied by a loss in energy and it is this energy loss which accounts for the major part of the friction of metals under most practical circumstances. Fracture must occur when the surface interactions are adhesive and can also take place due to relative motion of interlocking asperities. The formation of wear debris is of course evidence that fracture has taken place. However, the energy losses associated with fracture will, in most cases of sliding metals, be small in comparison with those due to plastic deformation. One reason for this is given in chapter 6 where it is shown that a wear particle is not formed at each asperity contact but that for most metals in normal atmospheric conditions, an asperity makes more than 1000 contacts before the formation of a wear particle.

Although energy is required to deform a metal elastically, most of this energy is recoverable and elastic energy losses are negligible compared with the energy losses associated with plastic deformation. However, some rubbers exhibit large irreversible energy losses due to their elastic deformation (elastic hysteresis) and in certain cases this is the major source of friction, (see section 4.8).

Summarising the above considerations shows that there are two sources of surface interaction, that is, adhesion and material displacement, and these can cause energy losses due to both elastic and plastic deformation and to fracture.

We will now consider various proposed mechanisms of friction in the light of the above considerations.

4.4 THE ADHESION THEORY OF FRICTION

4.4.1 Simple Theory

This theory due to Bowden and Tabor[2], uses as a starting point the fact that when metal surfaces are loaded against each other they make contact only at the tips of the asperities. Because the real contact area is small the pressure over the contacting asperities is assumed high enough to cause them to deform plastically. This plastic flow of the contacts causes an increase in the area of contact until the real area of contact is just sufficient to support the load, figure 4.6. Under these conditions, for an ideal elastic–plastic material

$$Ap_o = W$$

where A is the real area of contact, p_o is the yield pressure of the metal W is the normal load.

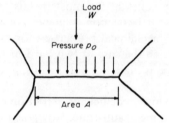

Figure 4.6   A single asperity contact—the asperities yield plastically until the area of contact has grown sufficiently to support the load

Over the regions of intimate metal-to-metal contact, Bowden and Tabor state that strong adhesion takes place, and that the junctions 'cold weld'. If s is the force per unit area of contact necessary to shear the junctions, that is, s is the shear stress necessary to cause plastic flow and final fracture and F is the friction force, then

$$F = As + p_e$$

where p_e is a term introduced by Bowden and Tabor to take account of the force required to 'plough' hard asperities through a softer surface, that is, a material displacement interaction causing plastic deformation. Bowden and Tabor state that for most situations involving unlubricated metals p_e is small compared with As and may be neglected. (The ploughing term is discussed in section 4.7.) Ignoring the ploughing term we can write

$$F = As = \frac{Ws}{p_o}$$

and

$$\mu = \frac{F}{W} = \frac{s}{p_o}$$

Thus this simple theory provides an explanation of the two laws of friction, that is, that the friction is independent of the apparent area of contact and the friction force is proportional to the load.

In the above analysis we have considered an ideal elastic–plastic material, and have ignored the effects of work-hardening. Therefore it is reasonable to take s equal to s_o the critical shear stress, and both p_o and s_o must refer to the softer of the two metals.

Now

$$\mu = \frac{s_o}{p_o} \tag{4.2}$$

This ratio s_o/p_o is fairly constant for most metals and the above analysis gives an indication of why the friction coefficient of a large range of metals varies little, while their mechanical properties, for example, hardness, vary by orders of magnitude. In the case of two hard metals in rubbing contact p_o is high, A is low and s_o is high. For soft metals, p_o is low, as is s_o, but A is large.

One way of obtaining lower coefficients of friction is to deposit a thin layer of soft metal onto a hard metal substrate. Now, the load carrying capacity is really due to the substrate and p_o is the yield pressure for the substrate. However, shearing takes place in the soft metal layer and the critical shear stress of the soft metal is the value we must use, and therefore

$$\mu \simeq \frac{s_o \text{ (soft)}}{p_o \text{ (hard)}} = \text{low}$$

4.4.2 Discussion of Simple Theory

There is no doubt that junction welding can take place during the rubbing of metals. For cleaned metal surfaces in high vacuum very high adhesion and friction coefficients have been recorded. For metals rubbed in normal atmospheric conditions, adhesion and transfer of metal fragments have been demonstrated using radioactive tracer techniques. However, the simple adhesion theory has been criticised for a number of reasons and it can be shown to be inadequate by a comparison of the absolute values of the friction coefficient predicted by the simple theory and those found experimentally. For most metals, s_o is about one-fifth of p_o and therefore the simple adhesion theory predicts that $\mu \simeq 0.2$.

Many metal combinations in air give a friction coefficient higher than 0.5, and metals in high vacuum give much higher values of μ. This led Bowden and Tabor to re-examine some of the assumptions in the simple theory, and to present a modified and more realistic description of friction in terms of adhesion[2].

4.5 MODIFIED ADHESION THEORY

The fact that very high values of friction are obtained for metals under high vacuum conditions, where adhesion is unimpeded by oxide films or other contaminants, indicates that the real contact area must be considerably larger than is indicated by the simple theory. In the simple theory it was assumed that A was defined by the yield pressure of the softer metal p_o and the normal load W. This is approximately true for static contact but in the case of friction, where a tangential force is also applied, yielding must take place as a result of the combined normal and shear stresses. To illustrate this let us look at the simplified two-dimensional stress system illustrated in figure 4.7a and assume that yielding occurs when the maximum shear stress attains a critical value[3].

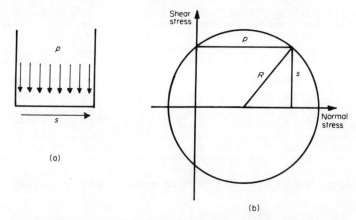

Figure 4.7   The Mohr's circle construction for finding the maximum shear stress for an idealised two-dimensional junction under normal and tangential stress

We can find the maximum shear stress in the system by using the Mohr's circle[4] construction, figure 4.7b. The maximum shear stress is the radius R of the circle therefore

$$\left(\frac{p}{2}\right)^2 + s^2 = R^2$$

When R reaches the critical resolved shear stress, yielding takes place. From this it is easy to see that yielding is dependent on the action of the combined stresses and not on p alone.

81

We will now examine how consideration of the combined stresses affects the value of the real area of contact in an asperity junction.

Consider first a single asperity contact under a normal load W. The area of contact will be A where $W/A = p_o$. If a tangential force is now gradually applied up to a value F, further plastic flow will take place. This flow causes an increase in the contact area, that is, junction growth is brought about by the superposition of the shear stress on the normal stress. The normal and shear stresses caused by the normal and shear forces must decrease as the area over which the forces act increases, and junction growth continues until the combined stresses obey a relationship of the form previously given for the two-dimensional system. The exact solution for the three-dimensional case is not known but we will assume that it is of the form

$$p^2 + \alpha s^2 = k^2$$

where α and k are constants yet to be determined, that is

$$\left(\frac{W}{A}\right)^2 + \alpha\left(\frac{F}{A}\right)^2 = k^2 \qquad (4.3)$$

where A is the area of contact of the junction. Now, if s is zero, the pressure over the junction must be p_o and therefore

$$k^2 = p_o^2$$

that is

$$p^2 + \alpha s^2 = p_o^2 \qquad (4.4)$$

If F increases to very large values then junction growth continues until W/A is small in comparison with F/A in which case we can write

$$\alpha s^2 \simeq p_o^2$$

In this case s must be approximately equal to s_o, the critical shear stress. Therefore

$$\alpha s_o^2 \simeq p_o^2$$

or

$$\alpha \simeq \frac{p_o^2}{s_o^2}$$

Now $p_o \simeq 5s_o$ therefore $\alpha \simeq 25$. However, experiment indicates that α should have a value somewhat lower than 25 and Bowden and Tabor assume that $\alpha = 9$. (This also implies that $p_o = 3s_o$.) We will see later that the exact value of α does not greatly affect the amount of junction growth taking place in many practically important cases.

From equations 4.3 and 4.4 we have

$$A^2 = \left(\frac{W}{p_o}\right)^2 + \alpha\left(\frac{F}{p_o}\right)^2$$

W/p_o is the area of contact derived from the simple theory in which only the effect of normal load is considered, and the additional term $\alpha(F/p_o)^2$ represents the increase caused by the shear or friction force.

It can be seen from the above discussion that for clean metal surfaces (that is, metals in high vacuum), large-scale junction growth is possible, resulting in very high friction coefficients. This has been confirmed experimentally[5]. In normal atmospheres metals are covered by an oxide or other contaminant film. We will now see how the above theory can be applied to such cases.

4.5.1 Adhesion Theory of Metals with Contaminant Films[2]

Consider an asperity junction under normal load W and a gradually increasing shear force F. We assume that at the junction there is a thin contaminating film with critical shear stress s_f. We also assume $s_f = cs_o$ where s_o is the critical shear stress for the metal and c is less than unity. While F and A, have values such that $F/A < s_f$, then junction growth will proceed as described previously for uncontaminated metals. However, when $F/A = s_f$ then the contaminating film will shear, junction growth will end, and gross sliding will occur.

Thus the condition for the start of gross sliding is

$$p^2 + \alpha s_f^2 = p_o^2$$

But it has already been shown that

$$p_o^2 = \alpha s_o^2$$

Therefore

$$p^2 + \alpha s_f^2 = \alpha s_o^2$$

or

$$p^2 + \alpha s_f^2 = \frac{\alpha}{c^2} s_f^2$$

Therefore

$$\frac{s_f}{p} = \frac{c}{[\alpha(1 - c^2)]^{1/2}}$$

The coefficient of friction $\mu = (F/W) = (s_f A/pA)$. Therefore

$$\mu = \frac{c}{[\alpha(1 - c^2)]^{1/2}} \tag{4.5}$$

As c tends to 1 then μ tends to infinity in agreement with results obtained for uncontaminated metals. We can plot μ against c for various values of α, figure 4.8, and we see that the value of μ drops rapidly as c reduces from unity.

83

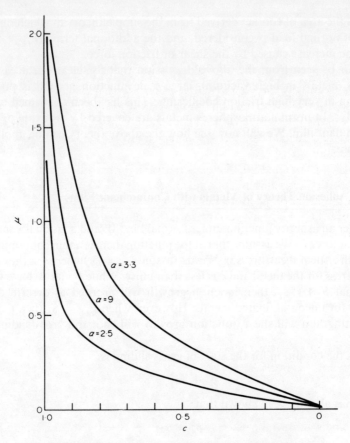

*Figure 4.8   The variation of μ with c for different values
of α. It can be seen that except at large values of c, the
exact value of α is not of major importance*

Thus a small amount of weakening at the interface produces a drastic reduction in μ.

When c is small then equation 4.5 can be written

$$\mu = \frac{c}{(\alpha)^{1/2}}$$

but

$$(\alpha)^{1/2} = \frac{p_o}{s_o}$$

Therefore

$$\mu = \frac{cs_o}{p_o} = \frac{s_f}{p_o}$$

84

that is

$$\mu = \frac{\text{critical shear stress of the interface}}{\text{yield pressure of the bulk metal}}$$

which is essentially the same result as obtained by the simple theory (compare discussion of soft metal films, or hard substrates). This is understandable since when the interface is weak in shear, appreciable junction growth does not occur and the real contact area depends only on the normal load and the yield pressure p_o.

This then is a more realistic theory. Although it is still based on a simple model and contains a number of assumptions, it is able to explain a remarkable range of friction phenomena.

However, it is no longer exclusively an adhesion theory. Let us examine what are the main points of the theory

(a) The real contact area is defined by plastic deformation.
(b) The two rubbing surfaces are separated by a film of shear strength which can vary from low values up to the bulk shear stress of the substrate material.
(c) The friction force is the force required to shear the separating film.

At one extreme, where metal-to-metal contact and junction welding take place, we can consider that either the separating film is of zero thickness, or that it is composed of the softer of the two metals. Under atmospheric conditions of dry sliding where the metals are covered by oxide films it is only at those points where the oxide film is broken that metal-to-metal contact and welding take place. In this case the effective shear strength of the interface will lie between the shear strength of the softer metal and that of the oxide film; its exact value depending on the relative amount of metal to metal and metal to oxide contact.

We will now discuss a number of criticisms of the theory that have been advanced.

(i) If two hard metals are rubbed together under atmospheric conditions, in general no adhesive component normal to the surfaces can be detected when the normal force has been released. Bowden and Tabor answer this by pointing out that the adhesion under rubbing conditions occurs while the normal force is exerted. To measure adhesion in the normal direction the normal load must first be released and elastic recovery will break many of the bonds during this process.
(ii) The adhesion theory of friction can explain the transfer of material from one rubbing surface to the other but offers no explanation of the formation of loose wear debris. Experimental work has shown that transfer occurs during single traversals but subsequent traversals on the same track produce loose wear particles. This suggests a possible change of mechanism during the rubbing process. It may be significant

that much of the experimental work carried out by Bowden and Tabor, giving results consistent with the adhesion theory, consisted of single rather than multiple pass experiments.

(iii) The adhesion theory is based on plastic deformation of asperities. On continued rubbing under conditions of low wear, work hardening will take place and a proportion of asperity contacts will now deform elastically. However, although the deformation of the asperities is now largely elastic, the area of contact is still related to the previous plastic deformation. This can be understood by considering an individual asperity contact under an initial normal load W. The deformation curve is as shown in figure 4.9. If the load is now released and reapplied up to a maximum value of W, the deformation is elastic but the total deformation or strain is exactly the same as in the plastic deformation case.

We will return to these criticisms in section 4.9 which contains a general discussion of friction theories.

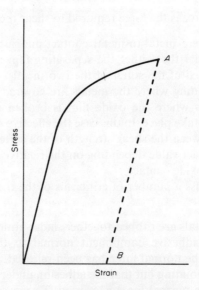

Figure 4.9 Typical stress–strain curve of metals. If the stress is relaxed at A and then reapplied the deformation is elastic along BA, but the resultant strain is equal to that caused by the initial plastic deformation

86

4.6 PLASTIC INTERACTION OF SURFACE ASPERITIES

In the Bowden and Tabor theory, the normal and yield stress on a single asperity were assumed to be representative of the stresses of all asperities.

No consideration was given to the possibility that at a single asperity, s and p could vary with time over large ranges so that s/p could also vary. This evolves as a feature of a theory based on the plastic interaction of asperities. This type of theory was first introduced by Green[6] and has been extended by Edwards and Halling[7] whose treatment is followed here. The basis of this analysis is that in the sliding of macroscopically flat surfaces the motion is parallel to the surface.

Consider two wedge-shaped asperities of equal angle interacting as shown in figure 4.10. Assuming that motion is in the direction shown and that

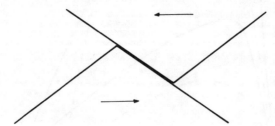

Figure 4.10   The idealised wedge-shaped as-perities studied in the 'plastic interaction theory'

deformation is plastic it is possible to calculate the instantaneous shearing force F_1 and normal force W_1 over the complete horizontal displacement l from the first contact until the asperities separate, both for the case where the asperities are in intimate contact and adhesion is possible and for the case where the asperities are separated by a film of shear strength cs_0 where c and s_0 have the same meanings as in the Bowden and Tabor theory. The development of the expressions for F_1 and W_1 is beyond the scope of this book but the form of the plot of normal and shear forces against displacement is shown in figure 4.11 for junction angles of $10°$ and for various values of c. The coefficient of friction is the sum of instantaneous shearing forces F_1 for all contacting asperities divided by the sum of instantaneous normal forces W_1 for all contacting asperities. Alternatively μ could be regarded as the mean F/W value for a single pair of interacting asperities, which are assumed to be typical of all the interacting asperities. Thus the coefficient of friction can be calculated from figure 4.11 and plots of μ against c, so obtained are shown in figure 4.12.

It is interesting to note that if the results are extrapolated the curve corresponding to junction angle θ equal to zero, corresponds identically with the

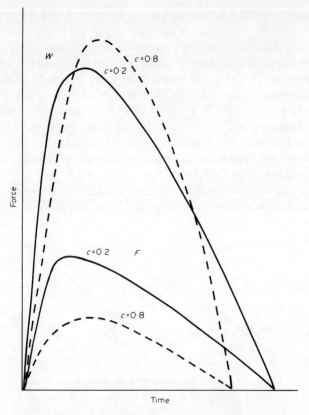

Figure 4.11 The variation of the normal force P and the friction force F throughout the junction life for a junction of angle 10°

Bowden and Tabor relationship

$$\mu = \frac{c}{\alpha(1 - c^2)^{1/2}}$$

The general relationship given by the Edwards and Halling theory is

$$\mu = \left[\frac{c}{\alpha(1 - c^2)^{1/2}} + \Phi\right] \Big/ \left[1 - \frac{c}{\alpha(1 - c^2)^{1/2}}\Phi\right] \qquad (4.6)$$

where Φ is a function of c and the geometry of the junctions and equal to zero when θ equals zero.

This then indicates that the Bowden and Tabor theory is a special case of the more general Edwards and Halling theory. The Edwards and Halling theory assumes plastic deformation and an array of similarly shaped asper-

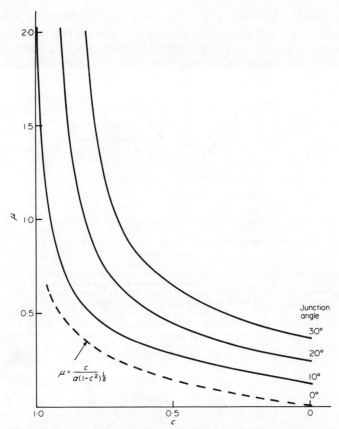

*Figure 4.12   The variation of μ with c for various junction
angles*

ities at the same height. However, using the methods developed in chapter 3 for normal load effects on rough surfaces, it is clearly possible to extend this type of theory to cover surfaces consisting of an array of asperities with a given height distribution which are interacting elastically and plastically depending on their height. The effects due to work hardening and dissimilar materials could also be included.

4.7 PLOUGHING EFFECT

Ploughing is that part of friction, included by Bowden and Tabor in their total friction force, caused by asperities on a hard metal penetrating into a softer metal and 'ploughing' out a groove by plastic flow in the softer metal. This is a major component of friction during abrasion processes and also it is probably important in cases where the adhesion term is small, for example, for

89

well-lubricated surfaces where the shear strength of the interfacial film is low.

Consider a hard material whose surface is composed of a large number of similar conical asperities of semi-angle θ in contact with a softer material whose surface is comparatively flat, figure 4.13. During rubbing only the

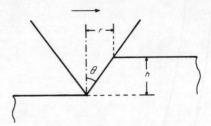

Figure 4.13 *The ploughing of a soft surface by a hard conical asperity*

front surface of each conical asperity is in contact with the opposing material and the vertically projected area of contact is given by

$$A = n\,\frac{\pi r^2}{2}$$

where n is the total number of asperities

$$W = Ap_o = n\,\frac{\pi r^2}{2}\,p_o$$

The friction force F is obtained in a similar manner by considering the total projected area of material which is being displaced by plastic flow, that is

$$F = nrhp_o$$

Therefore

$$\mu = \frac{F}{W} = \frac{2h}{\pi r}$$

but

$$\frac{h}{r} = \cot\theta$$

therefore

$$\mu = \frac{2}{\pi}\cot\theta \tag{4.7}$$

Similar expressions can be obtained for asperities of a different shape. Using this approach μ always equals one half the vertical projected area of the asperity divided by the horizontal projected area of the asperity. The above theory assumes that the yield pressure is the same in the vertical and horizontal direction. Kragelskii[8] introduces an extra multiplying factor which allows for a difference in work hardening in the two directions.

Bowden and Tabor state that the contribution of ploughing to friction is small for the following reason. The tangential resistance to sliding for a single asperity is made up of a shear term and a ploughing term, that is

$$F = A_V s_o + A_H p_o$$

where A_V and A_H are the vertical and horizontal projected areas.

For asperities on most metal surfaces, the angle θ is large and A_H/A_V is very small so that the ploughing term is negligible in comparison with the adhesion term. For rough surfaces where θ is larger, the ploughing term can be comparable to the adhesion term.

4.8 ELASTIC HYSTERESIS LOSSES

In chapter 8 it is shown that the major part of rolling friction of elastic (and visco-elastic) solids is accounted for by hysteresis losses. Also theoretical expressions for friction arising in this way are derived and shown to agree with experimentally determined values. When materials exhibiting elastic hysteresis are in sliding contact then a similar expression must appear in the friction term. If the materials are well lubricated then, as in rolling contact, elastic hysteresis is the major cause of friction.

Consider for example the sliding of a rigid hemisphere on a flat sheet of well-lubricated rubber. The deformation experienced by the rubber will sensibly be the same as if a rigid sphere of the same radius were rolled over it under the same load. Thus the friction force due to elastic hysteresis will be the same in each case and the theory developed in chapter 8 can be used for sliding contacts. It should be noted that this term is analogous to the ploughing term discussed earlier, but now the friction is due to elastic hysteresis rather than the force required to deform the material plastically. References are occasionally made to elastic hysteresis in metals. However, it is extremely doubtful whether this has any effect on friction, and the energy losses referred to are probably due to plastic deformation, analogous to the losses occurring during a fatigue cycle.

4.9 DISCUSSION OF THE VARIOUS FRICTION THEORIES

Initially, if we limit the discussion to the friction of metals, we can ignore any effects due to elastic hysteresis, which are negligibly small for metals. The 'ploughing' theory is nothing more than a particular case of the 'interaction of asperities' theory in which some simplifying assumptions have been made. Thus we need consider only the Bowden and Tabor 'adhesion' theory and the 'interacting asperity theory'.

Let us now consider those features of the two theories that are similar. The deformation in both cases is plastic, that is, the area of contact is defined by the flow stress, and the major sources of frictional energy losses are due to plastic deformation. Both theories include the effects due to a surface film of shear strength equal to or lower than the bulk shear strength of the softer of the two metals in contact. The only major difference between the theories appears to be that the Bowden and Tabor theory purports to rely solely on 'adhesion' as the surface interaction mechanism. However, as we have seen, the theory also covers those situations where an oxide film covers the metal surface in which case 'adhesion' is limited to those parts of the surface where the oxide film is penetrated. Where the surface films are not penetrated then the relevant shear strength in equation 4.5 is the shear strength of the film, but it is unlikely that the surface interaction is adhesive in these cases. Thus it would appear that the Bowden and Tabor theory must implicitly include the possibility of interlocking asperities. The interlocking asperity theory does not exclude the possibility of adhesion, in fact this possibility is explicitly included and leads to the conclusion that the stresses on individual asperities can be tensile over part of the contact, that is, that the contribution of an individual asperity to the total normal load can be negative, again over part of the contact. Thus the two theories are not in conflict and in confirmation of this conclusion we have the fact that the asperity interaction theory can give exactly the same expression for the friction coefficient as the Bowden and Tabor theory. If the 'ploughing' term is included in the Bowden and Tabor theory then good agreement between the two theories is obtained for all junction angles.

The 'interlocking asperity' theory appears to be the more satisfactory of the two and offers the possibility of further extension to include the rubbing of surfaces with a realistic asperity height distribution resulting in both elastic and plastic interactions, with work-hardening effects considered. However, although somewhat less realistic, the Bowden and Tabor theory does give friction coefficients of the correct magnitude over a wide range of conditions, and has the advantage of being analytically much simpler than the asperity interlocking theory.

When materials exhibiting elastic hysteresis losses are rubbed then the only difference between this situation and the metallic case is in the form of frictional energy losses. The theory presented in chapter 8 for rolling friction can be readily adapted to the case of a given height distribution of hemispherical asperities.

REFERENCES

1. G. Amontons. *Hist. Acad. R. Soc.*, (1699), (Paris), 206.
2. F. P. Bowden and D. Tabor. *The Friction and Lubrication of Solids.* Pt II. Oxford University Press, (1964).

3. S. Timoshenko. *Strength of Materials*. Pt II (3rd Edition). Van Nostrand–Reinhold, New York, (1956).
4. S. Timoshenko. *Strength of Materials*. Pt I (3rd Edition). Van Nostrand–Reinhold, New York, (1955).
5. D. H. Buckley. *Friction, Wear and Lubrication in Vacuum*. NASA, SP-277, (1971).
6. A. P. Green. Friction between unlubricated metals, a theoretical analysis of the junction model. *Proc. R. Soc.*, **228**, A, (1955), 191.
7. C. M. Edwards and J. Halling. An analysis of the plastic interaction of surface asperities and its relevance to the value of the coefficient of friction. *J. Mech. Engng Sci.*, **10**, (1968), 101.
8. I. V. Kragelskii. *Friction and Wear*. Butterworths, London, (1965).

5

Wear

5.1 INTRODUCTION

Wear occurs as a natural consequence when two surfaces with a relative motion interact with each other. Although our understanding of the various mechanisms of wear is now improving, no reliable and simple quantitative law comparable to that for friction has been evolved. This is not surprising since the wear process involves many diverse phenomena, interacting in a largely unpredictable manner. Furthermore, whereas the coefficients of friction of most materials lie between 0.1 and 1.0, corresponding wear rates can vary over many orders of magnitude.

Wear is often thought of as a wholly harmful phenomenon, but this is not so. For instance, the wear experienced in 'running-in' is beneficial and many forming operations, for example, machining and filing, rely on wear mechanisms. When wear is harmful then the rates of wear of sliding components can be reduced, but not eliminated, by lubrication, by careful design, and by material selection, and each of these aspects will be considered.

5.1.1 Definition

Wear is familiar to everyone and probably we all feel that we understand what is meant by wear. However, the formulation of a precise and all embracing definition of wear is difficult. A committee of the Institution of Mechanical Engineers decided on the following definition: 'the progressive loss of substance from the surface of a body brought about by mechanical action'.

Kragelskii[1] defines wear as: 'the destruction of material produced as a result of repeated disturbances of the frictional bonds'. Neither definition is perfect. For instance, the first appears to eliminate spark erosion as a form of wear, and the second perhaps places too much emphasis on fatigue effects in wear.

5.2 TYPES OF WEAR

In order to study and gain a better understanding of wear, it is essential to recognise that several distinct and independent mechanisms are involved. In his excellent 'Survey of Possible Wear Mechanisms', Burwell[2] lists four mechanisms

(1) Adhesive wear
(2) Abrasive wear
(3) Corrosive wear
(4) Surface fatigue

He also includes a fifth classification under the heading 'Minor Types' of wear, which covers erosion, cavitation and impact chipping. We will follow the same classifications, but will also separately consider phenomena such as fretting, which arise due to a combination of the above mechanisms.

5.2.1 Adhesive Wear

It has been shown earlier, (chapter 4), that macroscopically smooth surfaces are rough on an atomic scale, and that when two such surfaces are brought together contact is made at relatively few isolated asperities. As a normal load is applied the local pressure at the asperities becomes extremely high. The yield point stress is exceeded, and the asperities deform plastically, until the real contact area has increased sufficiently to support the applied load. In the absence of surface films the surfaces would adhere together, but very small amounts of contaminant prevent adhesion under purely normal loading. However, relative tangential motion at the interface acts to disperse the contaminant films at the points of contact, and cold welding of the junctions can take place. Continued sliding causes the junctions to be sheared and new junctions to be formed. As we have seen in the previous chapter, this naturally leads to a simple theory of friction with the coefficient of friction equal to the shear strength divided by the yield pressure.

This model also leads naturally to the formulation of a mechanism of wear, but not to a precise quantitative law of wear. The amount of wear depends on the position at which the junction is sheared. If shear takes place at the position of the interface then wear is zero. If shear takes place away from the interface then metal is transferred from one surface to the other. With further rubbing some of the transferred material is detached to form loose wear particles by one of the mechanisms described later.

The transfer of material from one surface to another by adhesion has been studied by several investigators. For example, Kerridge and Lancaster[3] used a radioactive pin of 60:40 brass, rubbing against a ring of tool steel. The transfer of material was demonstrated by placing a photographic film in contact with the tool steel after rubbing and obtaining an autoradiograph of the transferred brass from the wear track.

Law of Adhesive Wear

Having described the mechanism of adhesive wear we can now develop a simplified law of adhesive wear. Here we follow the approach of Archard[4], although other workers have derived similar expressions.

Assume that the contact is made up of a number of similar asperities each of radius a. The area of each contact is πa^2 and each contact supports a load of $p_o \pi a^2$, where p_o is the yield pressure. The surfaces will pass completely over each asperity in a sliding distance of $2a$ and we will assume that the wear fragment produced at each asperity is hemispherical in shape and of volume $2/3\pi a^3$.

Then the total wear volume Q, per unit distance of sliding is given by

$$Q = \Sigma \frac{\frac{2}{3}\pi a^3}{2a}$$

$$= \tfrac{1}{3}\Sigma \pi a^2 = \frac{\pi a^2}{3} \times n$$

where n is the total number of contacts. But each contact supports a load of $p_o \pi a^2$, therefore

$$\text{total load } W = p_o \pi a^2 n$$

or

$$n\pi a^2 = \frac{W}{p_o}$$

therefore

$$Q = \frac{W}{3p_o}$$

This equation has been derived assuming that all asperity encounters produce a wear particle. If only a fraction k of all encounters produce wear particles then the equation becomes

$$Q = k\frac{W}{3p_o} \qquad (5.1)$$

where k is the probability of an asperity contact producing a wear particle. All quantities in equation 5.1 are measurable except k, and it is this factor that represents the uncertainty in the equation. k must be found for different combinations of sliding materials and for different conditions of rubbing.

Relationships similar to equation 5.1 have also been developed by Holm[5] and by Burwell and Strang[6]. This equation leads to three 'laws' of wear.

(1) The volume of wear material is proportional to the distance of travel.
(2) The volume of wear material is proportional to the load.
(3) The volume of wear material is inversely proportional to the yield stress, or the hardness, of the softer material.

The first 'law' is found to be true for a wide range of conditions. A number of investigators have demonstrated the validity of the second law over a limited range of load. For instance, figure 5.1, taken from the results of Burwell

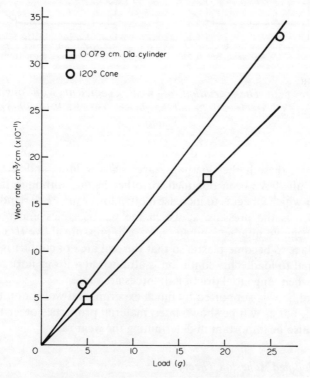

Figure 5.1   Wear rate as a function of load

and Strang[6], shows the rate of wear plotted against load for steel rubbing against steel. However, if the loading is increased further, plots of the form shown in figure 5.2 are obtained. These show k/H, the adhesive wear coefficient plotted against the average pressure (that is, load over apparent area of contact) for steels of different hardnesses. It can be seen that k remains constant up to a pressure of about $H/3$ where H is the hardness of the steel,

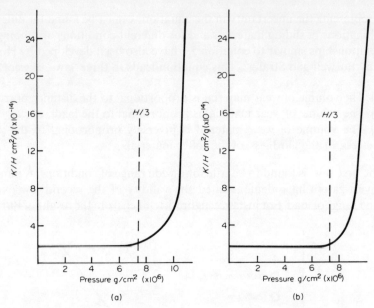

Figure 5.2 The variation of wear coefficient with apparent pressure for steel: (a) Brinell hardness 223; (b) Brinell hardness 430

and above this pressure k, and hence the wear rate, increase rapidly. It is found that at these higher loadings, large scale welding and seizure occur. Similar results have been obtained for other metals, although the average pressure at which k begins to increase is often lower than $H/3$. Under normal loading $H/3$ is the pressure at which the plastic zones under individual asperities begin to interact, and an increase in pressure above $H/3$ causes the whole surface to become plastic so that the real area of contact is no longer proportional to load. This condition is attained at a lower normal pressure than $H/3$ when tangential (frictional) forces are present.

The third 'law' is supported by much experimental work, notably that of Kruschov[7], but as will be shown later, material properties other than hardness, may also be important in determining the wear rate.

Rowe's Modified Adhesion Theory

Rowe[8] has modified the simple adhesion wear theory presented above to include the effect of surface films. Rowe starts with the Archard equation for wear

$$Q = k \frac{W}{3p_o}$$

or

$$Q = k'A$$

98

He then points out that in lubricated systems, including systems lubricated by surface films, k' must be related to both properties of the lubricant and those of the sliding metals. The volume of adhesive wear must be related to the metal–metal contact area A_m, that is

$$Q = k_m A_m$$

where k_m is a constant for the sliding metals and independent of lubricant properties or of surface films. A parameter β is now introduced such that

$$\beta = \frac{A_m}{A}$$

β is called the fractional surface film defect and is characteristic of the lubricant. For instance a poor lubricant would allow more metal–metal contact than a good lubricant, and therefore a poor lubricant would have a higher β than a good lubricant. Substituting for A_m we have

$$Q = k_m \beta A$$

but

$$A = \frac{W}{p_o}$$

Therefore

$$Q = k_m \beta \frac{W}{p_o} \qquad (5.2)$$

Now this equation contains k_m, characteristic of the sliding metals, and β, characteristic of the lubricant (or contaminating film). Rowe introduced one further refinement into this theory, similar to that introduced by Bowden and Tabor into the adhesive friction theory. He argued that the appropriate value for the yield pressure is that value obtained under combined shear and normal stresses rather than that under a static normal load. The relationship between these quantities is

$$p^2 + \alpha s^2 = p_o^2$$

where p = normal pressure
$\quad p_o$ = flow pressure under static load
$\quad s$ = shear stress
$\quad \alpha$ = a constant.

From friction theory (see chapter 4)

$$s = \mu p$$

Therefore

$$p = \frac{p_o}{(1 + \alpha\mu^2)^{1/2}}$$

99

and therefore the wear equation becomes

$$Q = k_m(1 + \alpha\mu^2)^{1/2}\beta \frac{W}{p_o} \tag{5.3}$$

5.2.2 Abrasive Wear

The term 'abrasive wear' covers two types of situation. In both cases wear is occasioned by the ploughing-out of softer material by a harder surface. In the first instance a rough hard surface slides against a softer surface. In the second case abrasion is caused by loose hard particles sliding between rubbing surfaces. The action of a file or emery paper against a softer metal is an example of the first type of abrasion. There are two requirements here— that one surface must be harder than the other, and that the hard surface must be rough. For instance, the wear rate of carbon against tungsten carbide is very low if the surface of the tungsten carbide has been smoothed sufficiently (C.L.A. 40 micron). This type of wear has been largely eliminated from modern machinery, due to a greater awareness of the importance of surface finish and the availability of instruments for the routine measurement of surface roughness.

However, the second mechanism is still of great importance. Not only do most mechanisms work in an environment containing much airborne dust and dirt, but also the products of corrosive wear are more often than not abrasive in character. Particles of hard metal produced by the adhesive wear mechanism can also cause abrasion. However, airborne dust and grit is the largest source of abrasive particles and the practical solution here is to exclude the dust by adequate sealing and filtration.

The fact that dust from the atmosphere is a major cause of abrasive wear probably explains the large variability in wear measurements. Thus, where attempts are being made to measure wear due to causes other than abrasion, Cattaneo and Starkman[9] have suggested that the lowest wear rates obtained are more appropriate than average values. They suggest that increases from the minimum values are due to insufficient precautions against ingress of abrasive particles. Statistical analysis appears to support their ideas.

Abrasive Wear—Quantitative Relationship

In order to obtain a quantitative expression for abrasive wear we shall assume a simplified model, in which one surface consists of an array of hard conical asperities all with the same semi-angle θ. The second surface is softer and flat.

Consider a single asperity creating a track through the softer surface as shown in figure 5.3. In traversing unit distance, the asperity displaces a

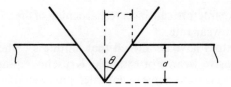

*Figure 5.3   Abrasive wear by a conical
indenter*

volume of material rd. But $d = r \cot \theta$, therefore the volume displaced by one asperity in unit distance $= r^2 \cot \theta$.

We assume that the material has yielded under the normal load and therefore the asperity supports a load of $\pi r^2 p_o / 2$ where p_o is the yield pressure of the softer material.

If there are n asperity contacts

$$\text{total normal load } W = \frac{n \pi r^2 p_o}{2}$$

and the total volume displaced in unit distance is Q where $Q = nr^2 \cot \theta$. Eliminating n

$$Q = \frac{2W \cot \theta}{\pi p_o} \tag{5.4}$$

It is interesting to note that this equation is of exactly the same form as the adhesive wear equation (equation 5.1) and therefore the 'laws' of wear listed under 'Law of Adhesive Wear' which were formulated from equation 5.1 apply equally well to abrasive wear.

The derivation of equation 5.4 has been based on an extremely simple model. No account has been taken of the distribution of asperity heights and shapes. Material will tend to build up in front of the asperities, altering the conditions as it does so. As discussed in the next section parameters such as Young's modulus will be important in certain situations, but are not considered in the simple theory. However, an equation of the form

$$Q = \frac{k_a W}{H} \tag{5.5}$$

where H is the hardness of the softer material and k_a the abrasive wear constant, is found to cover a wide range of abrasive situations.

The above derivation is appropriate to two-body abrasive wear. In the case of loose abrasive particles causing wear on rubbed surfaces, that is, three-body abrasion, the same form of equation will hold, but k_a will be lower, since in this case many of the particles will tend to roll rather than slide.

As wear due to abrasion proceeds, some blunting of the hard asperities or particles will occur, thus reducing the wear rate. However, an abrasive grit, which is brittle, can fracture causing a resharpening of the edges of the particle and an increase in wear rate.

Many investigators have confirmed that hardness is the most important parameter in abrasive wear. For example, Kruschov[7] plotted resistance to wear against hardness for a range of annealed pure metals, obtaining a linear relationship. He also found that prior work-hardening of the pure metals had no effect on the wear rate. These and other experiments have led to the conclusion that during abrasion a metal surface work-hardens to a maximum value, and it is this value of hardness which is appropriate when considering abrasion resistance.

Richardson[10] has shown that if wear resistance is studied as a function of the ratio of the hardness H_m/H_a, where H_m is the hardness of the metal surface and H_a the hardness of the abrasive, then the wear resistance increases rapidly as H_m/H_a becomes greater than 0.8. The region of abrasive wear where $H_m/H_a < 0.8$ is known as 'hard abrasive wear', and where $H_m/H_a > 0.8$ as 'soft abrasive wear'. In the 'soft' region the abrasive wear does not cease until H_m/H_a is much greater than unity. It has been suggested that the criterion for abrasive wear to be negligible is equality of the yield stresses of the metal and abrasive. Thus for contact between metals, abrasive wear should be negligibly low when the hardnesses are equal. However, if an abrasive non-metal is in rubbing contact with a metal, abrasive wear is appreciable until H_m/H_a is considerably higher than unity. This becomes apparent if we consider the relationship between hardness and yield stress, as measured by a diamond indenter

$$H = cp_o$$

where H is hardness, p_o is yield stress and c is a constant. For a metal, c is nearly 3 (see chapter 3). However, for a non-metal, with a much lower Young's modulus than a metal, elastic yielding around the indenter reduces the constraint around the plastic zone of the hardness indenter. Then c is less than 3, and can be as low as 1.2 to 1.3. Thus, equivalence of yield stress between metal and non-metallic abrasive, and hence negligible abrasive wear, requires H_m/H_a ratios of up to 2.5.

In the above discussion, based on hardness as the parameter of prime importance in abrasive wear, the value of Young's modulus has been introduced in connection with the relationship between hardness and yield stress. Young's modulus may have a more direct bearing on abrasive resistance. Oberle[11] has pointed out that if, in the presence of abrasive particles, a surface can elastically deform sufficiently to allow the particle to pass, then permanent damage to the surface will be avoided. As an example of this, water-lubricated rubber bearings for propeller shafts in ships operating in sandy waters have proved more abrasion resistant than materials with a higher value of Young's modulus, such as bronze. This indicates that the important parameter is the limit of elastic strain, that is, the yield stress divided by Young's modulus. Oberle has listed (table 5.1) various materials

TABLE 5.1   RATIO OF HARDNESS TO MODULUS OF ELASTICITY
FOR VARIOUS MATERIALS[11]

Material	Condition	Modulus of elasticity, E, lbf/in^2	Brinell hardness number, H	Hardness/elastic modulus, H/E (a)
Alundum (Al$_2$O$_3$)	Bonded	14×10^6	2000	143×10^{-6}
Chromium plate	Bright	12	1000	83
Grey iron	Hard	15	500	33
Tungsten carbide	9% Cobalt	81	1800	22
Steel	Hard	29	600	21
Titanium	Hard	17.5	300	17
Aluminium alloy	Hard	10.5	120	11
Grey iron	As cast	15	150	10
Structural steel	Soft	30	150	5
Malleable iron	Soft	25	125	5
Wrought iron	Soft	29	100	3.5
Chromium metal	As cast	36	125	3.5
Copper	Soft	16	40	2.5
Silver	Pure	11	25	2.3
Aluminium	Pure	10	20	2.0
Lead	Pure	2	4	2.0
Tin	Pure	6	4	0.7

in descending values of H/E. No controlled tests have yet been made to examine this relationship, but the trends indicated in the table appear to agree with general experience of wear resistance. It should be pointed out, however, that Spurr and Newcombe[12] interpreted their results to indicate that wear resistance increases with increasing elastic modulus, in direct disagreement with the work of Oberle.

5.2.3   Fatigue Wear

Rolling Contact

Adhesive and abrasive wear mechanisms depend on direct contact between solids and they produce a wear pattern that is progressive from the start of rubbing. If the surfaces can be separated by a lubricating film (and abrasive particles excluded), then these wear mechanisms cannot operate. This is the situation in well-designed rolling element bearings, where it is found that a fatigue mechanism of failure takes place. For this case, although direct contact does not occur, the opposing surfaces experience large stresses, transmitted through the lubricating film during the rolling motion. The nature and magnitude of the stresses can be found using the Hertz[13] equations. These show that the maximum compressive stresses occur at the surface, but the maximum shear stresses occur some distance below the surface as illustrated in figure 5.4. As rolling proceeds, the directions of the shear stresses for any

103

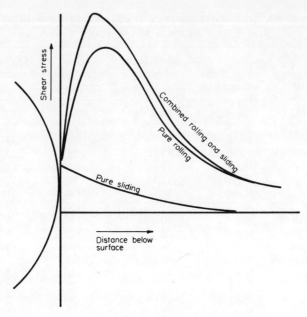

Figure 5.4   The variation of shear stress with distance
below the surface

element change sign. Fatigue failure is dependent on the amplitude of the reversed shear stresses and, if in rolling contact these are above the endurance limit, failure will eventually occur.

The position of failure in a perfect material subjected to rolling contact will be defined by the position of maximum reversed shear stress obtainable using the Hertz equations. If, superimposed on the rolling contact, there is also some sliding contact, then the position of failure moves nearer to the surface. However, materials are rarely perfect and the exact position of ultimate failure will be influenced by inclusions, porosity, microcracks and other factors.

Rolling contact fatigue wear is characterised by the formation of large wear fragments after a critical number of revolutions. Prior to this critical point, negligible wear takes place and a rolling contact bearing will operate normally until the wear particles are detached, when the useful life of the bearing is terminated. This is in marked contrast to the wear experienced in sliding bearings, where due to adhesion and/or abrasion, wear causes a gradual deterioration from the start of running. From this we can see that the amount of material removed by fatigue wear in rolling contact is not a very useful parameter. Much more relevant is the useful life in terms of number of revolutions or time at a given speed. The definition of life of a bearing used by manufacturers is the number of revolutions which will be reached or exceeded by 90 per cent of similar bearings. Testing of large numbers of

104

rolling bearings has shown that the life N, as defined above is inversely proportional to the cube of the applied load W. That is

$$W^3 \times N = \text{Constant} \tag{5.6}$$

The position (measured from the surface) of the maximum shear stress in pure rolling is proportional to $(WR)^{1/3}$ for a ball and to $(WR)^{1/2}$ for a cylinder[13], and the following figures are presented as a typical example. For a ball 1.0 cm in diameter, subjected to a load of 1000 MN/m^2 the maximum shear stresses occur at a depth of 0.12 mm from the surface of the ball.

The above discussion has been related to rolling elements with little surface traction. There are many mechanisms where rolling and sliding motions are combined, such as in hypoid gear teeth. Such mechanisms also fail by fatigue, but the increased traction causes the maximum shear stresses to be nearer the surface.

In practical applications the stresses transmitted by the oil films can be extremely high and EP lubricants are often necessary. In this case severe wear by adhesion is prevented at the expense of a low rate of wear due to corrosion, as shown later. This wear rate can be sufficiently low for eventual failure to be caused by fatigue. However, if some corrosive element such as water contaminates the lubricant, accelerated failure due to the combination of high stresses and corrosion, stress corrosion, can result.

Sliding Contact

When sliding surfaces make contact via asperities we have seen that wear by adhesion and abrasion can take place. However, it is conceivable that asperities can make contact without adhering or abrading and can pass each other, leaving one or both asperities plastically deformed. After a critical number of such contacts an asperity would fail due to fatigue, producing a wear fragment. It is difficult to prove directly that fatigue is a major cause of wear in any given set of conditions. Kragelskii attaches much importance to this mode of wear, as is implied by his definition of wear, given earlier in this chapter. He interprets many of his results and those of other investigators in terms of fatigue. Lancaster[14] has suggested that the low wear experienced when rubbing electrographite against metals may take place by fatigue, and Archard and Hirst[15] have suggested that the metal transferred by adhesion is finally detached by a fatigue process. A model of a low wear regime presented by workers at IBM[16] and, rather unfortunately, described as 'Zero Wear', invokes fatigue and corrosion as the only mechanisms responsible for wear in this regime.

The factor k in equation 5.1 has been interpreted as the probability of an asperity contact producing a wear particle, without any physical explanation as to the method of their production. Also, although the adhesive-wear theory can explain transferred wear particles, it does not explain how loose wear particles are formed. In particular, the occurrence of wear of the

105

harder of the two rubbing surfaces is difficult to understand in terms of the adhesion theory.

All these points can be explained on the assumption that wear is a fatigue process. The k factor is now interpreted by assuming that a wear particle is produced when an asperity has experienced a sufficient number of contacts and deformations to produce a fatigue fracture. When this occurs, a loose wear particle is produced and of course, this mechanism explains the production of wear particles from both the harder and the softer of the two rubbing surfaces. The fatigue mechanism does not exclude the possibility of transfer by an adhesive mechanism and therefore it appears that most of the wear phenomena can be explained at least qualitatively in terms of fatigue.

Kragelskii[1] has attempted to produce a quantitative fatigue theory of wear. He uses the results of Tavernelli and Coffin[17] who have shown that for a wide range of materials the plastic strain produced on each fatigue cycle is related to N, the number of cycles to failure, by the equation

$$\left[\frac{2\varepsilon_{\text{fail}}}{\varepsilon_{\text{p}}}\right]^2 = N$$

where ε_{p} is the plastic strain amplitude per cycle and $\varepsilon_{\text{fail}}$ is the plastic strain at failure in a tensile test. However, Kragelskii appears to have used the total strain rather than the plastic strain in his derivation, and in addition he makes a rather obscure estimate of the strain amplitude in terms of asperity geometry. Thus no satisfactory wear theory in terms of fatigue has yet been developed although the mechanism appears to be more realistic than that of the adhesion theory. In the opinion of the authors, it will be necessary to develop such a theory in order to advance our understanding of wear phenomena.

5.2.4 Corrosive Wear

When rubbing takes place in a corrosive environment, either gaseous or liquid, then surface reactions take place and reaction products are found on one or both surfaces. These reaction products are commonly poorly adherent to the surfaces, and further rubbing causes their removal. This process is then repeated. Thus corrosive wear requires both corrosion and rubbing. The rate of growth of, say, an oxide film on a steel will decrease exponentially with time, and therefore unless the oxide film is removed by rubbing, the metal-to-oxide reaction will rapidly become negligibly small. The detailed operation of corrosive wear is extremely complex. The reaction products depend on the exact composition of the environment. For instance, small quantities of water vapour in air cause the reaction product to be the hydroxide rather than the oxide. Mechanisms operating in an industrial environment or near the coast will generally corrode more rapidly than those operating in 'clean' air.

The most common liquid environments are aqueous and here small amounts of dissolved gases, commonly oxygen or carbon dioxide, influence corrosion. Corrosion is also influenced by the relative electropotential of rubbing metals.

The presence of a lubricant usually protects the surfaces from the corrosive environment. However, it is not uncommon for corrosive elements to be dissolved in lubricants, for example, water in oil, and also lubricants can be degraded in time and become progressively more corrosive.

Some mechanisms, notably hypoid gears, are required to work under conditions where severe adhesion is possible, that is, at a pressure greater than $H/3$. To prevent catastrophic metallic welding in these situations lubricants with mildly corrosive additives are used, that is, EP additives.

Where corrosion is a major cause of wear, there is usually a complex interaction between various mechanisms. The original wear of surface films can be due to either adhesion or abrasion. Since many commonly occurring films, notably iron oxides, are abrasive, the corrosion and abrasion will combine. High contact stresses can cause enhanced corrosion locally, leading to pitting. It is well known that internal stresses in metals, caused during forming operations, cause stress-corrosion cracking when in a corrosive atmosphere. This, combined with surface rubbing, can result in catastrophic wear, as can the corrosion of a single-phase in a two-phase alloy bearing.

Corrosion is not always a deleterious phenomenon. Oxide films and other corrosion products prevent adhesion of metal asperities and metallic wear in vacuum, where oxide films cannot form, is generally very high.

5.2.5 Fretting

The wear phenomenon known as fretting is not a distinct mechanism, but it is convenient to treat it separately.

Fretting occurs where low amplitude vibratory motion takes place between two metal surfaces loaded together. This is a common occurrence, since most machinery is subjected to vibration, both in transit and in operation. Examples of vulnerable components are shrink fits, bolted parts, and splines.

Basically fretting is a form of adhesive wear, the normal load causing adhesion between asperities and the vibrations causing rupture as described earlier. Most commonly, fretting is combined with corrosion, in which case the wear is known as fretting corrosion. In air the corrosion product is oxide, and the characteristic fine reddish-brown powder produced from steels is known as 'cocoa'. Here the initial wear debris is oxidised. The oxide particles are abrasive and because of the close fit of the surface, cannot readily escape. Further oscillatory motion causes abrasive wear and oxidisation, and so on.

Fretting wear can cause the formation of surface stress raisers and, if the vibratory stresses are sufficiently high, fatigue cracks are propagated leading to complete failure.

5.2.6 Erosion

The term erosion is usually taken to cover that form of damage experienced by a solid body, when a fluid, which may contain solid particles, impinges on to the surface of the body. Until fairly recently erosion has been considered a minor source of wear, and many reviews of wear refer to it only briefly, if at all. However, with the continuing development of mechanisms operating at high speed, together with the need for materials with high strength-to-density ratios, erosion has gained more prominence. This is exemplified by carbon-fibre reinforced plastic turbine blades, whose leading edges require special coatings to increase erosion resistance.

Erosion by Solid Particles

When a stream of solid particles is directed at a surface it is found that the wear rate is dependent on the angle of incidence of the particles, and that the wear rates for ductile and brittle solids follow different curves[18] as shown in figure 5.5. This suggests two different mechanisms of erosion, dependent on whether the eroded solid is ductile or brittle.

In the case of a ductile solid it is thought that the mechanism near to the peak is similar to that of abrasion. At angles close to 90° for a ductile material

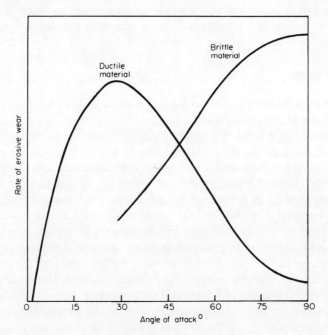

Figure 5.5   Dependence of rate of erosion on angle of attack of impinging particles

a fatigue mechanism is probably predominant. For brittle materials surface cracks are formed which link to give the wear particles. In both forms of wear, ductile and brittle, the wear rate is proportional to the kinetic energy of the impinging particles, that is, to the square of the velocity of the particles.

Fluid Erosion

When small drops of liquid are incident on the surface of a solid at high speeds (1000 m/s) very high pressures are experienced, exceeding the yield strength of most materials. Thus plastic deformation or fracture can result from single impact, and repeated impact leads to pitting and erosive wear.

Cavitation Erosion

Cavitation erosion arises when a solid and fluid are in relative motion, and bubbles formed in the fluid become unstable and implode against the surface of the solid. Damage by this process is found in such components as ships' propellers and centrifugal pumps.

The stability of a bubble is dependent on the difference in pressure between the inside and outside of the bubble, and the surface energy of the bubble. This last factor is a measure of the energy released by collapse of the bubble, and of the potential damage at collapse. Thus a reduction of surface tension of the liquid reduces damage, as does an increase in vapour pressure.

Spark Erosion

When an electric spark occurs between two surfaces permanent damage in the form of metal removal and deposition results. This is a well-known problem in the field of electrical contacts. When the contacts are rubbing, as in the case of a commutator, the sparking damage on the copper commutator can then cause excessive wear of the brush by abrasion. Damage can be cumulative since, if a particular commutator bar is recessed slightly and sparking takes place, then the brush-bar contact will deteriorate causing more severe sparking and even melting of the copper.

Spark erosion is used as a method of metal removal or cutting where normal machining is difficult, for example, for very hard metals, and also in cases where it is important to restrict the amount of subsurface damage, for example strain-free machining of metal single crystals.

5.2.7 Size of Wear Particles

We have now reviewed the various wear mechanisms but have not considered the factors influencing the size of the wear particles produced. Rabinowicz[19] has suggested that, if a wear particle is to be formed, the elastic energy stored prior to detachment must be greater than the energy for the new surface created. This approach leads to a criterion for whether or not an asperity contact produces a wear particle, and thus to a value of k. However, as is shown below, the theory is open to criticism.

Rabinowicz considers the energy stored in a hemispherical particle adhering to one surface, after yielding has taken place. The stored energy per unit volume of such a particle is $p_o^2/2E$, where p_o is the yield pressure, E is Young's modulus, and the volume of the particle is $\frac{2}{3}\pi r^3$. The surface energy of newly created surfaces, if the hemisphere is detached along its diametral plane, is $2\pi r^2\gamma$, where γ is the surface energy per unit area. Then according to Rabinowicz

$$\frac{2}{3}\pi r^3\left(\frac{p_o^2}{2E}\right) > 2\pi r^2\gamma$$

or

$$r > \frac{6E\gamma}{p_o^2}$$

If p_o is replaced by $H/3$ for a metal, then

$$r > \frac{54E\gamma}{H^2}$$

The exact value of the constant has been arrived at by assuming a hemispherical shape for the particle. For particles of other shapes this equation can be replaced by

$$r > \frac{KE\gamma}{H^2} \tag{5.7}$$

where K is $\simeq 54$ but depends on the particle shape.

Rabinowicz then goes on to say that H/E is roughly constant, and this leads to

$$r > \frac{K'\gamma}{H} \tag{5.8}$$

Rabinowicz interprets this to mean that only wear particles above a certain critical size, dependent on γ/H, can become detached. However, examination of table 5.1 indicates that H/E can vary over a wide range and that equation 5.7 is more realistic than equation 5.8. Furthermore, Levy, Lindford and Mitchell[20] argue that, in a rubbing situation, the value of r given in equation 5.7 should be regarded not as the minimum, but as the maximum value obtainable. It is certainly difficult to see how r could ever exceed the right-hand side of equation 5.7, and it is highly probable when other forms of energy, such as the kinetic energy of the rubbing surfaces, are available that wear particles will be detached before r has reached the critical size. Thus, the relevant relationship becomes

$$r < \frac{KE\gamma}{H^2} \tag{5.9}$$

This suggestion is supported by the experimental work of Aghan and Samuels[21], who show that under abrasive polishing conditions, the wear debris consists of particles several orders of magnitude smaller than the size predicted by Rabinowicz.

5.2.8 The IBM Engineering Model for Wear

The expressions presented in the previous sections have attempted analytically to relate wear to conditions of sliding, such as load and speed, and material properties, such as hardness and Young's modulus. None of these expressions has gained universal acceptance, and because of this, workers at the IBM Endicott Laboratory[16] have attempted to obtain empirically, expressions for wear which could be directly applied to the design of rubbing components. Such expressions have been obtained and appear to have wide applicability. The work at IBM can be divided into two parts. In the first, expressions are obtained which define the amount of rubbing which can be accommodated before the depth of wear is greater than the initial surface finish. In the second, expressions are presented which can be used to predict wear greater than the depth of the original surface finish. The two parts are entitled Zero Wear and Measurable Wear.

Zero Wear

First, a unit of sliding called a 'pass' is defined. This is the unit of sliding in which sliding has been experienced over a distance equal to the width of the apparent contact in the direction of sliding. For example, in the pin-on-cylinder wear apparatus shown in figure 5.6, a single pass would be of length

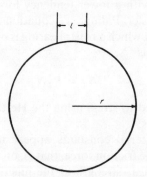

Figure 5.6 The concept of a pass in the IBM model of wear. In one complete revolution the cylinder experiences one pass and the pin experiences $2\pi r/l$ passes

111

l and for a complete revolution of the cylinder $2\pi r/l$ passes would be experienced.

It has been found that the condition for zero wear (wear less than the surface finish) in a single pass is

$$\tau_{max} < 0.54\,\tau_y \qquad (5.10)$$

where τ_{max} is the maximum shear stress experienced and τ_y is the shear stress necessary to cause yielding.

For zero wear in 2000 passes the condition is

$$\tau_{max} < \gamma_r \tau_y \qquad (5.11)$$

where γ_r is a constant which is experimentally determined and can have values of either 0.54 or 0.20.

The condition for zero wear for M passes is related to the condition for zero wear in 2000 passes by the expression

$$2000\,(\gamma_r \tau_y)^9 = M(\tau_{max})^9 \qquad (5.12)$$

The IBM workers have found that these expressions are valid for a large range of materials, both metals and plastics, for various geometries, for various lubricants and for different surface finishes.

The simplicity of the expressions is somewhat surprising. They appear to be independent of surface finish, but it should be remembered that zero wear is already defined in terms of the original surface finish. The fact that γ_r takes only two values 0.54 and 0.20 has no theoretical explanation as yet, but the choice of γ_r seems to be related to the tendency for material transfer, 0.54 being appropriate with a lower tendency for transfer and 0.20 with a higher tendency for transfer. Palmgren's equation 5.6 is similar to equation 5.12 relating the load P at which a roller bearing is operated to N, the number of cycles to failure. That is

$$P_1^3 N_1 = P_2^3 N_2$$

If the loads are converted to stress using the Hertz equation, then a ninth-power relationship results.

All the necessary material constants appear to be contained in τ_{max}, which is dependent on the friction force, that is, on the hardness (or Young's modulus) and on the shear strength at the junctions. Parameters such as speed and temperature do not appear to be taken into account in this model. However, these affect the material constants implicit in τ_{max}. For instance it was found that with certain plastics at very high rubbing speeds, the results did not agree with predictions. However, it was found also that the high speeds had caused melting of the plastics, thus modifying the material properties. Using the corrected values for τ_y and τ_{max} agreement between predicted and experimental results was obtained.

Measurable Wear

Here the amount of wear Q is related to the number of passes M, and also to the amount of energy E, dissipated in the form of wear during each pass. The appropriate relationship relating these quantities was found to be the differential relationship

$$dQ = \left(\frac{\partial Q}{\partial E}\right)_M dE + \left(\frac{\partial Q}{\partial M}\right)_E dM \qquad (5.13)$$

Two types of wear are distinguished, that involving severe transfer and that involving only a moderate degree of transfer. The second type is more common in practical systems and will be treated here.

After some simplifying assumptions, the following equation is obtained

$$d\left[\frac{Q}{(\tau_{max} l)^{9/2}}\right] = C\, dM \qquad (5.14)$$

where C is some constant for the system and l is the magnitude of sliding in a single pass that is, the length of the apparent contact area in the direction of sliding. Q is obtained by substituting the expression for τ_{max} and l into the equation and integrating. The constant C must in general be obtained experimentally by determining the amount of wear for a particular number of passes. However, Bayer[22] has indicated that the zero wear and measurable wear models can be combined to give an analytical expression for C. In this case the value of γ_r (either 0.54 or 0.20) must be found experimentally, as must the condition of transfer, high or low.

The IBM workers have presented, in a series of papers, methods of applying both the zero wear and measurable wear models to practical design problems.

5.3 VARIOUS FACTORS AFFECTING WEAR

In the following sections we will examine a number of factors which can cause considerable variation in the wear rates of rubbing surfaces. Although it is convenient to consider these factors under different headings, we will see that they interact and it is difficult to separate them one from another. For instance, high temperatures at the surface will be generated by high loads and speeds. The temperature influences surface film formation and can cause changes in the surface structure and hardness. A number of the factors considered below are treated in more detail in chapter 6.

5.3.1 Effect of Surface Films

Since both friction and wear are essentially surface phenomena the presence and properties of films covering rubbing surfaces have a critical effect. We will first consider wear under conditions where no surface films are present, that is, in high vacuum.

Wear Under Vacuum

A comprehensive study of friction and adhesion under a vacuum of 10^{-8}mmHg was carried out by Bowden and Rowe[23]. They found that under the action of normal and tangential forces, junction growth occurred, unimpeded by the presence of surface films. Thus welds were made over comparatively large areas of contact resulting in high friction coefficients. In order for tangential, that is, rubbing, motion to proceed these welded bonds must be broken. As discussed earlier, the wear taking place depends on the exact position of the fracture relative to the junction, but obviously, the larger the area of the junctions, the larger the wear rate. In fact most metals wear catastrophically under these conditions. These experiments of Bowden and Rowe were carried out largely in order to investigate fundamental properties. However, the high wear experienced under vacuo is now a problem of practical significance in high-flying aircraft, missiles and spacecraft. A great deal of work has now been accomplished aimed at understanding friction and wear under high vacuum, and also at providing practical bearings under these conditions. Much of this work is summarised by Bisson[24]. Conventional lubrication using oils and greases is of course impossible over long periods. Possible solutions to the problem are the use of low shear strength metal films, plastics, or solid lubricants. A fairly recent development has been the use of hexagonal close-packed metal as a bearing material in space environments, but this will be discussed more fully later.

Oxide Films

Most metals are covered by an oxide film and even after cleaning by machining or grinding acquire a film of oxide of between 5 and 50 molecular layers in five minutes or less. When an oxide film covering an asperity is removed by rubbing in a normal atmosphere the clean metallic surface will be covered by a monomolecular layer almost instantaneously. This must be taken into account in all considerations of metals rubbing under atmospheric conditions.

Unless the loads are very small, the oxide film does not prevent intermetallic contact, as can be seen from the results of Wilson[25] who measured the electrical contact resistance between various rubbing metal pairs, metallic contact being indicated by a sharp drop in the contact resistance. Where asperities make metallic contact, localised welding can and does take place, but as has been shown earlier (chapter 4), the oxide film prevents the junction growth which occurs in vacuo, and thus reduces both the friction and the wear rate.

Many investigators have found that the wear of metals can be divided into two regimes, mild wear and severe wear. In the mild wear region, occurring

114

at lower loads, the contact resistance is high, the wear debris is fine and consists mainly of metal oxide, and the rubbed surfaces become polished. In the severe wear region, at higher loads, the contact resistance is low, and wear debris includes coarse metallic particles and the rubbed surfaces are rough. This type of result is represented by figure 5.7, taken from Archard and Hirst[15].

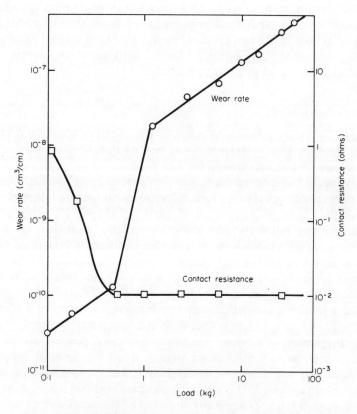

*Figure 5.7   The variation of wear rate and contact resis-
tance for 60/40 brass pin*

The properties of the oxide film are also important. For instance, the hard brittle oxide formed on aluminium provides poor protection against heavy wear. Barwell[26] presents an unexpected result as follows: steel was rubbed on steel in a vacuum of 10^{-4}mmHg; the wear was lower than when air was admitted to the apparatus. The interpretation of these results is that a thin, tough and tenacious oxide film was formed on the steel at 10^{-4}mmHg whereas when air was admitted the oxide film formed was thicker but less protective.

In studies of the wear of copper slip-rings it has been found that under the negative brush, the rate of copper oxidation was enhanced but the resultant film was poorly adherent, giving higher wear rates. Under the positive brush, the rate of copper oxidation was reduced, the film was then tough and adherent giving low wear rates. At higher currents, high wear rates were recorded under the positive brush, apparently as a result of complete cathodic reduction of the oxide film.

In this section we have considered the effect of oxide films on adhesive wear, but of course wear due to a combination of chemical attack and rubbing is defined as corrosive wear. Oxide films will also have a large influence on abrasive wear since many metal oxides are hard and when present in the form of wear debris act as abrasive particles.

Boundary Lubrication

The term 'boundary lubrication' refers to the situation where an oil film is present between two rubbing surfaces but its thickness is insufficient to prevent asperity contact through the film.

Many mechanisms operate totally under conditions of boundary lubrication. Others are designed to operate under full hydrodynamic lubrication, but as the oil film thickness is a function of speed, (see chapter 10) during starting up and running down the film thickness will be insufficient to provide complete separation of the surfaces, and the conditions of boundary lubrication will be obtained. We may initially consider the action of a boundary lubricant in much the same way as that of an oxide film, that is, it limits, but does not prevent, metallic contact at asperities, and inhibits junction growth. Further, a lubricant can reduce wear by limiting the access of a corrosive liquid or gas when a bearing is operating in a hostile environment.

The effectiveness of a boundary lubricant is less dependent on the rheological properties of the lubricant than it is on its chemical properties. It is found that liquid fatty acids are more effective than alcohols with a comparable chain length. It is generally accepted that this is due to the more reactive fatty acids being chemically absorbed on the metal surface, and it has been found that one or two molecular layers are effective in reducing wear by a factor as great as 1000. However, metals that do not react with the fatty acid are lubricated equally well by an alcohol of similar chain length. Thus we see that the effectiveness of boundary lubrication is increased in those cases where a solid metallic soap film is formed over the metal surface.

Under the very high pressures developed in hypoid gears, organic boundary lubricants are ineffective since they break down due to the high local temperatures at the contacts. To counteract this extreme pressure, additives, for example, organic chlorine or sulphur compounds, are added to the lubricant. These are stable at normal temperatures but react with the metal at the hot spots, forming metal chloride or sulphide films which inhibit welding of the asperities and reduce adhesive wear to acceptable levels.

116

Solid Lubricants

Solid lubricants can be applied to bearing surfaces by means of an adhesive such as a resin, or they may be introduced in powder form between two rubbing metals at low loads, in which case they adhere to the metal surfaces, eventually forming a fairly coherent covering. Also a solid lubricant in a fine powder form may be added to a liquid lubricant where again the rubbing surfaces eventually acquire a coating of the solid lubricant. The action of the solid lubricant is added to the oil in a motor car engine, 'running-in' will films, that is, it acts to reduce the number of metallic contacts and inhibits junction growth. Furthermore, it provides a low shear strength interface.

The effectiveness of reducing wear is demonstrated by the fact that if a solid lubricant is added to the oil in a motor car engine, 'running-in' will take almost twice as long as under conventional lubrication. 'Running-in' consists of removal by wear of the larger asperities produced by machining operations. Solid lubricants are used to reduce wear in those situations where a conventional lubricant (oil or grease) cannot be used—in high vacuum (space craft) or at high temperatures. They are also useful in preventing wear for short periods in emergencies, such as the complete loss of oil from a car engine.

Other Surface Layers

The wear between metals may be reduced by coating the surfaces with a thin layer of metal which has higher wear resistance than the substrate metal. This is obviously sound economic practice since many of the metals found to be wear resistant are expensive, but the cost of an effective film thickness can be very low.

Rhodium and chromium are both hard, and electroplated coatings of these metals have been used successfully in the protection of cylinder liners, crankshafts and similar applications. The wear resistant properties of these films have been demonstrated by Moore and Tabor[27]. Very successful hard facing and wear resistant coatings are deposited using spraying and fusing techniques. These coatings, based largely on cobalt, chromium and iron have good hot-hardness properties and are also highly corrosion resistant.

Films of soft metals such as indium and lead have been used on harder metal substrates. An example of their effective use has been in lubricating the dies used in deep drawing operations.

Besides these methods involving the deposition of a surface layer on to a substrate, there are a number of techniques where the surface of the substrate is subjected to chemical action to increase the wear resistance of the surface. Examples are the phosphating and sulfinuzing processes for steels. Ferrous surfaces can also be hardened by nitriding and carburising. Also hard surface layers can be produced by heat treatment. For instance, large cast-iron rolls for rolling mills are quenched so that the higher rate of cooling at the surface

117

produces a hard 'white iron' surface whereas the centre is composed of less brittle 'grey iron'.

Finally, mention should be made of the fact that the rubbing action itself produces changes in the surface layers. It has been found that prior work-hardening has no effect on abrasive wear resistance as the rubbing action work-hardens the surface to a maximum value. References to modified surfaces in rubbed ferrous materials are common, although the exact form of the modified material has rarely been analysed. A material referred to as a hard 'white etching' layer is found on rubbed cast-irons and steels. The formation of these layers is probably associated with hot spot temperatures being sufficiently high to take the carbides into solution, followed by rapid cooling.

5.3.2 Effect of Temperature

The temperature of rubbing surfaces influences the wear in three major ways. It can alter

(a) the properties of the rubbing materials
(b) the form of the surface contaminating film
(c) the lubricant properties.

We will consider these in turn.

In general the hardness of metals is temperature dependent, the higher the temperature, the lower the hardness. Thus as the tendency for asperities to adhere and the wear rate increase with decreasing hardness, in the absence of other effects, they also increase with increasing temperature. This effect has been demonstrated by M. J. Hordon[28]. In order to counteract this effect it is necessary to use metals with high hot-hardness for bearing materials operating at high temperatures. Metals commonly used at high temperatures include tool steels, and alloys with base composition of cobalt, chromium and molybdenum. However, at temperatures above about 850 °C, it is necessary to use cermets or ceramics.

A quite different effect of temperature is a temperature-induced phase change causing the properties of a bearing material to alter radically. An example of this is described in detail in chapter 6.

In normal atmospheres, most metals are covered by an oxide film. The exact form and thickness of film is dependent on the temperature of formation, and again this is discussed in chapter 6.

Another effect of frictional heating, when rubbing ferrous materials in normal atmospheres, is reported by Welsh[29]. He found that at low loads wear was high but at higher loads the wear rate dropped to a very low value. A metallographic study indicated that this fall in the wear rate was due to the formation of a hard surface layer, which Welsh attributed to interaction with the atmospheric nitrogen at the temperatures generated under the higher loads.

118

If the temperature of operation of a bearing lubricated by an oil is raised, deterioration is caused, first by oxidation of the oil and then by thermal degradation. This places a limit beyond which organic fluids are ineffective in reducing wear, and at high temperatures alternative lubricants must be employed, for example, solid lubricants, such as graphite and molybdenum disulphide.

Oxidation and thermal degradation cause irreversible changes in the lubricating properties of an oil. In general, under conditions of boundary lubrication, greatest protection is provided when the lubricant is solid. If the temperature of such a lubricant is raised beyond the melting point, wear rate increases but the change is reversible, since the lubricant again provides protection when the temperature is lowered. A fatty acid lubricating reactive metal surfaces will provide good wear resistance beyond its melting point due to the formation of the metallic soap.

5.3.3 Effect of Load

An increase in load causes an increase in the frictional force, and hence a temperature rise which produces the effects discussed earlier. Also, an increase of load can cause a transition from mild wear to severe wear. This occurs at about the point when

$$\frac{W}{A_a} = \frac{H}{3}$$

where W = load
A_a = apparent area of contact
H = hardness

and this has been demonstrated by Burwell and Strang[6] (see figure 5.2).

The transition from mild to severe wear is attributed to the interaction of the plastic zones beneath the contacting asperities[30]. During mild wear, the situation is as shown in figure 5.8, where there is no interaction between the plastic zones. As the load is increased, the plastic zones interact as in figure 5.9, and the subsurface regions become entirely plastic. Beyond this point Amonton's law is no longer obeyed, and severe wear occurs.

Figure 5.8   The independent plastic zones under the contacting asperities of a plane slider under relatively low loads

119

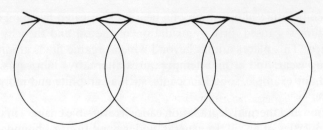

*Figure 5.9   The interation of the plastic zones of a
plane slider under high loads*

5.3.4   Effect of Compatibility

Rabinowicz[31] has suggested that metal pairs with low metallurgical compatibility will exhibit low friction and low wear. He defines 'metallurgically
compatible' metals as those which show a high degree of mutual solubility.
Thus compatibility ratings can be obtained from binary phase diagrams.
Rabinowicz has attempted to correlate adhesion, friction and wear results
from twelve different published studies, with compatibility of the metal pairs
used. He found zero correlation of compatibility with adhesion, a positive
correlation with friction and a greater correlation with wear.

This suggests that metal pairs for boundary lubricated or unlubricated
sliding should be chosen to have low mutual solubilities, in order for the wear
rate to be low. Certainly it has been generally accepted that in selecting metals
for bearings, similar metals, that is, metals with 100 per cent compatibility,
should be avoided.

5.3.5   Crystal Structure

Hexagonal metals give much lower friction coefficients when rubbing against
themselves than do face centred cubic and body centred cubic metals. This
is discussed in some detail in chapter 6 and will not be considered further here.

5.4   EXPERIMENTAL ASPECTS

Laboratory wear tests can be listed conveniently under two broad headings.
First, there are the tests in which specimens with convenient geometries are
run against each other; parameters which influence wear are varied one
at a time, and attempts are then made to correlate wear rate with the various
parameters. Tests under this heading range from simple comparative wear
studies to closely controlled experiments aimed at a better understanding of
fundamental wear mechanisms. Such simplified tests often produce results
which have little direct relevance to the complex wear situations met in
practice. Therefore in order to study the more practical problems of wear it is

120

more usual to perform a second type of test, which simulates the practical situation. Often in this type of test the conditions simulated are the worst wear conditions that are likely to be met in practice, and in some tests the conditions are greatly accentuated in order to reduce testing time. Results from tests of this type have limited applicability, and are unlikely to provide any fundamental information.

5.4.1 Wear Apparatus

There are a number of specimen geometries in common use for simple wear tests, and these are listed below.

The four-ball arrangement is one in which a single ball is rotated under load against a cluster of three balls. This is mainly used as a comparative test in assessing the performance of lubricants.

In the pin-on-disc type of machines, a stationary cylindrical pin, either hemispherically, conically or flat ended is loaded against a rotating flat disc or annulus. Alternatively two annuli, one stationary and one rotating, may be used.

The pin-on-cylinder geometry again has a stationary pin but here its end bears on the surface of a rotating cylinder. In this arrangement the cylinder may also have a linear movement along its axis so that the pin is always rubbing on virgin cylinder surface and the wear track is a spiral. The crossed-cylinder apparatus has a stationary cylinder loaded with its cylindrical surface against a rotating cylinder, with the cylinder axes at right angles. Again, the rotating cylinder may have a superimposed linear motion.

In order to simulate combinations of sliding and rolling two driven discs are loaded against each other. By varying the relative speeds of the discs pure rolling or any combination of sliding and rolling can be studied.

Each of the above arrangements has its own characteristic advantages and disadvantages. For example, the flat-ended pin-on-disc and the annulus-on-annulus retain their macroscopic geometry during a run, but the pin-on-disc is assymetric in that the disc material is open to the atmosphere over the major part of the run. In the annulus-on-annulus both surfaces are bearing continuously and exposure to the atmosphere is limited. Also, in this arrangement it is more difficult for wear debris to escape. The crossed-cylinder apparatus is probably the simplest apparatus to build. The macroscopic geometry changes as wear proceeds, the size of the wear scar being representative of the total wear. An example of a crossed-cylinder apparatus is shown in figure 4.2.

5.4.2 Precautions Necessary in Wear Measurement

Wear measurements are notoriously variable. Experiments on the same machine and in the same laboratory can give widely differing results at different times. Such variations are almost always due to lack of adequate

121

control of the conditions under which the surfaces are rubbing. For example, the ingress of extraneous grit may cause the wear rate to increase, small amounts of water vapour may alter the oxide film growth and handling of specimens can deposit contaminating films of grease. Thus surfaces must be prepared and cleaned in a reproducible way. Atmospheric conditions must be constant, and ingress of abrasive material must be prevented. An extreme example of these controlled conditions is in measuring wear under space environments. Here the specimens are prepared mechanically, electro-polished, and cleaned by electron bombardment and run under pressures of 10^{-12}mmHg. An example of such a rig is shown in figure 4.3b.

5.5 WEAR PREVENTION

When two surfaces in contact slide relative to each other, wear will inevitably take place. The only way to prevent wear is to separate the surfaces. This implies continuous running under hydrodynamic conditions. When a bearing is run down or started then some asperities will make contact. Under these conditions wear can be kept low by the use of adequate boundary lubricants or solid lubricants and by material selection. Hard materials wear less rapidly than the soft ones and the rubbing of similar materials should be avoided. Surfaces should be smooth, extraneous particles must be excluded from bearings by filtration of the air and lubricants, and corrosive atmospheres must be excluded where possible.

In practical situations wear is inevitable and in many cases bearings are so designed that one member has a very low wear rate whereas the mating member is considered replaceable, and has a higher wear rate. For example, the crank shaft in an internal combustion engine is costly to replace, and it is therefore made from a hard steel and is supported in relatively cheap bearing shells, of much softer metals (lead–tin, copper–lead, aluminium–tin alloys). Use of a softer metal bearing has additional advantages. It can deform easily to redistribute local high loads which might be caused by shaft distortion or misalignment, and extraneous abrasive particles can be absorbed. Even under extreme conditions, such as total loss of lubricant, because of the low melting point of many bearing materials damage to the shaft can be avoided over a short period of time.

5.6 APPLICATION OF WEAR RELATIONSHIPS TO DESIGN

One ultimate objective of wear studies is to be able to design components (in this context design includes material selection), that will run under a given set of conditions with acceptable known values of wear rate. None of the theories presented earlier enables bearing life to be predicted from a knowledge of load, speed, material properties and bearing geometry, but some ways of using present knowledge to give design criteria are described below.

5.6.1 Application of Archard's Equation

If we wished to apply Archard's[4] wear equation 5.1 to a bearing operating under conditions of boundary lubrication, then that specific bearing type would have to be run and the wear rate measured in order to find k for those particular conditions. If subsequently the conditions were changed, say by an increase of load, and k was found to be the same as before, then Archard's equation could be applied at all intermediate loads. In other words, having found k experimentally over a range of conditions, we can use Archard's equation to interpolate but not to extrapolate. Changes of geometry, speed or load could each cause variations in k. However, we can use this equation in a qualitative way, for example, an increase of load will generally cause an increase in wear rate, an increase in material hardness will generally reduce the wear rate.

5.6.2 Applications of the IBM Model

The IBM models for wear, described earlier, are an attempt to improve this situation. In a number of publications the IBM team have described how the wear models can be applied to practical situations. The following example is taken from one of these papers[32], and applies the 'zero wear' model to the design of the steel-on-steel cam and follower shown in figure 5.10. The problem in this case is to ensure a satisfactory wear performance, that is, depth of wear is less than the original surface finish, for 10^6 revolutions of the cam under a load of 1 oz. Using the Hertz theory the maximum shear stress is

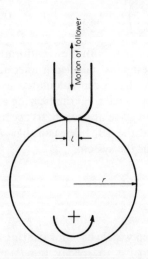

Figure 5.10   Cam and follower to be designed for 'zero' wear

123

found to be 7.95×10^3 lbf/in². The value of l, the dimension of the contact area in the direction of sliding is found, again from Hertz theory, to be 0.002 in.

Now because the follower is in constant contact, as opposed to the cam which is in contact only over $2\pi r/l$ of a revolution

$$M_{cam} = 10^6 \text{ passes}$$

$$M_{follower} = 6.28 \times 10^9 \text{ passes}$$

The value of γ_r must be determined experimentally, but once determined this value is appropriate for this combination of materials and lubricant for all loads. We now assume that γ_r has been determined and that in this case it had the value 0.54.

The values of yield points in shear are

$$\tau_{\gamma(cam)} = 4.0 \times 10^4 \text{ lbf/in}^2$$

$$\tau_{\gamma(fol)} = 1.5\hat{0} \times 10^5 \text{ lbf/in}^2$$

By substituting the above values in equation 5.12 the endurance limits can be calculated as

$$\left(\frac{2 \times 10^3}{M_{(fol)}}\right)^{1/9} \gamma\tau_{\gamma(fol)} = 1.552 \times 10^4 \text{ lbf/in}^2$$

for the follower, and

$$\left(\frac{2 \times 10^3}{M_{(cam)}}\right)^{1/9} \gamma\tau_{\gamma(cam)} = 1.09 \times 10^4 \text{ lbf/in}^2$$

for the cam. It can be seen that both these endurance limits are greater than τ_{max}, and therefore the design as specified is adequate.

If the endurance limits had been calculated as lower than τ_{max} then it would be necessary either to redesign the bearing and recalculate the endurance limits or, as an alternative approach, to accept a different criterion for acceptable wear and use the IBM Measurable Wear Model in order to calculate the wear rate for the system.

5.6.3 The *P–V* Factor

An approach that has been widely used for bearings operating under boundary lubrication conditions, is to specify the *P–V* factor for a particular bearing. This factor is the product of load per unit area and speed, and is found empirically for each type of bearing. The use of the *P–V* factor is considered more fully in chapter 6.

5.7 AN EXAMPLE OF WEAR IN PRACTICE—WEAR OF AN I.C. ENGINE

Prior to running, the various pairs of bearing surfaces in a new engine are not 'mated together'. There may be slight initial misalignments and there will certainly be 'high spots' on all surfaces. Bearing clearances will initially be small and therefore the cooling flow of oil is low and this, together with the initially higher friction, leads to higher than normal bearing temperatures. 'Running-in' is the process of removing high spots and mating the bearing surfaces; also the surfaces work-harden and become more wear resistant. During this period wear is higher than in normal running as the 'high spots' cause more asperity contacts, the misalignments reduce apparent contact areas and increase bearing pressures, and the higher temperatures usually cause higher wear rates.

If 'running-in' proceeds satisfactorily the 'high spots' are removed by adhesive wear and by plastic deformation. The wear gradually increases the real area of contact until it equals the design contact area, and the contact pressure drops. Frictional losses decrease during this period and bearing clearances increase, thus reducing the surface temperatures. The wear rate decreases due to all the above causes until it reaches the normal steady-state wear rate for the engine. However, if the initial misalignments are sufficiently high the apparent bearing surface area might be so low as to cause the pressure to be higher than one-third of the hardness. In this case, the wear rate is excessively high as shown by figure 5.2 and the real contact area may not grow sufficiently to reduce the pressure to less than $H/3$, that is, to reduce the wear rate to normal levels, before large-scale damage or even seizure has occurred.

The wear rate during 'running-in', even when misalignments are minimal, is higher than during normal running. Thus a greater quantity of wear debris is produced and this, together with small amounts of casting sand and machining swarf which are often present in new engines, is filtered out of the lubricating oil to avoid excessive abrasion.

After the 'running-in' period, the two major sources of wear are abrasion and corrosion. Abrasive wear is kept to a minimum by the exclusion of foreign particles, and by efficient oil filtration. Corrosion is thought to be a major mechanism of wear of cylinders and piston rings in car engines although modern oil formulations have reduced corrosion considerably. The presence of sulphur and additives in the petrol may be one cause of corrosion, but the major cause is carbonic acid formed from the combustion products, CO_2 and water. Burwell[2] describes the situation as follows:

A car is started from cold in the morning and driven to work, where it is parked and cools down, possibly to sub-zero temperatures. In the evening the process is repeated. Thus two cycles from cold to hot to cold are repeated daily even though the total distance driven might be small. Each time the engine is cooled the carbonic acid condenses on the cylinder walls, producing abrasive corrosion products (e.g., Fe_2O_3). The next start-up wears away these corrosion products producing additionally some abrasion. This type of mechanism has been confirmed by comparative wear tests in which two engines were run for equal times under similar conditions, other than that one was run continuously, and the other was run intermittently and allowed to cool after each run. The cylinder wear of the intermittently run engine was many times higher than that of the engine run continuously. Analysis of the oil from the engines also confirmed the importance of the corrosive mechanism during intermittent running.

During continuous running it has been found that hydrodynamic conditions exist between the piston and cylinder in all regions except those close to extreme piston positions, i.e., BDC and TDC where the rubbing speed is zero and lubrication is in the boundary region. This is confirmed by wear measurements on the cylinder wall from which it is found that wear is greater at the extreme ends of the stroke, and that wear at TDC is greater than that at BDC. This is due to the lower temperature and better availability of oil at the bottom of the stroke.

5.8 CONCLUSIONS

The above example of the piston/piston-ring interface in an internal-combustion engine illustrates the way in which the wear of a very common-place mechanism can be due to the combination, in varying degrees, of many of the different types of wear described in this chapter. Material at the interface is worn away by adhesion, abrasion, corrosion and fatigue. In addition, the relative importance of these wear mechanisms varies for different positions on the cylinder due to differences in environment, speed, load, temperature and efficiency of lubrication. This single example serves to show that the process of wear is indeed a very complex one, and perhaps explains why attempts to predict wear rates in practical situations have, so far, met with only limited success.

REFERENCES

1. I. V. Kragelskii. *Friction and Wear*. Butterworths, London, (1965).
2. J. T. Burwell. Survey of possible wear mechanisms. *Wear*, 1, (1957), 119–141.
3. M. Kerridge and J. K. Lancaster. The stages in a process of severe metallic wear. *Proc. R. Soc.*, 236A, (1956), 250–264.
4. J. F. Archard. Contact and rubbing of flat surfaces. *J. appl. Phys.*, 24, (1953), 981–988.
5. R. Holm. *Electric Contacts*. Almgvist & Wicksells, Uppsala, (1946).
6. J. T. Burwell and C. D. Strang. Metallic wear. *Proc. R. Soc.*, 212A, (1952), 470–77.
7. M. M. Kruschov. Resistance of metals to wear by abrasion; related to hardness. Instn mech. Engrs Conf. Lubrication and Wear, London, (1957), 655–59.

8. C. N. Rowe. Some aspects of the heat of absorption in the function of a boundary lubricant. *Trans. Am. Soc. Lubric. Engrs*, **9**, (1966), 100–111.
9. A. G. Cattaneo and E. S. Starkman. A.S.M. Symposium on Mechanical Wear, Cleveland, Chapter IV, 1950.
10. R. C. D. Richardson. The wear of metals by relatively soft abrasives. *Wear*, **11**, (1968), 245–75.
11. T. L. Oberle. Properties influencing wear of metals. *J. Metals*, **3**, (1951), 438–9.
12. R. T. Spurr and T. P. Newcombe. The friction and wear of various materials sliding against unlubricated surfaces of different types and degrees of roughness. Instn. mech. Engrs. Conf. Lubrication and Wear, London, (1957), 269.
13. H. Hertz. *Z. reine angew. Math.* **92**, (1881), 155.
   or see
   S. P. Timoshenko and J. N. Goodier. *Theory of Elasticity*. McGraw-Hill, New York, (1970), p. 409.
14. J. K. Lancaster. The relationship between the wear of carbon brush materials and their elastic moduli. *Br. J. appl. Phys.* **14**, (1963), 497–505.
15. J. F. Archard and W. Hirst. The wear of metals under unlubricated conditions. *Proc. R. Soc.*, **236A**, (1956), 397–410.
16. R. G. Bayer, W. C. Clinton, C. W. Nelson and R. A. Schumacher. Engineering model for wear. *Wear*, **5**, (1962), 378–91.
17. J. F. Tavernelli and L. F. Coffin. A compilation and interpretation of cyclic-strain fatigue tests. *Trans. Am. Soc. Metals*, **51**, (1959), 438–53.
18. J. G. A. Bitter. A study of erosion phenomena. *Wear*, **6**, (1963), 51–69.
19. E. Rabinowicz. *Friction and Wear of Materials*. Wiley, New York, (1965).
20. G. Levy, R. G. Linford and L. A. Mitchell. Wear behaviour and mechanical properties; the similarity of seemingly unrelated approaches. C.E.G.B. Report No. RD/B/N2250, (1972).
21. R. L. Aghan and L. E. Samuels. Mechanisms of abrasive polishing. *Wear*, **16**, (1970), 293–301.
22. R. G. Bayer. Prediction of wear in a sliding system. *Wear*, **11**, (1968), 319–31.
23. F. P. Bowden and G. W. Rowe. The adhesion of clean metals. *Proc. R. Soc.*, **233A**, (1956), 429–42.
24. E. E. Bisson, Friction and bearing problems in the vacuum and radiation environments of space. *Advanced Bearing Technology*, NASA, SP-38, (1964), 259–287.
25. R. W. Wilson. Influence of oxide films on metallic friction. *Proc. R. Soc.*, **212A**, (1952), 450–52.
26. F. T. Barwell. Wear of metals. *Wear*, **1**, (1957–8), 317–32.
27. A. J. W. Moore and D. Tabor. C.S.I.R. (Australia), Tribophysics Report A46, (1942).
28. M. J. Hordon. Adhesion of metals in high vacuum. Adhesion or Cold Welding of Materials in Space Environment. A.S.T.M., STP 431, (1967), 109–127.
29. N. C. Welsh. Frictional heating and its influence of wear of steel. *J. appl. Phys.*, **28**, (1957), 960–68.
30. M. C. Shaw. Lubrication, friction and wear under gross plastic deformation. Friction and Wear Interdisciplinary Workshop, NASA Lewis Research Center, Cleveland, Ohio, TM-X-52748, (Nov. 1968).
31. E. Rabinowicz. Compatibility criteria for sliding metals. Friction and Lubrication in Deformation Processing. Am. Soc. mech. Engrs, (1966), N.Y., 90–102.
32. R. G. Bayer, W. C. Clinton, T. C. Ku, C. W. Nelson, R. A. Schumacher, J. L. Sirico and A. R. Wayson. Applying the wear model to design problems. A.S.M.E. Design Eng. Conf., Paper No. 66 MD-12, (May, 1966).

6

Tribological Properties of Solid Materials

6.1 INTRODUCTION

In previous chapters on friction and wear theories we have described the basic interactions between moving surfaces, and the effects of these interactions on friction coefficients and wear rates. In the present chapter the aim is to explain the effects of material properties and environment on the observed behaviour and, in particular, to describe the behaviour of materials which have good tribological properties.

Undoubtedly the most effective way to reduce friction and wear is to completely separate the solid surfaces by means of a film of a viscous fluid, as described in the chapters on hydrodynamic and hydrostatic lubrication. As long as such a fluid film can be maintained the coefficient of friction will be very low, of the order of 0.003 or less, and wear will be entirely eliminated. Under these circumstances the properties of the solid surfaces are only of secondary importance. However the properties of the surfaces do become very important under the two sets of circumstances described below.

(1) If a film of fluid lubricant cannot effectively separate the solid surfaces, as during the stopping and starting of hydrodynamically lubricated equipment, then conditions of boundary lubrication obtain and, as this implies, metal-to-metal contact can occur at contacting asperities. We therefore need to use materials which will give acceptable tribological

128

behaviour during these periods of metal-to-metal contact. In the first part of this chapter we therefore describe the tribological properties of clean metal surfaces, and of surfaces covered by naturally occurring films such as oxides, and discuss the criteria which give such surfaces acceptable tribological properties.

(2) If for some reason—which may be technical or economic—fluid lubrication is not considered to be suitable for a particular application, we must use solid surface materials which have inherently good tribological properties. In the second part of this chapter we shall discuss the properties and uses of such materials.

Throughout the chapter we shall consider the effects of environmental variables, such as temperature and atmosphere. We shall indicate the limits of usefulness of the various materials in terms of these variables, and the operational parameters such as load and speed.

6.2   TRIBOLOGICAL PROPERTIES OF METALS

6.2.1   Clean Metal Surfaces

We have already seen, in chapters 4 and 5, that the friction coefficient and wear rate on rubbing together clean metal surfaces can become very high, and in many cases gross seizure of the opposing surfaces can occur. This has been shown by the modified Bowden and Tabor theory of friction to be the result of the plastic deformation of the contacting asperities under the combined normal and tangential stresses.

Under most normal conditions steps are taken to minimise the amount of metal-to-metal contact but, as we have seen, this can rarely be eliminated and care must therefore be taken in material selection to reduce the ensuing friction and wear. Some of the material properties which must be considered are described below.

Hardness

Many metals show a transition from mild to severe wear when the nominal contact pressure, that is, the load divided by the apparent area of contact, becomes greater than some fraction of the hardness. For many metals this fraction has a value of about one-third. This transition is generally attributed to the interaction of the plastic zones beneath contacting asperities, as shown in figure 6.1, so that gross plastic deformation can take place. To avoid this effect it is clearly desirable to choose materials which have a hardness several times greater than the apparent contact pressure.

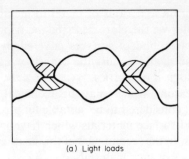

(a) Light loads

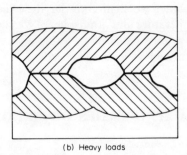

(b) Heavy loads

Figure 6.1   Plastic zones be-
low contacting asperities

Mutual Solubility

There is some evidence to suggest that metals exhibiting a high degree of mutual solubility will have poor tribological properties. One analysis by Rabinowicz[1] of published friction, adhesion and wear results for different metal couples has shown zero correlation of solubility with adhesion, a positive correlation with friction, and a greater correlation with wear.

This suggests that metal pairs for boundary lubricated or unlubricated sliding should be chosen to have low mutual solubilities. Certainly it is generally accepted that similar metals, that is, metals with 100 per cent mutual solubility, should be avoided.

Crystal Structure

The Bowden and Tabor theory indicates that perfectly clean surfaces which are able to deform plastically will exhibit seizure before gross sliding can occur. There is however one important class of metals which does not behave in this way and which will give reasonable friction coefficients, ($\mu \simeq 0.2$–0.4) and low wear rates even when absolutely clean. These are the metals which have hexagonal close packed crystal structures and which plastically deform by slip on a single slip plane, the basal plane. It is generally thought that the good tribological properties of these materials are explained by this limited slip behaviour, as the plastic deformation of the type invoked in the Bowden and Tabor theory requires slip on several different slip planes. Without this it is not possible to get the continuous junction growth which is the basis of the theory.

Although this type of behaviour was of no more than theoretical interest a few years ago, recent developments have created a demand for strong materials which can be relied upon to run together under completely clean conditions. Materials of this type have recently been used as hinges on the doors of spacecraft, and they are also likely to find increasing use in inaccessible areas such as the interior of nuclear reactors.

Before leaving the subject of clean metals it is again worth mentioning the use of soft metal films on hard substrates which has already been described in chapter 5. These films are also finding increasing use in vacuum chambers and space environments. An interesting metal in this context is lead, which as a film on a hard substrate, is an excellent lubricant under vacuum conditions, although poor in normal atmospheres. This difference is thought to be due to the continuous conversion of lead to lead oxide under atmospheric conditions.

6.2.2 Surfaces in Normal Atmospheres

Under normal atmospheric conditions most metals are covered by oxide films and clean metal surfaces will acquire such a film 5 to 50 molecules thick within a few seconds. Thus if the oxide is removed from an asperity by rubbing it will almost certainly have been reformed before the asperity makes another contact with the opposing surface.

The presence of an oxide film can be very beneficial, but unless the loads are small the oxide cannot completely prevent intermetallic contact. As a result of this, it has been found that the wear behaviour of many metals under normal atmospheric conditions, can be divided into two regimes. In the mild wear regime, which occurs at lower loads, the electrical contact resistance is high, the wear debris is fine and consists mainly of oxide, and the metal surfaces have a burnished appearance. In the severe wear regime, which occurs at higher loads, the contact resistance is low, the wear debris includes coarse metallic particles, and the surfaces are rough. The transition between these two regimes is clearly at the load at which large scale penetration of the oxide films, and consequently metallic adhesion, take place.

The properties of the particular oxide films are important. For example the oxide formed on aluminium is hard and brittle and offers little protection whereas, under suitable conditions, the films formed on steels are tough and tenacious and offer excellent protection. This latter fact has been utilised in nuclear reactors, where plain stainless steel bearings are used at elevated temperatures under which conditions they form suitable oxide films.

6.2.3 Effects of Temperature

It is clear that the temperature of the rubbing surfaces can have a major influence on all the properties described above. We can consider them in turn.

Hardness

The hardness of metals generally decreases with increasing temperature and it can be seen that, in the absence of other effects, wear rates will increase with increasing temperature. To counteract this effect it is necessary to use materials of high hot-hardness, such as tool steel and alloys based on cobalt, chromium, and molybdenum, for high temperature bearings.

Mutual Solubility

The mutual solubility of metal pairs is a function of temperature, and although this effect has not been investigated, it is therefore quite possible that increases in solubility with increasing temperature could adversely effect tribological properties.

Crystal Structure

Temperature-induced phase changes can have a profound effect on tribological properties. An excellent example of this is cobalt which at 417 °C changes from its low temperature hexagonal close packed structure to a face centred cubic structure. Wear experiments by Buckley[2] on cobalt, the results of which are illustrated in figure 6.2, show that the wear rate at 350 °C is one hundred times greater than that at 280 °C. This is undoubtedly due to the combined ambient and frictional heating transforming the interface material into the face centred cubic regime. This enables junction growth to take place, due to the increased number of operative slip systems, as mentioned in the foregoing section headed 'crystal structure'.

Changes of temperature have a marked effect both on rates of oxidation and on the types of oxide formed. This can clearly have a marked influence on tribological properties. This has been strikingly demonstrated by Kragelskii[3] by rubbing pure Armco iron against itself at different speeds (and hence different interface temperatures). At low speeds the wear rate was high, with evidence of much adhesion, but at higher speeds the wear rate fell by almost three orders of magnitude and the surfaces became smooth and polished. The temperature of the interface at the transition speed was calculated to be about 1000 °C. To prove that this was a temperature rather than a speed effect the experiment was repeated at very low speed while high current pulses were passed through the contacts to raise them to the same temperature. The wear rate was again very low and the surfaces were polished.

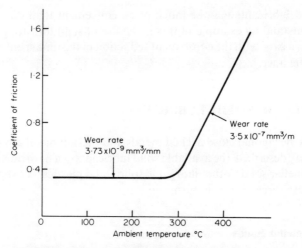

Figure 6.2 Friction coefficient and wear rate for cobalt on cobalt in vacuum at various temperatures— pressure 10^{-7} N/m^2, sliding speed 2 m/s, load 9.81 N

6.3 SELF-LUBRICATING MATERIALS

In this section we shall describe materials which have intrinsically good tribological properties, but before describing specific examples it is worth considering why we use self-lubricating materials.

6.3.1 Advantages of Self-lubricating Materials

(a) Solid lubricants operate over a greater temperature range than fluids. Oils become thin and decompose or oxidise at high temperatures; at low temperatures they become very viscous, and eventually may solidify.

(b) Solid lubricants give better surface separation than liquid boundary lubricants at high loads and low speeds.

(c) Many solid lubricants are much more chemically stable than liquids and they can be used in environments such as strong acids, solvents, and liquid gases.

(d) Solid lubricants are usually very clean and can be used in environments where cleanliness is essential, such as in food-processing equipment.

(e) Solids can often be used to provide permanent lubrication for parts of equipment which are inaccessible after assembly.

(f) By using solid lubricants, designs can be simplified by eliminating complicated passageways and oil-circulating equipment.

(g) Solid lubricants are quite stable in very radioactive environments, whereas oils and greases are degraded.

133

(h) Solid lubricants may be much more convenient than oils or greases. An outstanding example of this is the use of self-lubricating bushes on modern cars, with the consequent reduction in the time spent on periodic maintenance.

6.4 TYPES OF SOLID LUBRICANT

Besides the hexagonal close packed metals and the soft metal films which we have already described, the available solid lubricants can be divided into three groups, lamellar solids, other inorganic solids, and plastics. We shall describe each of these three groups in turn.

6.4.1 Lamellar Solids

A lamellar solid is one in which the atoms are bonded together in parallel and comparatively widely spaced sheets. The two best known and most widely used examples are graphite and molybdenum disulphide, which have the crystal structures shown in figures 6.3 and 6.4. Under many circumstances these are both excellent lubricants, as are other lamellar solids such as tungsten disulphide, cadmium chloride and cadmium iodide. However not all lamellar solids act as lubricants, and there is as yet no theory which will tell us whether or not a particular lamellar solid will be suitable.

The lamellar solids which can lubricate effectively do have certain characteristics in common. The most important characteristic seems to be the ability to form a strongly adherent transferred film on to the surface being

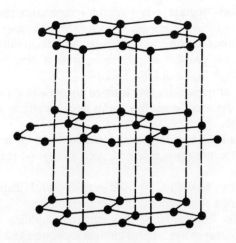

Figure 6.3 The crystal structure of graphite

134

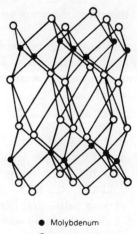

● Molybdenum
○ Sulphur

Figure 6.4 The crystal structure of molybdenum disulphide

lubricated. After the initial running in period, during which this film is formed, the interface therefore consists of lubricant on lubricant.

A second characteristic is that the material at both surfaces develops a preferred orientation which is shown schematically in figure 6.5. This orientation reduces the mechanical interaction between the surfaces, as can

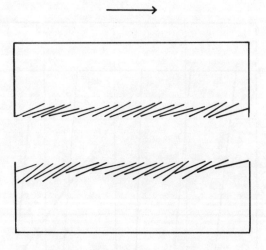

Figure 6.5 Schematic illustration of the surface orientation developed during the rubbing of lamellar solid lubricants

be shown by reversing the direction of motion when the friction coefficient increases significantly.

Although a great amount of experimental work has been performed on graphite and molybdenum disulphide there is as yet no universally accepted explanation of their behaviour. In the following paragraphs we shall briefly describe some of the suggested explanations.

Graphite

The earliest suggested explanation for the properties of graphite was that due to Bragg[4], who suggested that the shear strength parallel to the atomic sheets was very low. This would allow the sheets of atoms to slide over each other rather like a pack of playing cards, and hence enable the graphite to act as a boundary lubricant. This explanation was accepted until the Second World War, when it was found that graphite brushes in electrical generators of high-flying aircraft wore out very rapidly. A systematic investigation of this effect by Savage[5] showed that graphite had very poor tribological properties in the absence of condensable vapours. This effect is very marked and the amount of vapour needed for effective lubrication can be very low: for example exposing dry graphite to a pressure of 400 N/m^2 of water vapour decreases the wear rate by three orders of magnitude, and the friction coefficient by a factor of five. The effectiveness of different vapours varies enormously, and while nitrogen has no lubricating effect at atmospheric pressure many organic vapours are effective at very low pressures, as shown in figure 6.6.

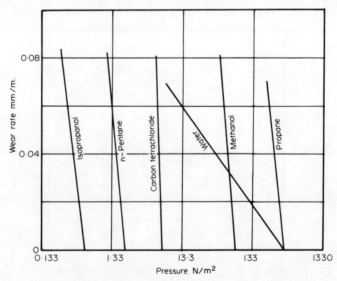

Figure 6.6   Effects of condensable vapours on the wear of graphite[9]

136

Alternative explanations of this effect have been advanced. The first explanation suggests that the vapours penetrate between the layers and reduce the shear strength to a very low level. The second, and more widely supported, suggestion is that the vapours saturate the surface forces, and particularly the edge forces, on individual crystallites. The adhesion between neighbouring crystallites is therefore very low so that they can easily slide over each other. The surface layer of graphite, rather than the individual crystals, can thus be sheared by low forces.

This second explanation has been supported by the work of Midgely and Teer[6] and the work of Arnell, Midgely and Teer[7], who have shown that even nearly amorphous mechanical carbons and brittle pyrolytic carbons exhibit the same low friction behaviour as graphite. Such behaviour is not compatible with a low shear strength theory.

Molybdenum Disulphide

Unlike graphite, molybdenum disulphide, MoS_2, appears to be an intrinsically good lubricant and in fact its friction coefficient is lower under high vacuum conditions than in the presence of water vapour. Again there are alternative explanations of this inherent low friction. The first explanation suggests that molybdenum disulphide is a true low shear strength solid, with shear taking place between the adjacent layers of the sulphur atoms shown in figure 6.4. The second explanation, which is similar to that advanced for graphite, is that the edge forces on the crystals are saturated by oxygen atoms. Unlike the vapours on graphite, however, the oxygen is not volatile, but persists up to the decomposition temperature of the solid.

Lamellar solids can be used in several different forms, the most common ones being described below:

(1) As dry powders or dispersions in fluids. The oldest method of using these lubricants, and one which is still used extensively with MoS_2 is to rub the surfaces to be lubricated with dry powders. Burnished films of MoS_2 tend to adhere better than graphite, and the former is therefore often applied as a dispersion in a solvent which subsequently evaporates to leave a film of dry powder. Dispersions and dry powders are used extensively to facilitate the assembly of close fitting parts, to lubricate metal working components such as wire drawing dies, and as parting agents for screw threads.

(2) Solid Blocks. Graphite and graphite carbons are often bonded together into solid blocks which can, for example, be made into solid thrust bearings. Such carbon bearings were used for many years as the clutch release bearings of cars. Solid carbon bearings are generally made from a mixture of finely divided coke and a carbon binder such as pitch, which is then fired to a very high temperature. This heat treatment graphitises the entire mixture, and by varying the heat-treatment temperature and time, a large range of materials can be made. These range

from highly crystalline electrographite, which is generally used for low load applications such as electrical generator brushes, to almost amorphous mechanical carbons, which have high strength and are commonly used for thrust bearings.

(3) Bonded Films. Solid lubricants are frequently bonded to metal surfaces by using organic resin binders, and such films have wear lives two orders of magnitude higher than those of dry powder films. Because of their excellent tribological properties bonded MoS_2 films are used extensively throughout industry. The main drawback of these films, compared with a dry powder film, is that the decomposition of the resin imposes a limit to the maximum running temperature, and therefore to the load and speed.

(4) Metal Composites. The thermal limit mentioned in the previous paragraph can be overcome by incorporating the lubricant in a metal matrix. The relative proportions of lubricant and metal in these films are quite critical as can be seen from figure 6.7. Too much lubricant weakens the composite, whereas too little provides an insufficient supply of lubricant.

These composites provide an excellent illustration of the fundamental principle which lies behind the design of bearings, that is, to provide high load-carrying capacity with a low shear strength. If the composites contain insufficient lubricant they have excellent load-carrying capacity, but the shear strength is close to that of the pure metal. Such

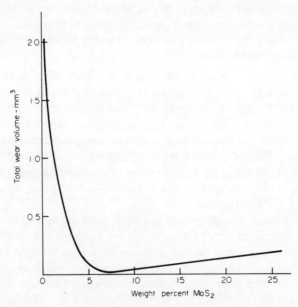

Figure 6.7   Wear of powder metallurgy specimen containing molybdenum disulphide[10]

138

composites therefore have friction coefficients close to those of the pure metals. Furthermore the lubricant cannot provide adequate surface coverage and the wear rate is also high. On the other hand, if we have excess lubricant, the surface shear strength is very low, but the composite structure is weakened, and the load-carrying capacity is also low.

(5) Grease Additions. The major use of MoS_2 is as an additive to greases and oils. Such filled greases are very effective during running-in operations and increase the effectiveness of lubrication under heavy loads. They are also used to prevent seizure of threaded components.

6.4.2 Inorganic Solids for High Temperature Lubrication

Apart from the lamellar solids already described many inorganic solids have been tried as solid lubricants with varying degrees of success. It is not the purpose of this chapter to present an exhaustive list of the successful lubricants, but two of the inorganic solids which are very promising as high temperature lubricants are described below.

(a) Lead monoxide. This is a poor lubricant at temperatures below approximately 250 °C, but above this temperature it is a better lubricant than MoS_2 and it retains excellent properties at temperatures up to approximately 650 °C. Lead monoxide can be used in ceramic bonded films which have excellent wear properties, and such a film at 650 °C has better properties than a resin bonded MoS_2 film at room temperature. This material has an upper temperature limit of approximately 700 °C, as the coating softens at higher temperatures.

(b) Calcium Fluoride. For lubrication at temperatures above 700 °C one of the most effective lubricants so far discovered is a ceramic bonded film of calcium fluoride. This retains excellent properties at temperatures above 1000 °C, and it has better wear properties at high temperatures than either molybdenum disulphide or lead oxide at their optimum temperatures.

6.4.3 Plastics

Plastics are used successfully in many tribological applications, the most widely used being nylon, fluorocarbons such as polytetrafluoroethylene (PTFE), and phenolic laminates. In addition to the advantages already cited for self lubricating materials, plastic bearings have the following advantages

(1) They absorb vibrations well and are quiet in operation.

(2) They readily deform to conform to mating parts: machining tolerances and accuracy of alignment are therefore less critical than for metal parts.

139

(3) They are easily formed into complicated shapes, either by machining or moulding.
(4) They are very cheap.

Plastics are used in many forms, such as solid plastic, resin bonded films, and as composites impregnated with other substances to give improved tribological, physical and mechanical properties. There is an enormous range of materials available and the aim here is simply to describe the properties of the base materials and to indicate the relative advantages of the various forms.

6.5 TRIBOLOGICAL PROPERTIES OF PLASTICS

6.5.1 Friction of Plastics

It has been shown by King and Tabor[8] that, for most plastics, the friction force between a plastic surface and a steel slider is as predicted by the simple Bowden and Tabor theory of friction, that is, the friction force is approximately equal to the bulk shear stress of the plastic multiplied by the effective area of contact. It is not, however, correct to state specific values of friction coefficients for plastics. Such materials are usually viscoelastic and their deformation is therefore dependent on strain rate. As this implies, the friction coefficients of plastics can vary very significantly with such parameters as sliding speed and surface roughness. However, the coefficients of friction of most plastics, on metals and on themselves, are generally in the range 0.2–0.4. PTFE is an outstanding exception to this generalisation in that the coefficient of friction for PTFE sliding on itself can be as low as 0.05 which is the lowest known value for any solid. There is as yet no firm explanation of this low friction coefficient, although it is generally ascribed to the intrinsically low adhesion between PTFE molecules. PTFE has long rigid molecules in which carbon atoms are effectively shielded by the surrounding fluorine atoms. The actual adhesion forces between molecules are therefore thought to be very low, so that the surface molecules can roll or slide over each other. On the other hand the mechanical interlocking of the molecules in the bulk material gives it a relatively higher hardness and bulk shear strength.

6.5.2 Wear of Plastics

With the exception of PTFE, the friction coefficients of plastics are not particularly low, but the main advantage of plastics is that they wear at low and reasonably predictable rates. This enables a designer to select, with some accuracy, a bearing material which will give the desired lifetime under specified conditions of load and speed. The rate of wear of a plastic bearing is obviously a function of load and speed, and it is found that we can define a design criterion, the $P–V$ factor, for plastic bearings as described below.

6.5.3 The P–V Factor

The derivation of the P–V factor is based on the reasonable assumption that the rate of wear will be proportional to the rate of energy dissipation at the sliding interface. Starting from this assumption we shall derive the relationship between wear rate and the P–V factor for the two basic bearing configurations.

(i) Flat Bearings. If we have a flat bearing surface of area A, as shown in figure 6.8, subjected to a normal load W, then the amount of energy

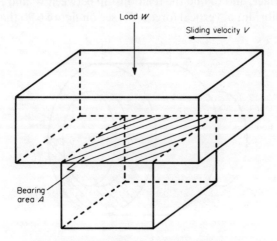

Figure 6.8   Schematic diagram of flat bearing for derivation of P–V factor

dissipated on sliding a distance $\mathrm{d}x$ is given by $\mu W\,\mathrm{d}x$, where μ is the coefficient of sliding friction. The rate of energy dissipation is therefore given by

$$\mu \times W \times \frac{\mathrm{d}x}{\mathrm{d}t} = \mu W V$$

where V is the sliding speed at the interface. If we then assume that the volume wear rate $\dot{Q}$ is proportional to the rate of energy dissipation we have, for constant μ

$$\dot{Q} \propto W V$$

The factors we are normally interested in are the depth of wear, or the rate of linear wear normal to the sliding surface. We can see that the depth of wear at any time is simply equal to the volume of wear divided by the total contact area. Therefore

$$\text{rate of linear wear} = \frac{\dot{Q}}{A}$$

141

and

$$\text{rate of linear wear} \propto \frac{W}{A} V$$

or

$$\text{rate of linear wear} \propto PV$$

(ii) Sleeve Bearings. Figure 6.9a shows a circular bearing subjected to a normal load W, which is supported over the shaded bottom half of the bearing. We assume that we have a normal reaction R per unit area at this interface, and to find the relationship between W and R we consider the equilibrium of vertical forces. We see on figure 6.9b that the reactive

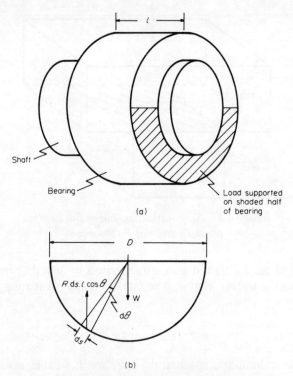

Shaft

Bearing

Load supported
on shaded half
of bearing

(a)

(b)

*Figure 6.9   Schematic diagram of solid lubricated
journal bearing and forces acting*

force over an element of surface of length ds will be given by $R\,\mathrm{d}sl$, where l is the axial length of the bearing, and the vertical component of this force will be

$$R\,\mathrm{d}sl\cos\theta$$

but

$$\mathrm{d}s = \frac{D}{2}\,\mathrm{d}\theta$$

142

Therefore the total vertical reaction is given by

$$\frac{RlD}{2} \int_{-\pi/2}^{+\pi/2} \cos \theta \, d\theta = RlD$$

This is equal to the applied vertical load W so that

$$R = \frac{W}{Dl}$$

But the rate of energy dissipation per unit area at the interface $= \mu R V$. Therefore the total rate of energy dissipation at the interface

$$= \mu R V \times \frac{\pi D}{2} l$$

$$= \mu \frac{W}{Dl} V \frac{\pi D}{2} l$$

$$= \mu W V \frac{\pi}{2}$$

Therefore

$$\text{volume wear rate} \propto \mu W V \pi$$

We are normally interested in the radial wear rate and we can see from figure 6.10b that this is given by

$$\frac{\text{volume wear rate}}{\pi D l}$$

Therefore

$$\text{radial wear rate} \propto \frac{\mu W V \pi}{\pi \, dl}$$

or

$$\text{radial wear rate} \propto \mu P V$$

where P is the load per unit projected area.

We can see from the above derivations that the same P–V factor applies to each type of bearing, provided that P is taken as the load per unit *projected* area.

To a good approximation it is possible to specify a certain P–V factor which must not be exceeded for a certain life expectancy. The use of a single factor is open to some criticism since it assumes the same sensitivity of wear rate to changes of P and V. This is frequently not the case and a more exact practice is to give plots of acceptable wear rates on pressure velocity diagrams

143

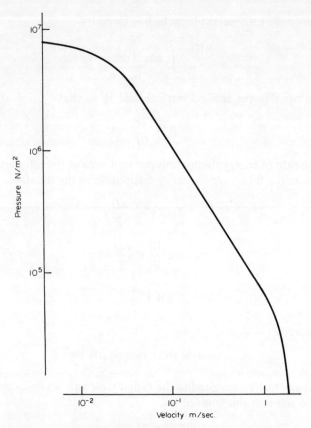

Figure 6.10 Typical limiting P–V curve of PTFE
based material for wear rates of 25 μm in 100 hours

as shown in figure 6.10. It can be seen from these diagrams that the P–V factor for a particular wear rate is reasonably constant except at the extremes of load and speed.

The P–V limit of a bearing is that P–V factor at which the bearing would fail rapidly due to melting or thermal decomposition.

It is also found that the steady state wear rate, $\dot{Q}$, for a plastic bearing is proportional to the P–V factor over much of the usable P–V range. We can therefore define a wear coefficient K by the equation

$$\dot{Q} = KPV$$

It can be seen from this equation that, if the wear rate is known for one P–V value, K can be calculated and the wear rates for other P–V factors determined.

It should be pointed out that the use of the P–V factor is not confined to plastic bearings; similar factors can be given for other types of bearing and, in particular, for resin bonded solid lubricant films.

144

6.5.4 Factors Influencing the Wear of Plastic Bearings

(1) Lubrication. The performance of plastic bearings can be improved by lubrication. Periodic lubrication can raise the $P-V$ limit by several hundred per cent, and under continuous lubrication conditions the operation is limited only by the mechanical strength of the plastic.

(2) Temperature and Heat Dissipation. It has already been stated that the $P-V$ limit of a bearing is reached when the interface starts to melt or thermally decompose. The $P-V$ limit is therefore clearly affected by the ambient temperature. It is also affected by the temperature rise at the sliding interface and hence by the rate of heat dissipation from the interface. In general the thermal conductivities of plastics are low and some technique has to be adopted to assist the dissipation of heat. Such techniques include the addition of thermally conducting fillers, such as metal powders, and the use of the plastic as thin liners inside metal sleeves. In addition it is possible to use bearings intermittently at increased $P-V$ levels, allowing the bearing to cool between operating periods.

6.5.5 Filled and Reinforced Plastics

All types of plastic bearings are used either by themselves or with a large variety of fillers or reinforcements. The additives are used for several different purposes, as listed below.

(1) Improvement of Mechanical Properties. One disadvantage of plastic bearings, in comparison with metals, is their lack of rigidity and strength. These properties can be appreciably improved by various additions; by far the most common being chopped glass-fibre. The addition of glass fibre increases tensile strength, rigidity, and creep resistance, and hence enables reinforced bearings to operate to higher $P-V$ values.

In the case of phenolic laminate bearings the properties of the bearing are of course affected by those of the laminating material. Typical laminating materials are paper, linen, canvas, and woven glass-fibre. In general the coarser materials such as canvas give the highest strength and toughness, whereas the finer materials have better machining properties and can be used for precision components.

(2) Improvement of Thermal Properties. We have already discussed the use of metal fillers to improve thermal conductivity and heat dissipation. An additional disadvantage of plastic materials is that they generally have high thermal expansion coefficients compared with metals. To allow for this discrepancy, plastic bearings have often to be made to much looser fits than other bearings. This discrepancy is again reduced by filling or reinforcing the plastic, thus enabling the bearings to be made to finer tolerances.

145

(3) Improved Friction and Wear. The most common additives to improve the friction and wear properties of plastic bearings are graphite and molybdenum disulphide. These can improve $P-V$ values by an order of magnitude, and wear rates by two or three orders of magnitude. They also have the additional advantage of improving, to some extent, the thermal and mechanical properties of the materials.

REFERENCES

1. E. Rabinowicz. *Proc. Am. Soc. mech. Engrs/Am. Soc. Lubric. Engrs Lubrication Conference*,(October 1970).
2. D. H. Buckley. *Cobalt*, **38**, (1968), 20–28.
3. I. V. Kragelskii. Friction and Wear. Butterworths, London, (1965), p. 91–2.
4. W. L. Bragg. Introduction to Crystal Analysis. Bell, London, (1924).
5. R. H. Savage. *J. appl. Phys.* **19**, (1948), 1.
6. H. W. Midgely and D. G. Teer. *Nature*, **189**, (1961), 735.
7. R. D. Arnell, J. W. Midgely and D. G. Teer. *Proc. Instn. mech. Engrs*, **179**, 3j, (1966), 115.
8. R. F. King and D. Tabor. *Proc. Phys. Soc.*, **65B**, (1953), 728.
9. R. H. Savage and D. L. Schaefer. *J. appl. Phys.*, **27**, (1956), 136.
10. R. L. Johnson, M. A. Swikert and E. E. Bisson. NACA TN.2027, (1950).

FURTHER READING

E. R. Braithwaite. *Solid Lubricants and Surfaces*. Pergamon Press, Oxford, (1964).
E. E. Bisson and W. J. Anderson. *Advanced Bearing Technology*. NASA SP-38, (1964).
F. J. Clauss. *Solid Lubricants and Self Lubricating Solids*. Academic Press, London, (1972).

7

Friction Instability

7.1 INTRODUCTION

Unwanted vibrations which may arise during the operation of machines are costly in terms of reduction of performance and service life, sometimes endangering equipment and personnel. This chapter is concerned with those vibrations occurring through the agency of friction forces at the sliding parts. It introduces methods of study, examines the characteristics, and considers the prevention or alleviation of the vibrations.

The reader will have encountered such vibrations manifested, for instance, by the squealing of brakes, the creaking of hinges and the ringing of wine glasses. He may also have witnessed the jerkiness which may occur when a large mass is driven at slow speeds along slideways. All these involve friction in a dominant role. They are most prevalent at low sliding speeds and often they produce not only noise, but also severe wear and dimensional inaccuracy. The indirect effects may be serious, as in textile machinery where oscillations can lead to objectionable patterning in the cloth[1].

The engineer concerned generally with vibrations requires a comprehensive knowledge of theory; for our purpose, the study of frictional vibrations at an elementary level, a short review of the basic ideas is a useful preliminary.

A system is said to be in a state of vibration when one or more of the variables describing its state changes in a periodic manner. The most simple example is the linear oscillator, figure 7.1, which may be represented, when free from external influences, by an ordinary differential equation of the form

$$\frac{d^2x}{dt^2} + \omega_n^2 x = 0 \tag{7.1}$$

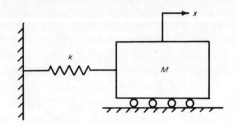

Figure 7.1   Linear oscillator

where x is a system variable and ω_n is a parameter of the system. The solution of equation 7.1, representing the unforced motion, is

$$x = A \sin(\omega_n t + \varepsilon) \tag{7.2}$$

where A and ε are constants determined by the initial state of the system. Equation 7.2 shows that the motion is sinusoidal, of amplitude A and frequency ω_n rad/s.

In those situations where account has to be taken of energy-dissipative forces, a more realistic linear model is of the type

$$\frac{\mathrm{d}^2 x}{\mathrm{d} t^2} + 2\zeta\omega_n \frac{\mathrm{d} x}{\mathrm{d} t} + \omega_n^2 x = 0 \tag{7.3}$$

where ζ is a constant parameter of the system. The solution of equation 7.3 appears in three forms. For $|\zeta| < 1$

$$x = A\mathrm{e}^{-\zeta\omega_n t} \sin[\omega_n(1 - \zeta^2)^{1/2}t + \varepsilon] \tag{7.4}$$

for $|\zeta| > 1$

$$x = A_1 \mathrm{e}^{[-\zeta\omega_n + \omega_n(\zeta^2 - 1)^{1/2}]t} + A_2 \mathrm{e}^{[-\zeta\omega_n - \omega_n(\zeta^2 - 1)^{1/2}]t} \tag{7.5}$$

and for $|\zeta| = 1$

$$x = (A_1 + A_2 t)\mathrm{e}^{-\zeta\omega_n t} \tag{7.6}$$

where A, A_1, and A_2 are constants determined by the initial state of the system.

The parameters ω_n and ζ are often called the undamped natural frequency and damping ratio respectively. If the damping ratio is negative then the system is unstable, displaying divergent motion whenever disturbed.

More complex linear systems, such as those having several degrees of freedom, are represented by differential equations of higher order whose solutions are combinations of the expressions 7.4, 7.5 and 7.6.

In the case of forced motion, the equation 7.3 is replaced by an equation of the type

$$\frac{\mathrm{d}^2 x}{\mathrm{d} t^2} + 2\zeta\omega_n \frac{\mathrm{d} x}{\mathrm{d} t} + \omega_n^2 x = f(t) \tag{7.7}$$

148

where $f(t)$ represents an external disturbance independent of the motion. The solution of equation 7.7 is the sum of the complementary function and the particular integral. The complementary function is the solution of the homogeneous equation 7.3, takes the form 7.4, 7.5 or 7.6 and, for $\zeta > 0$, represents a decaying, or transient motion. The particular integral is a solution of equation 7.7 which does not depend on the initial conditions and represents the steady-state motion. When $f(t)$ is a periodic function the particular integral will be periodic also. Often $f(t)$ is sinusoidal and applying standard mathematical procedures the particular integral is found to be sinusoidal also, of the same frequency but differing in amplitude and phase from the forcing function. The basic features of the transient and forced oscillations are shown in figures 7.2a and b.

The mathematical models that have been discussed are representative of many types of linear systems under varying conditions and it is worthwhile reviewing them now from a more physical viewpoint. We shall consider a

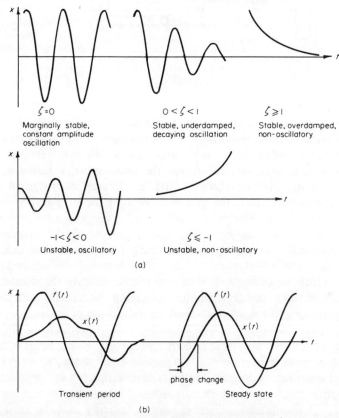

Figure 7.2 System responses: (a) transient motions, (b) forced oscillations

149

specific example, the mass–spring–dashpot system subjected to an external force $F(t)$ and a displacement $Y(t)$ as indicated in figure 7.3.

If $F(t)$ and $Y(t)$ are identically zero, then the element is a passive one with no energy source to sustain oscillations. If $F(t)$ or $Y(t)$ are made to vary in a prescribed way by an external source, then the response to these disturbances consists of transient and forced motions. The former is of the type (7.4), (7.5) or (7.6), and the latter is represented by the particular integral solution of an equation similar to equation 7.7. If $F(t)$ or $Y(t)$ is periodic, then the system will settle into a state of forced oscillation, energy being drawn from the source from which originates $F(t)$ or $Y(t)$. A common example is foundation vibration caused by unbalance of an engine.

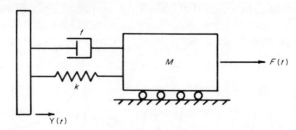

Figure 7.3 Damped linear system subjected to
disturbances

If $Y(t)$ again varies with time and $F(t)$ is related to the motion of the mass, then it is possible that oscillations occur even though the displacement $Y(t)$ is non-periodic. The system is said then to be in a state of self-excited oscillation. The energy necessary to sustain the oscillations is drawn from the source providing the displacement $Y(t)$, and energy is dissipated by the dashpot and possibly, as in the case of friction, by the force $F(t)$.

In both forced and self-excited oscillations, an energy source is available; in the latter case the frequency is greatly influenced by the parameters of the system itself. One example of self-excited vibrations is aircraft wing flutter where the aerodynamic forces are affected by wing deflections. Another, which we shall consider in detail later, involves the stick-slip of a mass on slideways, the drive motor supplying the energy and slideway friction providing forces which depend on the motion of the mass.

To conclude this review of vibration theory it is pertinent to emphasise that many systems are nonlinear and are represented by non-linear equations. Because the principle of superposition used in linear theory no longer holds, analytical treatment is difficult. Methods of investigation appropriate to the non-linearity resulting from the frictional forces will be considered in section 7.3. A non-linear oscillation may be so jerky that the variation of velocity, say, with time, is more like a saw-tooth than a sinusoidal wave, and then it may be called a relaxation oscillation.

150

7.2 CHARACTERISTICS OF FRICTIONAL VIBRATIONS

In this study, we shall concentrate attention on systems having a single degree of freedom and exclude those having several degrees of freedom where friction may act as a coupling agent.

To gain a basic understanding we shall consider the system represented by the model shown in figure 7.4. Suppose that the support A is driven at constant speed. Then if the bearing resistance F is zero the mass M will attain eventually the same speed as A although there may be a period of transient oscillation influenced by the dashpot. If F is not zero and varies with position or velocity, for instance, then under some conditions oscillations persist. We shall make the simplifying assumption that the friction varies with speed either as shown in figure 7.5a or as shown in figure 7.5b and these will be referred to as the stiction and negative gradient cases respectively.

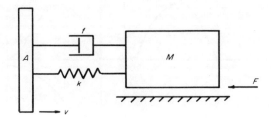

Figure 7.4   Linear drive system

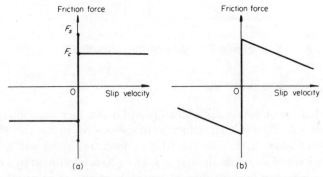

Figure 7.5   Friction–velocity characteristics: (a) stiction case; (b) negative gradient case

7.2.1 Stiction Case

Let the initial state of the system be such that the spring is uncompressed and M is stationary. When A moves there will be no movement of M until the drive force, that is the spring force plus the dashpot force, reaches a value equal to the stiction force F_s. As soon as M starts to move, the friction

151

falls rapidly to the value F_c so that there is an unbalanced force $F_s - F_c$ causing a sudden acceleration. The velocity of M increases until the drive force has fallen to the value F_c. If the mass velocity is then greater than that of the support A, the spring force continues to fall and the mass decelerates. If the damping is low, then eventually M comes to rest and the cycle of events is repeated. Acceleration, velocity and displacement curves for the mass are shown in figure 7.6, and it is clear that the motion is of the stick–slip type.

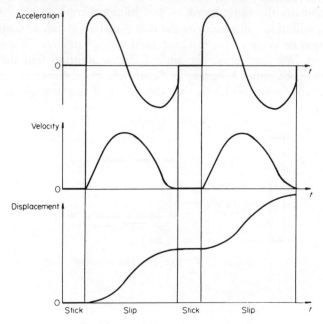

Figure 7.6   Stick–slip oscillation

7.2.2   Negative Gradient Case

Suppose that the mass is moving steadily and that it then encounters a small disturbance causing a sudden reduction in speed. This causes the drive force to grow and when it exceeds the friction force the mass will accelerate, possibly to a speed above the drive speed. This phase is followed by a deceleration phase, which in turn is followed by acceleration and so on. If the drive damping coefficient is low compared with the friction–velocity gradient, then the amplitude of the velocity swings increases at least until either reversal occurs or velocity ranges having positive friction gradients are entered.

The analysis for the stiction case will be covered in a later section. The analysis for the negative gradient case is straightforward providing that we consider velocity variations small enough to allow linearisation of the friction–velocity curve.

152

The equation of motion for the mass M is

$$M\frac{d^2X}{dt^2} + f\left(\frac{dX}{dt} - v\right) + k(X - vt) = -F \tag{7.8}$$

where X is the displacement of the mass M
   v is the drive velocity
   f is the dashpot coefficient
   k is the spring constant
   F is the frictional resistance.

Providing that v is constant, equation 7.8 may be written in terms of the spring extension, x, as

$$M\frac{d^2x}{dt^2} + f\frac{dx}{dt} + kx = -F \tag{7.9}$$

For small variations about the velocity v, the frictional force may be expressed as

$$F = F_v - \lambda\left(\frac{dX}{dt} - v\right) \tag{7.10}$$

From equations 7.9 and 7.10 it follows that

$$M\frac{d^2x}{dt^2} + f\frac{dx}{dt} + kx = -\left(F_v - \lambda\frac{dx}{dt}\right)$$

that is

$$M\frac{d^2x}{dt^2} + (f - \lambda)\frac{dx}{dt} + kx = -F_v \tag{7.11}$$

The complementary function solution of equation 7.11 is of the form of equation 7.4, 7.5 or 7.6 with the damping ratio given by

$$\zeta = \frac{f - \lambda}{2(Mk)^{1/2}}. \tag{7.12}$$

If the drive damping is low and the friction–velocity gradient is high then $f - \lambda$ is negative, the damping ratio is negative and the motion following a disturbance is divergent, being oscillatory with increasing amplitude if

$$-1 < \zeta < 0$$

7.2.3 Oscillation Characteristics

We shall anticipate some of the results of later analyses to give a qualitative appreciation of the problem.

Effect of Drive Speed

Many experimenters have observed that oscillations are usually worst at low speed. If the drive velocity is high then the oscillation decays. If the velocity is low, the oscillations become more severe and saw-toothed. The lowest drive speed for which the oscillation decays is often called the critical speed.

Oscillation Amplitude

In the sinusoidal oscillations the velocity–time curve is symmetrical about the mean velocity. This is not true when sticking occurs, as in case 7.2.1, and at speeds well below the critical speed, peak velocities many times the average velocity are reached during the slip period. This influences the choice of friction model parameters for use in analysis.

Oscillation Frequency

The oscillation frequency cannot be equated to the natural frequency of the drive system itself. However, if the oscillations are not violent the drive frequency is a useful guide, and may be used when selecting the friction model.

7.3 REVIEW OF ANALYTICAL METHODS

Linear systems are those which may be represented by linear differential equations. A consequence of linearity is the superposition property, namely that linear combinations of the homogeneous equation of the form of equation 7.3, are also solutions. Also the responses $r_1(t), r_2(t)$, of the system to disturbances $f_1(t)$, $f_2(t)$, may be combined in the form $a_1 r_1(t) + a_2 r_2(t)$ to generate the response to a disturbance $a_1 f_1(t) + a_2 f_2(t)$. Thus the form of the response is unaffected by the magnitude of the disturbance. In particular a linear system when disturbed from a steady operating state will display either a transient response always or a divergent response always, irrespective of its initial state and the magnitude of the disturbance. Moreover, in the case of an unstable oscillatory linear system, the vibration amplitude increases indefinitely with time. Non-linear systems are often difficult to analyse and understand because of the absence of the superposition property and the fact that the whole character of the response may be dependent upon the size of the disturbance and the initial operating state. Also, an unstable non-linear system may oscillate with a constant amplitude. Methods of analysis have been developed which can be applied to any linear system, but it is necessary to select the method of attack appropriate to a given non-linear problem. We shall now discuss briefly the methods which are in common use.

7.3.1 Perturbation Methods

Useful information can often be obtained by studying the behaviour of the system in the neighbourhood of the operating state. Providing that the non-linearities are continuous, tangent approximations may be used to

154

yield linear equations. Such an analysis may be sufficient to confirm that a system is stable or unstable. In the latter instance, the analysis may suggest design modifications for stabilising the system. This approach is feasible when studying the effect of negative gradients of the friction–velocity curve as we did in section 7.2.2. Where the character of the non-linearity is given in graphical form, a tangent may be drawn at the operating point to give a straight-line approximation; where it is expressed analytically, a Taylor series for the non-linear function, with only the linear terms retained, is used to represent the non-linearity. This method is not suitable for discontinuous non-linearities such as the stiction characteristics of the example of section 7.2.1. Instead the method outlined next may be tried.

7.3.2   Piecewise Linear Approximation

We can often represent the non-linear characteristic with sufficient accuracy by a series of linear expressions, each one valid within a finite region of the input variable. The idea of this approximation is illustrated graphically in figure 7.7. The calculation of the transient response to a disturbance proceeds in a step-by-step manner, each step being of finite size. The final state of one step is used as the initial state for the solution of the linear equations of the next step interval. Sometimes it may be possible to draw conclusions about not only the transient behaviour but also the steady state. This approach is used in section 7.5.

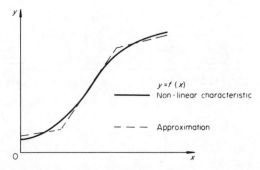

Figure 7.7   Piecewise linear approximation

7.3.3   Phase–Plane Analysis

The phase–plane method is applicable to the study of the transient response of unforced systems described by ordinary second-order differential equations with constant coefficients. So far we have illustrated the system response by plotting displacement versus time. For a positioning system governed by a second-order differential equation, the velocity and displacement together specify the state and we can show how the state varies by plotting curves,

155

known as trajectories, in the velocity–displacement plane. In terms of the velocity and displacement variables, the second-order equation is transformed to a first-order equation which may then be solved either directly or graphically to yield the trajectories. The method is particularly useful when the non-linearity is a function of either velocity or displacement. Here it will be demonstrated by a simple linear example.

Suppose that the drive system is equipped with low friction slideways so that equation 7.9 is replaced by

$$M \frac{d^2x}{dt^2} + f \frac{dx}{dt} + kx = 0 \qquad (7.13)$$

Writing $y = dx/dt$ equation 7.13 becomes

$$My \frac{dy}{dx} + fy + kx = 0 \qquad (7.14)$$

Thus at any point (x, y) of the phase plane, the slope, dy/dx, of the trajectory may be calculated and the trajectory constructed graphically. The equation for the line joining points of equal trajectory slope s, called an isocline, is given by

$$(Ms + f)y + kx = 0$$

which is a straight line passing through the origin. Once the isoclines have been drawn the approximate trajectory, starting from a given initial state, can be rapidly constructed. Figure 7.8a shows some isoclines and a typical trajectory for the case where the drive system is underdamped. The trajectory spirals into the origin. In the case $f = 0$, the response of the system is continuous oscillation and the trajectory is closed as shown in figure 7.8b. For comparison, the corresponding displacement–time curves are shown. If we need to calculate the time to travel from state A to state B, we can use the relation

$$t = \int_A^B \frac{1}{y} dx \qquad (7.15)$$

7.3.4 Describing Function Methods

The stability of linear systems may be studied using frequency response methods[2], developed by Nyquist and others. The methods are extremely powerful and hinge on the fact that sinusoidal inputs, such as would occur during steady oscillation, produce sinusoidal responses when applied to linear elements. Non-linear effects, however, cause distortion such as when a saturating amplifier clips the peaks of a sinusoidal input voltage. Consequently the Nyquist method cannot be applied directly to non-linear

156

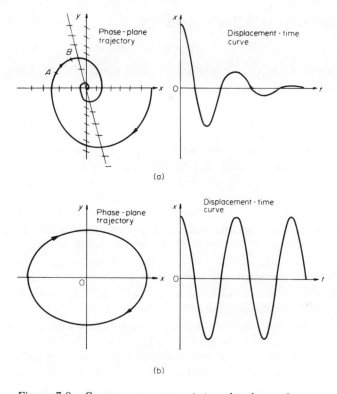

Figure 7.8 *System responses: (a) under-damped case,*
(b) zero-damping case

systems such as those involving friction. If appreciable inertia is present, the higher frequency harmonics representing the distortion may be attenuated leaving an almost sinusoidal response, and the frequency response method can be modified, as in the describing function approach[3], to yield useful information about the approximate amplitude and frequency of the oscillation. The reader who wishes to consider the use of this method should consult the references quoted[2,3].

7.3.5 Simulation

Analogue computers have been used in the study of systems involving friction[4,5]. The non-linearities present can be represented with reasonable accuracy by standard computer elements. Precise data regarding the friction characteristics, both static and dynamic, are unlikely to be available to the system designer, but simulation permits rapid examination of the performance for a range of likely frictional conditions.

157

7.4 FRICTIONAL FORCE MODELS

The analysis of frictional oscillations requires models for the system elements and the frictional forces. In electrical, mechanical or fluid mechanisms, substantial progress has been made in meeting this need. Frictional forces on the other hand, depend on many factors in ways which are not yet quantified and which, in many cases, are subject to large variations in the industrial environment. Consequently the analyst must select a model which he believes is representative and of sufficient detail to meet the object of his study, whether it be the elimination of troublesome oscillations or the prediction of critical velocities and frequency and amplitude characteristics.

Since frictional force varies considerably with normal pressure between the mating surfaces, geometry changes in some mechanisms may produce significant changes in frictional force, and successful analysis may depend more on taking into account the normal pressure variations than including changes in the coefficient of friction. Here we confine our attention to those systems where the normal force is constant, and are concerned with the stability of steady sliding motion.

It is evident that the coefficient of friction will vary from one point of a slideway to another. However, it is unlikely that such changes will be other than small and irregular. Consequently, although they may cause a succession of small disturbances to the motion, they are unlikely to be a major cause of prolonged oscillation. Of course, the disturbances in velocity which they initiate may be followed by oscillatory motion if the system is unstable.

Changes in frictional forces which are related to velocity or acceleration changes are important. It is worthwhile emphasising that frictional vibration usually occurs at low sliding speeds although less severe oscillations have been observed at high speeds[6]. Where hydrodynamic lubrication exists, frictional forces often increase the damping. So we concern ourselves with models valid in the neighbourhood of zero velocity.

We have already met the basic models in section 7.2, figures 7.5a and b. In the first model, the friction force is assumed to drop instantaneously as slip starts and then remain constant. In the second, the friction decrease is gradual. In both cases, the friction during stick is supposed to match instantaneously the drive force tending to cause motion until this force increases to a value just exceeding F_s, at which instant slip commences. Also the frictional force during slip depends on the instantaneous value of velocity.

A simple extension is to combine these models in various ways as shown in figures 7.9a and b. At greater speeds the force resisting motion is likely to be greater so the model may be modified to that shown in figure 7.9c. Of course the discontinuities present in these models will not occur in practice, but their piecewise linear nature eases the analysis. Alternatively, we select an f–v curve with a continuous gradient during slip and represented by an

158

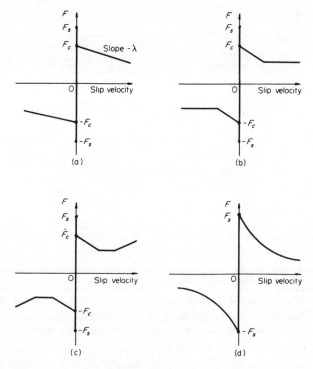

Figure 7.9   Simplified models for the friction–velocity characteristic

expression such as

$$F = F_s - A(1 - e^{-\lambda v_{\text{slip}}}) \tag{7.16}$$

Such a model is sketched in figure 7.9d. It need not be put into analytical form if the phase–plane method is used.

A prolonged stick phase normally results in a greater value of the friction, F_s, at the start of slip than is the case when the stick time is short[7]. The effect varies considerably with the slide conditions, particularly with the lubricant. A typical variation is shown in figure 7.10, and some researchers have represented the effect by an expression such as

$$F_s = F_1 + (F_2 - F_1)(1 - e^{-\alpha t_s}) \tag{7.17}$$

At this stage in our discussion, the position looks favourable since there is an amount of data describing the variation of frictional force with velocity. Unfortunately most of this data has been obtained from steady state measurements, under conditions of constant velocity for each friction force recorded. During an oscillation, however, the velocity is not constant and it would be preferable to use data obtained from experiments made under conditions of velocity variations and oscillation frequencies similar to those occurring

159

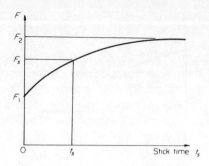

*Figure 7.10   Effect of stick time
on static friction*

in the vibration study. Studies by Bell and Burdekin[8], have shown that for typical machine-tool slideway conditions the slope of the $f-v$ curve for dynamic conditions, that is during oscillation, differs from that recorded under steady-state conditions. However the concept of a model such as that in figure 7.5b may still be of use in the analysis of frictional oscillations providing that the slope is measured under the appropriate dynamic conditions; this approach has been used with some success by Bell and Burdekin[9]. Another dynamic effect is hysteresis, for in some conditions it has been observed that the $f-v$ curve is not single-valued[1]. The slope during acceleration may differ from that during deceleration, as is indicated by the simplified model shown in figure 7.11. Another method which might account for dynamic effects is to regard the friction as a function of both velocity v and acceleration a in the simplified form

$$F = F_s + \lambda_1 v + \lambda_2 a \qquad (7.18)$$

where λ_1 and λ_2 are constants.

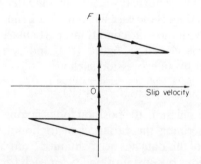

*Figure 7.11   Simplified hystere-
sis effect in the friction–velocity
model*

160

7.5 ANALYSIS OF STICK–SLIP OSCILLATIONS

We have seen in section 7.2 that a friction-velocity characteristic having a negative gradient will give rise to oscillations if the drive system damping is insufficient, and the analysis was straightforward. Here we wish to study the behaviour of the same drive systems, figure 7.4, when the friction–velocity relationship displays a stiction characteristic, besides the negative gradient, as shown in figure 7.9a. Derjaguin *et al*[10] have presented an extensive treatment of this situation, including the effect on the oscillations of the dependence of friction on the time of stationary contact.

7.5.1 Description of the Motion

Before the analysis, we will discuss the main features of the motion, starting from a condition where the mass is at rest, the drive spring is uncompressed and the drive end commences to move with constant velocity v. The friction force will be assumed independent of the time of stationary contact.

The spring is compressed until its compressive force plus the dashpot force reaches a value equal to the stiction force F_s. The mass starts to slip and it is assumed that instantaneously the friction force drops to the value F_c. Let us measure the time, t, from that instant. Then the unbalanced force acting on the mass at $t = 0$ is $F_s - F_c$ and hence the acceleration at $t = 0$ is $(F_s - F_c)/M$. The slip velocity increases until eventually the spring force drops sufficiently for deceleration to commence. If the deceleration phase ends without the velocity falling to zero, then sticking does not reoccur, and subsequently the mass velocity tends to the drive velocity, v, in the manner shown in figure 7.12c.

If the velocity falls to zero during the deceleration phase, then there are two possibilities for the next phase, either reversal or sticking. In this treatment we shall exclude the former, since it does not occur[10] if $F_s < 3F_c$ which is normally the case. If sticking occurs then again the drive spring is compressed until the mass slips, with the consequent drop in friction, and the system is then in the same state as at the beginning of the first slip phase. Thus the cycle is repeated and so on.

Summing up, the criterion for this theory is that stick–slip occurs if the velocity during slip falls to zero following an initial sudden acceleration from rest caused by the drop in friction $F_s - F_c$.

7.5.2 Analysis

Let x denote the spring extension. Then since the drive velocity v is constant

$$\frac{dx}{dt} = \frac{dX}{dt} - v \tag{7.19}$$

161

and

$$\frac{d^2x}{dt^2} = \frac{d^2X}{dt^2} \tag{7.20}$$

The equation of motion for the mass is

$$M\frac{d^2X}{dt^2} = -f\frac{dx}{dt} - kx - \left[F_c - \lambda\frac{dX}{dt}\right] \tag{7.21}$$

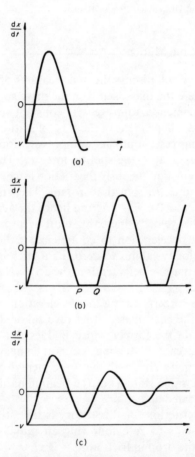

Figure 7.12 Velocity variations (velocity = v + (dx/dt)): (a) velocity during slip falls below zero; (b) stick–slip oscillation ensues; (c) increased damping, velocity remains positive and oscillation decays

162

that is

$$M \frac{\mathrm{d}^2 x}{\mathrm{d}t^2} + f \frac{\mathrm{d}x}{\mathrm{d}t} + kx = - \left[F_c - \lambda \left(\frac{\mathrm{d}x}{\mathrm{d}t} + v \right) \right] \qquad (7.22)$$

that is

$$M \frac{\mathrm{d}^2 x}{\mathrm{d}t^2} + (f - \lambda) \frac{\mathrm{d}x}{\mathrm{d}t} + kx = - [F_c - \lambda v] \qquad (7.23)$$

In terms of the damping ratio, ζ, and the undamped natural frequency, ω_n, given respectively by

$$\zeta = \frac{f - \lambda}{2(Mk)^{1/2}} \qquad (7.24a)$$

and

$$\omega_n = \left(\frac{k}{M} \right)^{1/2} \qquad (7.25b)$$

the equation of motion becomes

$$\frac{1}{\omega_n^2} \frac{\mathrm{d}^2 x}{\mathrm{d}t^2} + \frac{2\zeta}{\omega_n} \frac{\mathrm{d}x}{\mathrm{d}t} + x = -S \qquad (7.26)$$

where

$$S = \frac{F_c - \lambda v}{M \omega_n^2} \qquad (7.27)$$

The solution of equation 7.26 is found by adding the particular integral solution, $x = -S$ to the complementary function solution which is of the form of equation 7.4. It may be written as

$$x = -S + e^{-\zeta \omega_n t} \{ A \cos[\omega_n (1 - \zeta^2)^{1/2} t] + B \sin[\omega_n (1 - \zeta^2)^{1/2} t] \} \qquad (7.28)$$

from which we find

$$\frac{\mathrm{d}x}{\mathrm{d}t} = e^{-\zeta \omega_n t} \{ [-\zeta \omega_n A + \omega_n (1 - \zeta^2)^{1/2} B] \cos[\omega_n (1 - \zeta^2)^{1/2} t]$$

$$- [\omega_n (1 - \zeta^2)^{1/2} A + \zeta \omega_n B] \sin[\omega_n (1 - \zeta^2)^{1/2} t] \} \qquad (7.29)$$

At $t = 0$ the following conditions apply

$$kx + f \frac{\mathrm{d}x}{\mathrm{d}t} = -F_s \qquad (7.30)$$

and

$$\frac{\mathrm{d}x}{\mathrm{d}t} + v = 0 \qquad (7.31)$$

From equations 7.28, 7.29, 7.30, 7.31, we obtain

$$x = -S + e^{-\zeta\omega_n t}\left[(2\zeta v_1 - \delta)\cos(\omega_n\gamma t) + \frac{1}{\gamma}(2\zeta^2 v_1 - \zeta\delta - v_1)\sin(\omega_n\gamma t)\right]$$

(7.32)

and

$$\frac{dx}{dt} = \omega_n e^{-\zeta\omega_n t}\left[-v_1\cos(\omega_n\gamma t) - \frac{\zeta v_1 - \delta}{\gamma}\sin(\omega_n\gamma t)\right]$$

(7.33)

where $\delta = (F_s - F_c)/k$, $\gamma = (1 - \zeta)^{1/2}$ and $v_1 = v/\omega_n$. The acceleration during slip is

$$\frac{d^2x}{dt^2} = \omega_n^2 e^{-\zeta\omega_n t}\left[\delta\cos(\omega_n\gamma t) + \frac{v_1 - \zeta\delta}{\gamma}\sin(\omega_n\gamma t)\right]$$

(7.34)

A plot of dx/dt versus t, from equation 7.33, takes the form shown in figure 7.12a. If it intersects the line $(dx/dt) = -v$ then sticking occurs and the actual motion is stick–slip as shown in figure 7.12b, where PQ represents the stick phase. If there is no intersection, the motion is as shown in figure 7.12c.

Examining the curves in figure 7.12 we see that dx/dt overshoots zero, which would be its steady value in the absence of stick–slip, and then under-shoots. It is the severity of this undershoot which is important. The under-shoot is increased if the initial acceleration is increased.

For a given sliding velocity, v, the critical acceleration, and hence the critical value of the difference $F_s - F_c$, is determined by the condition that the curve of undershoot touches the line $(dx/dt) = -v$. In other words, (d^2x/dt^2) is zero when $(dx/dt) = -v$. From equation 7.34, (d^2x/dt^2) is zero when

$$\tan(\omega_n\gamma t) = \frac{\gamma}{\zeta - v_1/\delta}$$

(7.35)

Note that the lowest value of t satisfying equation 7.35 occurs at the first overshoot, whereas we require the second lowest value, say T_1.

A second relation between T_1 and δ/v_1 is obtained from equation 7.33 and the condition $(dx/dt) = -v$. We shall denote the important parameter δ/v_1 by P. In terms of the system parameters

$$P = \frac{F_s - F_c}{Mv\omega_n}$$

(7.36)

These relations may be solved numerically, for a given value of ζ, to yield the critical value of P. The report by Derjaguin et al[10] presents a curve showing the relation between ζ and the critical value of P, and also examines the duration of the slip and stick stages.

164

7.5.3 Phase–Plane Analysis

The phase–plane is useful for a graphical study and shows clearly the importance of the relative values of the initial acceleration at the start of slip and the sliding velocity. By differentiating equation 7.26 with respect to t we find

$$\frac{1}{\omega_n^2}\frac{d^2}{dt^2}\left(\frac{dx}{dt}\right) + \frac{2\zeta}{\omega_n}\frac{d}{dt}\left(\frac{dx}{dt}\right) + \left(\frac{dx}{dt}\right) = 0 \qquad (7.37)$$

In terms of non-dimensional phase–plane coordinates

$$u = \frac{1}{v}\frac{dx}{dt}$$

$$w = \frac{1}{v\omega_n}\frac{d^2x}{dt^2}$$

equation 7.37 becomes

$$w\frac{dw}{du} + 2\zeta w + u = 0 \qquad (7.38)$$

A trajectory satisfying equation 7.38 and the critical condition of touching the line AA, where $u = -1$, is sketched in figure 7.13a. It is drawn for a value of ζ which is positive and less than unity.

More generally the initial state is represented by a point on the line AA. If it lies below P, at Q say, then for the same value of ζ, the trajectory will not touch AA and stick–slip is absent. If the initial state is above P, say at R, then the trajectory intersects AA and stick–slip is present. Hence, for a given value of ζ, it is the value of the parameter $w_{t=0}$ which determines whether or not stick–slip occurs. In terms of the system parameters

$$w_{t=0} = \frac{F_s - F_c}{Mv\omega_n}$$

$$= P$$

Now consider the motion following a transient disturbance from a state of steady sliding. If the disturbance causes the point (u, w), representing the state, to pass into the cross-hatched region, then a period of sticking follows and stick–slip will ensue if the steady sliding speed is less than the critical speed.

An alternative use of the phase–plane method is to plot trajectories in the $(x, dx/dt)$ plane as in figure 7.13b.

Note that if $\zeta > 1$ the trajectory is as shown in figure 7.13c and stick–slip does not occur.

165

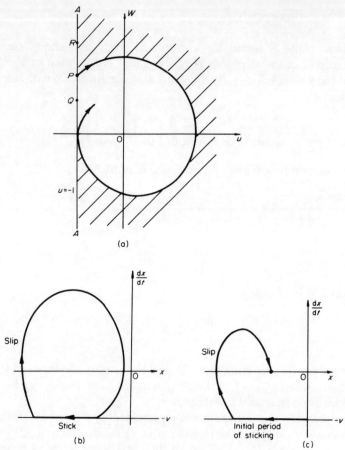

Figure 7.13 Phase-plane trajectories: (a) critical velocity condition (u, w) plane; (b) stick-slip trajectory (x, dx/dt) plane; (c) over-damped case

7.5.4 Geometry Effects

The analysis given above is concerned with situations where the friction force varies with speed. In other situations, friction forces may vary also with geometrical changes. These will not be discussed in this book except to say that they feature prominently in investigations of brake squeal and associated topics[11,12].

7.5.5 Zero Damping Case

To conclude this chapter with an example, we use the theory presented earlier to study the motion of the drive system when both the driving damping and the friction force variation with velocity are negligible with the consequence that the damping ratio ζ is zero.

166

Equation 7.26 becomes

$$\frac{1}{\omega_n^2} \frac{d^2x}{dt^2} + x = -\frac{F_c}{k} \qquad (7.39)$$

At $t = 0$, the conditions are

$$kx = -F_s \qquad (7.40)$$

and

$$\frac{dx}{dt} + v = 0 \qquad (7.41)$$

The solution of equation 7.39, satisfying conditions 7.40 and 7.41 is

$$x = -\frac{F_c}{k} - \delta \cos(\omega_n t) - v_1 \sin(\omega_n t) \qquad (7.42)$$

which leads to

$$\frac{dx}{dt} = \omega_n[\delta \sin(\omega_n t) - v_1 \cos(\omega_n t)] \qquad (7.43)$$

and

$$\frac{d^2x}{dt^2} = \omega_n^2[\delta \cos(\omega_n t) + v_1 \sin(\omega_n t)] \qquad (7.44)$$

The maximum slip velocity is reached at a time τ given by

$$\tan(\omega_n \tau) = -\frac{\delta}{v_1}$$

that is

$$\tan(\omega_n \tau) = -P \qquad (7.45)$$

From equation 7.45 we obtain

$$\sin(\omega_n \tau) = \frac{P}{(1 + P^2)^{1/2}}$$

and

$$\cos(\omega_n \tau) = -\frac{1}{(1 + P^2)^{1/2}}$$

Substituting these expression in equation 7.43 gives the maximum slip velocity v_{sm} as

$$v_{sm} = v + \left(\frac{dx}{dt}\right)_{max}$$
$$= v[1 + (1 + P^2)^{1/2}] \qquad (7.46)$$

167

The spring deflection, x, at the instant of maximum slip velocity is $-F_c/k$. From equation 7.43 the duration of slip τ_{sl} is obtained as

$$\tau_{sl} = 2\tau \qquad (7.47)$$

and equation 7.42 establishes that the spring deflection at the end of slip is $(F_s - 2F_o)/k$. If $F_s - 2F_c > 0$ then there is a tendency for velocity reversal to occur. If we assume that this would result in an immediate reversal of the friction force to at least a magnitude of F_c then velocity reversal will not actually start unless $F_s > 3F_c$. We shall ignore such an improbable case. Hence at the end of slip, sticking occurs, and during this stage

$$\frac{dx}{dt} = -v$$

Consequently if the duration of sticking is τ_1, then the spring deflection at the end of sticking will be given by

$$x = \frac{F_s - 2F_c}{k} - v\tau_1 \qquad (7.48)$$

Friction Independent of Stick Time

Assuming that the friction force can rise instantaneously from F_c to F_s then the duration of the stick stage satisfies

$$F_s = -k\left[\frac{F_s - 2F_c}{k} - v\tau_1\right]$$

that is

$$\tau_1 = \frac{2(F_s - F_c)}{kv} \qquad (7.49)$$

The total period is

$$\tau_{sl} + \tau_1 = \frac{2}{\omega_n}(\pi - \tan^{-1} P + P)$$

The principal features of the oscillation are shown in figure 7.14a.

Friction Dependent on Stick Time

The above analysis requires modification if the friction force at the end of the stick period grows with the duration of sticking, t, according to some relation

$$F = F(t) \qquad (7.50)$$

If the duration τ_1 of the stick period is low then

$$F(\tau_1) < F_s$$

168

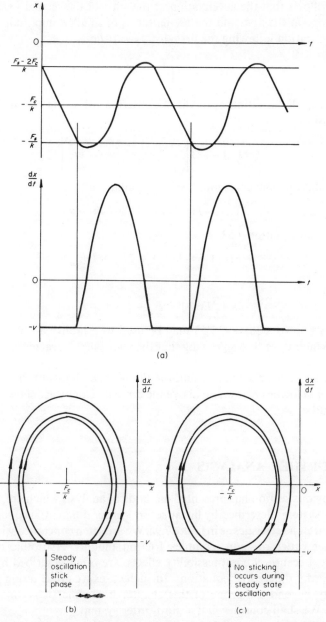

Figure 7.14 Oscillations of drive system, zero damping case: (a) stick–slip oscillations, F_s independent of stick time; (b) stick–slip oscillations, F_s dependent on stick time, stick phase tends to finite size; (c) raised velocity, F_s dependent on stick time, stick phase decreases and is absent after several cycles

169

and it follows that the acceleration at the start of the second period of slip is less than at the first, and the oscillation is of smaller amplitude. The stick period duration following the first slip, τ_1, satisfies

$$F(\tau_1) = -k\left[\frac{F_s - 2F_c}{k} - v\tau_1\right] \tag{7.51}$$

In general, the duration τ_n of the stick period following the nth period of slip satisfies

$$F(\tau_n) = -k\left[\frac{F(\tau_{n-1}) - 2F_c}{k} - v\tau_n\right] \tag{7.52}$$

The oscillation converges into a steady oscillation if

$$F(\tau_n) - F(\tau_{n-1}) \to 0 \quad \text{as} \quad n \to \infty$$

Hence from equation 7.52

$$F(\tau_\infty) = -F(\tau_\infty) + 2F_c + kv\tau_\infty$$

that is

$$F(\tau_\infty) = F_c + \tfrac{1}{2}kv\tau_\infty \tag{7.53}$$

If $F(t)$ is known, equation 7.53 may be used to establish a critical velocity, above which τ_∞ is zero signifying that the stick–slip character of the oscillation disappears after sufficient cycles of alternate slipping and sticking. The mass velocity then varies sinusoidally between limits $(0, 2v)$. The trajectories for velocities below and above the critical velocity are shown in figures 7.14b and c.

7.6 FURTHER ANALYSIS

The previous section has presented methods of analysis which are applicable to drive systems described by linear second-order differential equations with frictional characteristics as in figure 7.9a. We are often concerned with systems of higher order or more complex friction models. Hydraulic drives, for example, because of compressibility effects, are often described by third or higher order differential equations. In such cases we are entering regions of difficulty beyond the scope of this chapter but some suggestions may be helpful. We shall consider first a third-order system.

In section 7.5, we used for the stiction case the criterion that stick–slip persists if sticking reoccurs following slip from an initial sticking state. Now for a third-order system the state is dependent on three variables, not the mass velocity and spring deflection only as was the case for the second-order drive system. Consequently the state at the end of the stick period following the first step is not necessarily identical with the initial stick state.

170

In such a case, one possible procedure is to approximate the system by a second-order equation for a preliminary attack, establishing the approximate critical velocity before refining the value using analogue or digital computers.

If the problem is to study stability in a small neighbourhood of a steady sliding velocity, and the friction–velocity curve is linear within this region, then the Routh–Hurwitz stability criterion may be applied. The governing equation should be arranged in the form

$$a_0 \frac{d^3 z}{dt^3} + a_1 \frac{d^2 z}{dt^2} + a_2 \frac{dz}{dt} + a_3 z = A \tag{7.54}$$

where z is the dependent variable, A is constant, and a_0, a_1, a_2 and a_3 are coefficients which depend on the drive parameters and the local gradient of the f–v curve. Then by the Routh–Hurwitz criterion the system will be stable providing a_0, a_1, a_2 and a_3 are all positive and satisfy the inequality

$$a_1 a_2 > a_0 a_3 \tag{7.55}$$

If the criterion is satisfied, then a small disturbance to the steady sliding motion will not result in prolonged oscillation. For a statement of the Routh–Hurwitz criterion applicable to higher order systems the reader should consult, for instance, reference 2.

In situations where it is desirable to take into account the varying gradient of the f–v curve, then the phase–plane method may be used[13], provided that the drive system is described by a second-order differential equation similar to equation 7.7. The combination of friction non-linearity and higher-order drive system equations usually requires the application of simulation methods, describing function methods or advanced theory of non-linear vibrations.

7.7 ELIMINATION OF STICK–SLIP

In this section guide lines are suggested for the redesign of the system to free it from oscillation troubles. First, note that the trouble stems from the interaction of the drive dynamics with the friction characteristics of the bearing surfaces. Consequently we may consider altering one or both of the drive and bearing systems.

One approach is to make the frictional forces so small that their variation with velocity is of no importance. This may be achieved by using hydrostatic bearings, ones specially coated with low-friction material, or those using rolling elements. Such bearings are described elsewhere in this book. Although there are applications where the designer is already considering their use for other reasons, in others he may wish to avoid them on grounds of expense and complication. Also, low-friction bearings do not provide the damping that plain bearings may furnish outside the troublesome velocity region.

A second approach notes that for the situations we have considered the undesirable feature of the $f-v$ curve is a negative gradient or stiction effect. One way which has been used successfully with machine-tool drives[14], is to select the slideway lubricant from among those mineral oils with polar additives so that the $f-v$ curve gradient is approximately zero or even slightly positive for low velocities.

Looking now at the effect of the drive design parameters it is seen immediately that one way of ensuring stability, or reducing oscillation amplitude, is to increase the drive damping. This may be achieved by a direct method such as adding a dashpot device, or indirectly as in servo-mechanisms practice where velocity feedback improves stability[15]. Added damping may make the system sluggish in velocity regions where the friction gradient is positive. If the designer is prepared to add complexity he may introduce non-linear damping so that the damping coefficient is high in the dangerous speed regions and low elsewhere.

Normally, a very stiff transmission system is chosen so that excessive deflections under load are avoided. Decreasing the stiffness usually causes the stick–slip amplitude to increase but it should be remembered that the drive damping ratio, for a fixed amount of damping is reduced when the stiffness is increased. Probably the major effect of increasing transmission stiffness results from the consequent increase in natural frequency. When the speed fluctuates slowly the $f-v$ curve, at low speeds, may display a strong negative gradient characteristic whereas at high frequencies of speed fluctuation the gradient may be effectively zero[9]. Also, in situations where the growth of friction forces during the stick stage of an oscillation is important, the increased stiffness is beneficial, since it decreases the stick time.

REFERENCES

1. H. Catling. Stick–slip friction as a cause of torsional vibration in textile drafting rollers. *Proc. Instn mech. Engrs*, **174**, No. 17, (1960), 575–86.
2. B. Porter. *Stability Criteria for Linear Dynamical Systems*. Oliver & Boyd, Edinburgh, (1967).
3. H. E. Merritt. *Hydraulic Control Systems*. Wiley, New York, (1967).
4. A. J. Healey and J. D. Stringer. Dynamic characteristics of an oil hydraulic constant speed drive. *Proc. Instn mech. Engrs*, **183**, Pt 1, No. 34, (1968–9), 683–93.
5. J. Parnaby. Electronic analogue computer study of non-linear effects in a class of hydraulic servomechanisms. *J. mech. Engng Sci.*, **10**, No. 4, (1968), 346–59.
6. J. B. Hunt, I. Torbe and G. C. Spencer. The phase-plane analysis of sliding motion. *Wear*, **8**, (1965), 455–65.
7. S. Kato, N. Sato and T. Matsubayashi. Some considerations on characteristics of static friction of machine tool slideway. *Trans. Am. Soc. mech. Engrs*, Series F., *J. Lubric. Tech.*, **94**, No. 3, (July, 1973), 234–47.
8. R. Bell and M. Burdekin. Dynamic behaviour of plain sideways, *Proc. Instn mech. Engrs*, **181**, Pt 1, No. 8, (1966–7), 169–84.
9. R. Bell and M. Burdekin. A study of the stick-slip motion of machine tool feed drives. *Proc. Instn mech. Engrs*, **184**, Pt 1, (1969–70).

172

10. B. V. Derjaguin, V. E. Push and D. M. Tolstoi. A theory of stick-slip sliding of solids. *Proc. Conf. Lubric. and Wear.*, *Instn mech. Engrs*, (1957), 257–68.
11. R. T. Spurr. A theory of brake squeal. *Proc. Instn mech. Engrs*, No. 1, (1961–62), 33–52.
12. R. P. Jarvis and B. Mills. Vibrations induced by dry friction. *Proc. Instn mech. Engrs*, **178**, Pt 1, No. 32, (1963–64), 847–66.
13. B. R. Dudley and H. W. Swift. Friction relaxation oscillations. *Proc. R. Soc.*, (1949), 849–61.
14. M. E. Merchant. Characteristics of typical polar and non-polar lubricant additives under stick-slip conditions. *J. Lubric. Engrs*, **2**, (June 1946), 56–61.
15. P. L. Taylor. *Servomechanisms*. Longman, Harlow, (1969).

8

The Mechanics of Rolling Motion

8.1 INTRODUCTION

It is not without significance to society that the principle of rolling as a method of translation dates from antiquity. None the less, this method of translation may be properly claimed to be the product of man's ingenuity, and indeed it is known that such an advanced society as the Incas did not discover this principle.

In view of the very long history of usage of this method of translation, the lack of useful experimental data concerning it is somewhat surprising. There is undoubtedly a widespread awareness among engineers of the convenience and efficiency exhibited by rolling systems. Thus we find a whole catalogue of engineering situations to which rolling motion has been applied, namely, wheels of all types, ball- and roller-bearings of many designs, variable speed friction drives, belt drives (which may be considered as the rolling of a pulley along a continuous belt), knife edges which constitute a small radius rolling contact, gear teeth which function by a process of combined rolling and sliding, and others.

These developments suggest that any rolling situation should be analysed from three standpoints where any combination of the following effects may be contributory.

174

8.1.1 Pure Rolling or Free Rolling

This represents the most elemental form of rolling motion and is probably most nearly approached in the case of a cylinder or ball rolling without constraint in a straight line along a plane.

8.1.2 Rolling with Combined Applied Surface Tractions

These conditions obtain in the case of a driven or braked wheel, where the applied forces induce normal and tangential tractions in the contact zone between the wheel and its track.

8.1.3 Rolling with Regions of Slip in the Contact Zone Necessitated by the Geometry of the System

A ball rolling around the inner race of a ball-bearing may, to a first approximation, be considered to rotate about some instantaneous axis of rotation necessitated by the geometric conformity between the ball and its track, figure 8.1. From elementary kinematic considerations it then follows that surface particles in the contact zone remote from this axis of rotation must

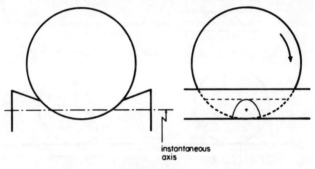

instantaneous
axis

Figure 8.1

necessarily slip. None the less, it should be recognised that such slip may in part be suppressed because of the ability of the materials to sustain elastic deformation.

In a similar way, a ball rolling around a curved track is subjected to a twisting moment which induces a degree of slip between the ball and its track.

8.2 FREE ROLLING

Since all bodies are made from deformable materials some deformation must ensue when they come into contact under load. For bodies having a geometry defined by circular arcs, such contact occurs over a relatively restricted area, but it is imperative that we appreciate that 'line contact'—the contact between two cylinders or a cylinder and a plane—and 'point contact'—the contact

175

between two spheres or a sphere and a plane—still necessitate some finite area of contact. The shape and size of the area of contact will depend upon such factors as the individual geometry, the load, and the deformation characteristics of the materials. Thus a cylinder rolling along a plane results in a rectangular area of contact of width $2a$, figure 8.2a, whereas a sphere rolling along a plane results in a circular contact area, figure 8.2b. Both these solutions are special cases of the well known hertzian analysis from which we might postulate the general case shown in figure 8.2c. For this case the contact would be elliptical in shape defined by the two semi-axes a and b, (see chapter 3).

As body 1 in figure 8.2b rolls along body 2, it will be recognised that

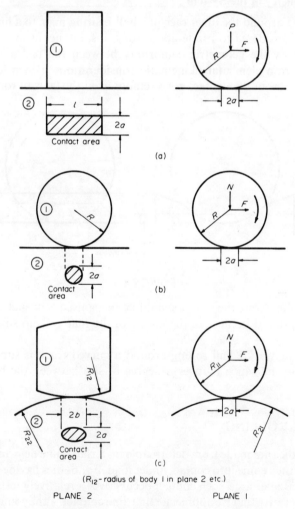

(R_{12} - radius of body 1 in plane 2 etc.)

PLANE 2   PLANE 1

Figure 8.2   Types of contact areas

176

material ahead of the line of centre in both bodies is being compressed, while material to the rear of this line is having this stress state relieved. For these conditions hertzian analysis applied to the elastic deformation of such bodies results in the following observations. (See chapter 3 for an elementary treatment of hertzian contact).

8.2.1 Cylinder on a Plane

The local pressure p in the contact area due to an applied load P is

$$p = \frac{2P}{\pi a l}\left(1 - \frac{x^2}{a^2}\right)^{1/2}$$

that is, a semicircular pressure distribution with a maximum pressure at the centre of the contact zone of value $2P/\pi a l$ (figure 8.3). The value of the contact semi-width a is given by

$$a = \left[\frac{4PR}{\pi l}\left(\frac{1 - v_1^2}{E_1} + \frac{1 - v_2^2}{E_2}\right)\right]^{1/2}$$

where v is Poisson's ratio and E is Young's modulus, suffixes 1 and 2 relate to the bodies shown in figure 8.3.

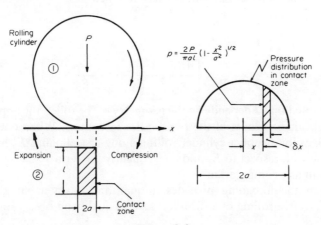

Figure 8.3

Consider an elemental strip δx at a distance x from the centre of the contact zone. During the forward compression of the material in bodies 1 and 2, this pressure gives rise to a resisting moment M about the centre of the contact zone where

$$M = \int_0^a p l x \, dx = \frac{2Pa}{3\pi}$$

177

Under ideal conditions the cylinders would be subjected to an equal and opposite moment arising from the pressure distribution in the rear part of the contact region. For the present, however, we will neglect this effect so that if the cylinder rolls a distance x we note that the elastic work done due to the forward compression alone would be

$$\phi = \frac{Mx}{R} = \frac{2Pax}{3\pi R}$$

Let us now postulate that the rearward recovery does not replace the whole of this elastic work of forward compression. This might be explained by an energy dissipation due to the elastic hysteresis loss occurring in the complex straining of the material which must occur during the rolling process. If this loss is defined by a coefficient ε we can see that the work dissipation during rolling a distance x will be $\varepsilon\phi$ and, if this loss represents the rolling resistance, it follows that

$$Fx = \varepsilon\phi = \varepsilon\frac{2Pax}{3\pi R}$$

where F is the force required to overcome this resistance. The ratio F/P may be specified as the coefficient of rolling resistance λ, having an analogous meaning to the coefficient of sliding friction μ, hence

$$\lambda = \frac{F}{P} = \frac{2}{3}\frac{\varepsilon a}{\pi R}$$

$$= \frac{2}{3} \times \frac{\varepsilon}{\pi R} \times \left[\frac{4PR}{\pi l}\left(\frac{1 - v_1^2}{E_1} + \frac{1 - v_2^2}{E_2}\right)\right]^{1/2} \tag{8.1}$$

It will be noted that λ depends on the geometry of the roller, the applied load and the elastic constants of the two bodies in contact. When one body is very rigid, as for a rubber cylinder rolling along a steel plane, E_2 has a very large value with respect to E_1 and consequently the term $1 - v_2^2/E_2$ may be neglected in any computation.

Although the foregoing provides a quantitative statement about the resistance to free rolling of a cylinder, several qualifying observations must be made.

(a) The hysteresis loss factor ε used in the expression for λ, is not the same as that which would be obtained from a simple tensile test. This arises because in the rolling process, the elastic deformation of any particular element of material passing through the contact region is a very complex process in relation to the relatively simple uniform deformation pattern in simple tension. Consideration of these two processes of deformation suggests that the hysteresis loss factor for a rolling cylinder would be about three times the loss factor obtained in a simple tension test.

(b) The hysteresis loss factor has here been assumed constant, whereas

178

some evidence exists to suggest that at larger strains it may vary with the degree of straining. Furthermore, little is known about the effect of strain rate on hysteresis, so that the foregoing analysis would suggest that the free rolling resistance is independent of velocity.

(c) For a cylinder rolling along a plane it is well established that the maximum stress condition exists below the surface of contact, (chapter 3). Thus at high loads it is probable that this high stress region could reach the condition for plastic flow giving the contact as shown in figure 8.4. As rolling proceeds, any element of material just below the

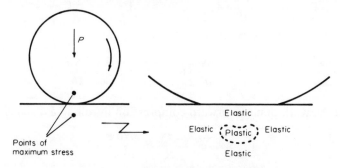

Figure 8.4

surface passes through an elastic–plastic–elastic condition. This process will lead to an appreciable plastic energy dissipation, even though the actual contact surfaces are still at an elastic stress level. It has been shown that, under certain loading conditions, such a situation induces residual stresses so that after several cycles of rolling a condition of shake-down to a purely elastic state may be achieved. Thus, in situations where any plastic deformation occurs, the resistance to rolling will be increased by the plastic work dissipation, in addition to any elastic hysteresis effects.

(d) In this analysis it has been assumed that the materials in contact are perfectly elastic and provide a finite, time independent, hysteresis loss. With many materials this is not the case, since there exists a relaxation time and recovery to the rear of the contact zone may be incomplete. Furthermore, such effects are known to be both time and temperature dependent, but the results of analyses based on such visco-elastic behaviour cannot be treated in this volume.

8.2.2 Sphere on a Plane

For this case the pressure distribution in the contact region is given by

$$p = \frac{3N}{2\pi a^2}\left(1 - \frac{r^2}{a^2}\right)^{1/2}$$

and the radius of the contact circle by

$$a = \left[\frac{3}{4} NR\left(\frac{1 - v_1^2}{E_1} + \frac{1 - v_2^2}{E_2}\right)\right]^{1/3}$$

To derive the resisting moment arising from the forward compression of material, we again consider an elemental strip δx, figure 8.5. The total vertical force on such a strip due to the pressure distribution is then given by

$$\frac{3N}{2\pi a^2} dx \int_{-(a^2-x^2)^{1/2}}^{+(a^2-x^2)^{1/2}} \left[1 - \frac{(x^2 + y^2)}{a^2}\right]^{1/2} dy = \frac{3N}{4a^3}(a^2 - x^2) dx$$

Thus the resultant resisting moment M is

$$M = \frac{3N}{4a^3} \int_0^a (a^2 - x^2)x \, dx = \frac{3Na}{16}$$

Again the work done in forward compression in rolling a distance x is

$$\phi = \frac{Mx}{R} = \frac{3Nax}{16R}$$

and introducing a hysteresis loss factor for this condition ε one obtains

$$Fx = \varepsilon\phi = \varepsilon\frac{3Nax}{16R}$$

where

$$\lambda = \frac{F}{N} = \frac{3}{16}\frac{\varepsilon a}{R} \tag{8.2}$$

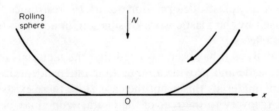

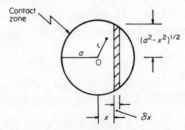

Figure 8.5

180

In this equation the contact size a is defined by

$$a = \left[\frac{3}{4} NR\left(\frac{1 - v_1^2}{E_1} + \frac{1 - v_2^2}{E_2}\right)\right]^{1/3} \qquad (8.3)$$

The same general observations which were made for the cylinder rolling along a plane still apply, except that the hysteresis loss factor for this condition is only about twice that which will be obtained for simple tension.

8.2.3 General Elliptic Contact

For these conditions the pressure distribution over the ellipse of contact is given by

$$p = \frac{3N}{2\pi ab}\left[1 - \frac{x^2}{a^2} - \frac{y^2}{b^2}\right]^{1/2}$$

For the special case of $a = b$ this reduces to the circular contact condition of the preceding case. The size of the contact ellipse is defined by the semi-axes a and b given by

$$a = k_a\left[\frac{3N}{\Sigma_\rho}\left(\frac{1 - v_1^2}{E_1} + \frac{1 - v_2^2}{E_2}\right)\right]^{1/3}$$

$$b = k_b\left[\frac{3N}{\Sigma_\rho}\left(\frac{1 - v_1^2}{E_1} + \frac{1 - v_2^2}{E_2}\right)\right]^{1/3}$$

where

$$\Sigma_\rho = \frac{1}{R_{11}} + \frac{1}{R_{12}} + \frac{1}{R_{21}} + \frac{1}{R_{22}}$$

In these formulae k_a and k_b are functions of the four principle radii of curvature of the contacting bodies and are discussed in chapter 3.

In this case the total vertical force on an elemental strip δx, figure 8.6, is given by

$$f = \frac{3N}{2\pi ab}\,dx \int_{-b(1 - x^2/a^2)^{1/2}}^{b(1 - x^2/a^2)^{1/2}} \left(1 - \frac{x^2}{a^2} - \frac{y^2}{b^2}\right)^{1/2} dy$$

Thus

$$M = \int_0^a fx\,dx = \frac{3Na}{16}$$

and

$$\phi = \frac{Mx}{R_{12}} = \frac{3Nax}{16R_{12}}$$

181

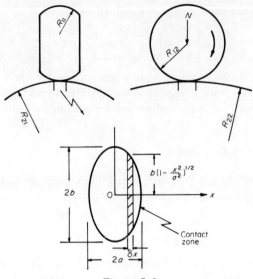

Figure 8.6

Hence

$$\lambda = \frac{F}{N} = \frac{3}{16} \frac{\varepsilon a}{R_{12}} \qquad (8.4)$$

where a is given by

$$a = k_a \left[\frac{3N}{\Sigma_\rho} \left(\frac{1 - v_1^2}{E_1} + \frac{1 - v_2^2}{E_2} \right) \right]^{1/3} \qquad (8.5)$$

In this case ε would be in the range two to three times the value which will be obtained from a simple tension test. When a is approximately equal to b the value will be near to 2 whereas when b is very much greater than a the conditions will approach those of the rolling cylinders and a value nearer to 3 will be more appropriate.

8.2.4 Heathcote Slip

In the foregoing arguments the contact conditions have been such as to allow the assumption that the contact zone lies substantially within a single plane. In many engineering situations, such as ball-bearings of all types, this assumption is no longer tenable. Consider a ball rolling along a grooved track as shown in figure 8.7. The contact area will be an ellipse and provided the subtended angle of the contact at the centre of the rolling element β is less than 30° it is found that the dimensions a and b are still predicted with reasonable accuracy by hertzian formulae. For this situation, as rolling proceeds, we must consider the rotation to occur about an instantaneous axis of rotation such as AA. As far as the contact ellipse is concerned we then

182

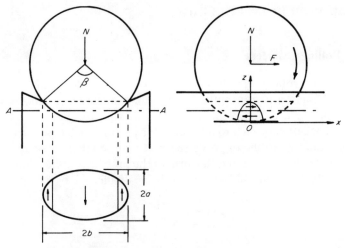

Figure 8.7   Heathcote-type slip

see that three regimes of relative slip must occur; the magnitude of slip being defined by the product of the angular velocity and the distance of the particular contact point in the xz plane from the axis AA.

The force F producing motion is now seen to consist of two components. One component accommodates the hysteresis loss, using the arguments discussed above, and the other component exists to accommodate the sliding friction losses. The latter term may be considered as a resisting moment about the axis AA, arising from the summation of these frictional effects. The method of carrying out such a summation is discussed in a later section, since this effect is complicated by one further factor.

In most engineering applications dealing with steel bodies the size of the contact ellipse and the height of the axis AA above O will be very small, that is, several orders of magnitude less than the radius of the ball. Thus the actual sliding velocities occurring in the contact zone during rolling will be very small and furthermore, because of the rolling motion, the sliding material rapidly passes through the contact region. This means that the magnitude of relative displacement between particles in the ball and the track is also very small (typically in the region of 10^{-2} mm/s for steels). Such small relative displacements can, to some degree, be accommodated by a tangential surface straining of the ball and track materials without the requirement for any slip. Thus one finds that in problems of this type some of the sliding indicated in the simplified picture of figure 8.7 does not occur, and that within the contact zone there are definite areas within which no slip occurs. These areas are naturally most prone to occur in the regions where the slip in figure 8.7 would tend to be very small, that is, adjacent to the instantaneous axis of rotation AA. This somewhat complicated situation, giving rise to areas of sticking and areas of slipping, is covered by the subject of microslip and will now be dealt with in greater detail.

183

8.3 MICROSLIP IN ROLLING

8.3.1 Rolling Cylinders

Consider two rolling cylinders subject to a normal load and in which a tangential traction is transmitted across the contact zone, figure 8.8a. This could apply to a cylinder subjected to a driving torque, as in a railway driving wheel, or subjected to a braking torque, in which case the tangential traction would be in the opposite sense. It will be tacitly assumed that the width of the contact zone and the form of the pressure distribution are those predicted by hertzian analysis and will be unaffected by the presence of

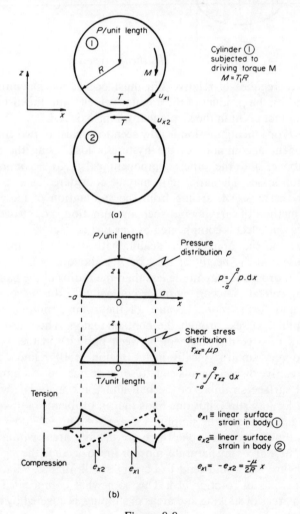

Figure 8.8

184

tangential tractions. If full sliding occurs over the contact region, the total surface traction T will be μP and at any point in the contact region the tangential surface shear stress τ_{xz} will be given by μp where μ is the appropriate coefficient of sliding friction, figure 8.8b.

In the following solution the cylinders will be considered as having unit axial length, so that the tractions T and P are in the form of load per unit length. From contact theory it can readily be shown that within the contact region such a situation produces a linear strain e_x given by

$$e_x = -\frac{\mu}{2R} x \qquad (8.6)$$

It will be noted that within the contact region e_x is a linear function of x with a slope of $\mu/2R$. Outside the contact region a more complicated distribution occurs as shown in figure 8.8b, but its exact form need not concern us in the following solutions. This represents the solution for a rolling cylinder where $T = \mu P$.

We now consider the case when $T < \mu P$ and to do so we must consider the effect of the resultant surface strains e_x on the probability of relative motion between surface particles on the two cylinders. To do this we must obtain an expression for the velocity of discrete particles incorporating the displacements due to surface straining.

Consider a particle distance X from the origin of coordinates in the unstrained state and which, due to straining, is displaced a further distance u in the x direction. The coordinate position of the particle will then be defined by x, that is

$$x = X + u$$

whence its velocity dx/dt is

$$\frac{dx}{dt} = \frac{dX}{dt} + \frac{du}{dx} \times \frac{dx}{dt}$$

if we assume that u is time invariant, that is, that steady state conditions prevail.

Since strains are small, x is approximately equal to X, so that $(dx/dt) \approx (dX/dt)$. But dX/dt is the velocity of the particle if straining is neglected, that is, the rigid body velocity of the particle U_x. The strain in the x direction e_x is given by du/dx so that we may write the resultant velocity of the particle, say V_x as

$$V_x = U_x + U_x e_x = U_x(1 + e_x) \qquad (8.7)$$

If we now consider coincident particles in the surfaces of the two cylinders we recognise from Newton's third law that the traction on the two bodies must be in opposition but equal in magnitude so that at any point

$$e_{x1} = -e_{x2}$$

185

The strain pattern in figure 8.8 is seen to satisfy this criterion. Thus the velocities of the two coincident particles on the surfaces of the bodies in contact may be written

$$V_{x1} = U_{x1}(1 + e_{x1}) \tag{8.8}$$

$$V_{x2} = U_{x2}(1 + e_{x2}) = U_{x2}(1 - e_{x1})$$

Self-evidently these two velocities can only be identical in areas where e_{x1} and hence e_{x2} are constant, since by definition U_{x1} and U_{x2} are constant rigid body approach velocities. This therefore defines the basic condition for areas in which the tendency to slip may be accommodated by surface straining of the bodies. When $V_{x1} = V_{x2}$ equations 8.8 result in the relationship

$$U_{x1} - U_{x2} = -e_{x1}(U_{x1} + U_{x2})$$

or

$$\frac{\Delta U}{U} = -2e_{x1} \tag{8.9}$$

where ΔU is the difference between the two rigid body circumferential linear speeds and U is the mean circumferential linear speed.

$$U = \frac{U_{x1} + U_{x2}}{2} \tag{8.10}$$

To summarise we note that when T is less than μP there may exist regions of contact where no relative slip occurs, provided that within these regions the surface strain due to the tangential tractions are of a constant value, and that the circumferential velocities of the two cylinders will be different at points remote from the contact zone, that is, where surface strains are not present.

An acceptable solution to this problem may now be provided by the argument shown diagramatically in figure 8.9. Figure 8.9a shows the normal pressure distribution according to hertzian theory, while figure 8.9b shows the tangential traction distribution when $T' = \mu P$ and $\tau_{xz} = \mu p$ throughout the contact region. For this condition the strain e'_x is given by equation 8.6 throughout the contact region and is shown in figure 8.9c. The strains outside the contact region are also shown but are not essential to the argument. Now suppose a traction T'' is applied over a contact zone of width 2α, the distribution and resultant strains are obviously as shown in figures 8.9d and 8.9e, when T'' is in the opposite sense to T' for each body. If we now add the traction in figures 8.9b and 8.9d we get a resultant traction $T = T' + T''$ for each body, and this is distributed as shown in figure 8.9f. Adding the resultant strains gives the result shown in figure 8.9g; the constant strain arising because the slope of the strain distribution is the same in both figures 8.9c and 8.9e. This must be so, since equation 8.6 shows that the slope of the strain distribution $\mu/2R$ is independent of the value of either the applied traction or

186

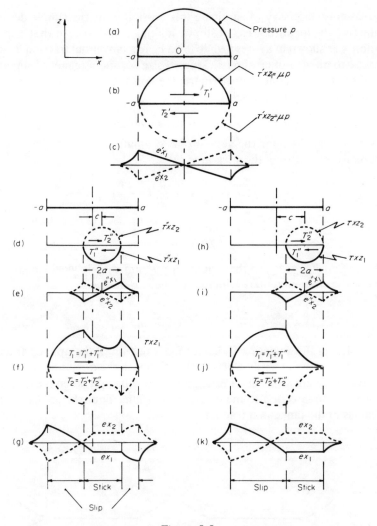

Figure 8.9

the size of the contact area. Examining figures 8.9f and 8.9g shows that we now have a possible solution to our problem. The resultant value of T is obviously less than μP. In the region where the strain is constant, that is, the region in which no slip occurs according to equation 8.9, we see that τ_{xz} is less than μp, whereas outside this region $\tau_{xz} = \mu p$ and in these areas slip must occur.

This solution therefore appears to satisfy the essential features of the problem. However, if we examine the outcome in greater detail we find that in the forward region of slip the inconsistency arises that the slip is in the same direction as the traction μp, thereby contravening the basic laws of friction.

187

It is therefore necessary to eliminate this possibility by the simple device of eliminating the forward areas of slip by making $c = a - \alpha$, so that our final solution is as shown in figures 8.9h to 8.9k. Here a tangential traction $T < \mu P$ gives rise to an area of no slip (sticking) of size 2α, and a region of slip at the rear of the contact zone. Furthermore, the constant strain in the region of sticking is given by Figure 8.9k

$$e_{x1} = e'_{x1} + e''_{x1} = -e_{x2} = -(e'_{x2} + e''_{x2})$$

but since $x'' = x - c$ for the system of coordinates used and c denotes the location of the centre of the stick area

$$e_{x1} = -\frac{\mu}{2R}x + \frac{\mu}{2R}(x - c)$$

$$= -\frac{\mu}{2R}c$$

In this region $V_{x1} = V_{x2}$ so that equation 8.9 immediately yields the relation between the circumferential velocities of the two bodies

$$\frac{\Delta U}{U} = -2e_{x1} = \pm \frac{\mu}{R}c \qquad (8.11)$$

The $\pm$ sign in this equation indicates the direction of the applied traction. The positive sign will apply when the traction is in the direction of rolling as for body 1 in figure 8.10a, and the negative sign will apply when the traction is in the opposite direction to the direction of rolling, figure 8.10b. Combining the values of the tangential traction

$$T_1 = T'_1 - T''_{11}$$

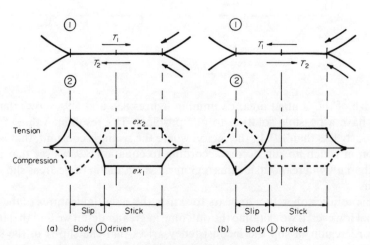

Figure 8.10

188

but $T'_1 = \mu P$ = area of figure 8.9b. Since T'' is the same form as μP but over a contact width 2α rather than $2a$, it follows that $T'' = \mu P(\alpha/a)^2$ = area of figure 8.9h. Hence

$$T_1 = \mu P \left(1 - \left(\frac{\alpha}{a} \right)^2 \right) \tag{8.12}$$

that is

$$T_1 = \mu P \left(2\frac{c}{a} - \frac{c^2}{a^2} \right) \tag{8.13}$$

since $\alpha = a - c$.

Substituting c from equation 8.11 in equation 8.13 then produces the relationship between the tangential traction and the so called creep velocity ΔU

$$\frac{\Delta U}{U} = \frac{\mu a}{R} \left[1 - \left(1 - \frac{T}{\mu P} \right)^{1/2} \right] \tag{8.14}$$

or in non-dimensional form

$$\frac{\Delta U R}{U \mu a} = 1 - \left(1 - \frac{T}{\mu P} \right)^{1/2} \tag{8.15}$$

We therefore see that by considering the linear strain pattern which arises from the applied tangential traction, we may identify areas of sticking and areas of slipping in the contact zone between two rolling cylinders. We note that due to the requirements of the elementary laws of friction this sticking zone must always lie at the front part of the contact zone. This latter requirement is only true for the contact of materials having similar elastic properties[2]. For materials having dissimilar elastic properties a solution of rather greater complexity has been proposed which reduces to the foregoing solution when the elastic constants are the same[3].

The validity of equation (8.15) has been proved by experiments in which the creep velocity ΔU has been measured for a range of values of T_1, P, U and R[4]. Typical results are shown in figure 8.11 where the non-dimensional groups $\Delta U R / U \mu a$ and $T/\mu P$ are seen to agree with equation 8.15 when a steel cylinder rolls along a steel track.

8.3.2 Rolling Spheres

Consider two spheres rolling together while subject to a normal load and a transmitted tangential surface traction. The solution for the rolling cylinder may be applied approximately to this case by considering the sphere to be represented by a row of elemental cylinders of axial length δy embracing the

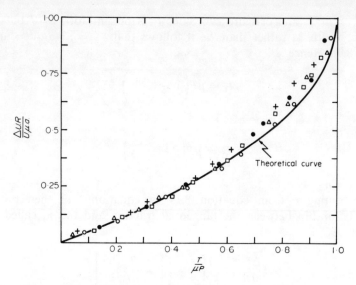

*Figure 8.11   An experimental and theoretical comparison
of the microslip between rolling cylinders*

contact circle of the sphere, figure 8.12. The pressure distribution for a
sphere is (chapter 3)

$$P = \frac{3N}{2\pi a^2} \left[1 - \frac{x^2}{a^2} - \frac{y^2}{a^2} \right]^{1/2} \tag{8.16}$$

At any point y the contact width will extend from $-(a^2 - y^2)^{1/2}$ to
$+(a^2 - y^2)^{1/2}$ whence the normal load on the cylinder element δy is then

$$f_z = \frac{3N}{2\pi a^2} \delta y \int_{-(a^2 - y^2)^{1/2}}^{(a^2 - y^2)^{1/2}} \left| 1 - \frac{x^2}{a^2} - \frac{y^2}{a^2} \right|^{1/2} dx$$

$$= \frac{3N}{4a^3} (a^2 - y^2)\, dy$$

that is

$$f_z = \frac{3N}{4a^3} X^2\, dy$$

The normal load per unit length of each of the elemental cylinders is then
given by

$$P = \frac{f_z}{\delta y} = \frac{3N}{4a^3} X^2$$

which is distributed over a contact width of $2X$ where

$$X^2 = a^2 - y^2$$

190

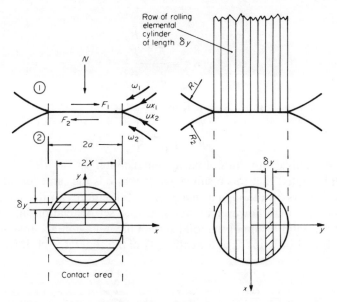

Figure 8.12 Method of utilising the 'strip theory of microslip'

The total transmitted tangential traction F may obviously be considered as being the summation of the individual tangential tractions per unit length T, which are applied to each of the cylindrical elements δy. Except for the case when $F = \mu N$, and thus $T = \mu P$ for each element δy, we have as yet no method for defining the value of T acting on any given element δy for the conditions when F is less than μN. We may resolve this difficulty by considering the velocity pattern in such a situation. Provided that the contact zone is small with respect to the ball size, it is seen that the rigid body approach velocity for every one of the elemental slips δy will be the same, that is,

for body 1

$$U_1 = \omega_1 R_1$$

for body 2

$$U_2 = \omega_2 R_2$$

Substituting these values in equation 8.11 which relates to the velocity difference and the constant strain in the stick area in a rolling cylinder, defines the location of the centre of the stick area as follows

$$\pm Kc = \frac{U_1 - U_2}{U_1 + U_2} = \frac{\omega_1 R_1 - \omega_2 R_2}{\omega_1 R_1 + \omega_2 R_2}$$

191

Therefore

$$\pm Kc = \frac{\phi - \dfrac{R_2}{R_1}}{\phi + \dfrac{R_2}{R_1}} \qquad (8.17)$$

where $\phi = \omega_1/\omega_2$ and $K = \mu/R$ using

$$\frac{1}{R} = \frac{1}{R_1} + \frac{1}{R_2}$$

that is, the equivalent radius of the system (chapter 3).

It will be noted that this result suggests that c is the same for all the elemental cylinders δy although its magnitude still requires a value for ϕ, as yet undefined. We may however obtain a further set of relationships by considering the equilibrium of the total system having a contact area as shown in figure 8.13. For strips δy at the extremity of the contact circle it is seen that

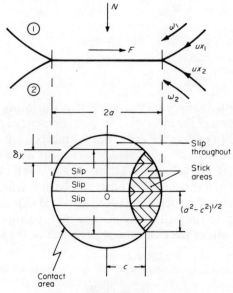

Figure 8.13

no-stick areas exist so that in these regions the tractions per unit length on the elemental slip δy, T_{SL}, is given by

$$T_{SL} = \mu P = \mu \frac{3N}{4a^3} X^2$$

$$T_{SL} = \mu \frac{3N}{4a^3} (a^2 - y^2) \qquad (8.18)$$

From this equation it is seen that the value of T_{SL} is a function of y having a zero value at $y = a$. For other strips when c has a real physical value, the

192

stick area is at the leading edge of the contact zone and the traction, T_{ST}, over such an elemental strip ΔY is, using equation 8.13

$$T_{ST} = \mu P \left[2\frac{c}{X} - \frac{c^2}{X^2} \right]$$

substituting for P and X then gives

$$T_{ST} = \frac{3\mu Nc}{4a^3} \left[2(a^2 - y^2)^{1/2} - c \right] \tag{8.19}$$

Hence in the elemental strips having stick areas defined by c the tangential traction per unit length is again a function of the y coordinate of the particular strip.

Referring to figure 8.13 we can easily identify the y values where equations for T_{SL} and T_{ST} should apply. Self-evidently, the line $x = c$ will intersect the circular contact boundaries at values $Y = \pm(a^2 - c^2)^{1/2}$ so that for y values outside these limits the equation for T_{SL} applies while within these limits T_{ST} must be used.

Returning to our original problem of two rolling spheres subjected to a normal load N and a tangential traction F, we see that equilibrium requires that

$$F = 2 \int_{(a^2-c^2)^{1/2}}^{a} T_{SL}\, dy + 2 \int_{0}^{(a^2-c^2)^{1/2}} T_{ST}\, dy$$

$$= 2\frac{3\mu N}{4a^3} \int_{(a^2-c^2)^{1/2}}^{a} (a^2 - y^2)\, dy + 2 \times \frac{3Nc}{4a^3} \int_{0}^{(a^2-c^2)^{1/2}} [2(a^2 - y^2)^{1/2} - c]\, dy$$

These integrations are straightforward and may be used to replace c in equation 8.17 thus producing a relationship between the total transmitted traction F and the ensuing angular velocity ratio ϕ. The pattern of stick areas is of course that already shown in figure 8.13.

Experimental results for steel balls rolling on steel tracks show good agreement with the predictions of this theory[5,6]. Further evidence for the validity of these arguments has been provided by photoelastic studies of rolling spheres which demonstrate the form of contact stress distribution shown in figure 8.9j[7].

8.3.3 Sphere Rolling in a Grooved Track

Heathcote Slip

This topic has been mentioned previously (section 8.2.4) and it will be recalled that it arises due to the geometric conformity between the ball and its track. We may, however, now consider the implications of the surface straining in mitigating such slip requirement by again utilising the simple solution for rolling cylinders.

193

Consider the elliptic contact area to be represented by a row of cylindrical elements of axial length δy and let the angular velocities of the ball and track be ω_1 and ω_2 respectively. Because of the curvature of the contact zone we now see that the rigid-body approach velocities will be given by

$$U_1 = \omega_1(R_1 - z)$$
$$U_2 = \omega_2(R_2 + z)$$

where z is the vertical coordinate position of the particular element δy at a distance y from the origin of coordinates O, (figure 8.14). If R_c is the actual

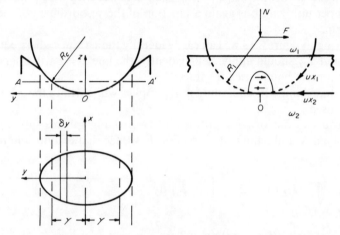

Figure 8.14   The geometry and kinematics of conforming contacts

conforming radius of the ball and its track after deformation (see Chapter 3), we may write

$$z = R_c - (R_c^2 - y^2)^{1/2}$$

and as y and z are small with respect to R_c in most practical cases this may be reduced since

$$z = R_c - R_c\left(1 - \frac{y^2}{R_c}\right)^{1/2}$$

$$= R_c - R_c + \frac{y^2}{2R_c} - \cdots = \frac{y^2}{2R_c}$$

For any given elemental cylinder δy we may then substitute in equation 8.11 and thus obtain

$$\pm Kc = \frac{\omega_1\left(R_1 - \dfrac{y^2}{2R_c}\right) - \omega_2\left(R_2 + \dfrac{y^2}{2R_c}\right)}{\omega_1\left(R_1 - \dfrac{y^2}{2R_c}\right) + \omega_2\left(R_2 + \dfrac{y^2}{2R_c}\right)}$$

194

Dividing this equation by $\omega_2 R_1$ and putting $\omega_1/\omega_2 = \phi$ gives

$$\pm Kc = \frac{\left(\phi - \dfrac{R_2}{R_1}\right) - \dfrac{y^2}{2R_cR_2}(1 + \phi)}{\left(\phi + \dfrac{R_2}{R_1}\right) + \dfrac{y^2}{2R_cR_2}(1 - \phi)} \qquad (8.20)$$

It will be apparent that ϕ will have a value near to R_2/R_1 so that the two terms in the numerator of equation 8.20 are comparable, whereas in most practical situations y^2 is very much less than R_c, so that the second term in the denominator may be neglected with respect to the first term. Equation 8.20 then becomes

$$\pm Kc = \frac{\left(\phi - \dfrac{R_2}{R_1}\right) - \dfrac{y^2}{2R_cR_1}(1 + \phi)}{\left(\phi + \dfrac{R_2}{R_1}\right)} \qquad (8.21)$$

This equation shows that for any elemental cylinder δy at a coordinate value y the instantaneous value of c is a quadratic function of y since ϕ is independent of y. Furthermore it is seen that when $c = 0$, that is, sticking occurs throughout the contact zone at that value of y, the values of y must be given by

$$y^2 = 2R_cR_1\left(\frac{\phi - \dfrac{R_2}{R_1}}{1 + \phi}\right)$$

Let these particular values of y be identified by γ.

The patterns of sticking and slipping in the contact zone must then follow the form shown in figure 8.15, which shows some typical patterns for high and low geometric conformity between the ball and its track. It will be noted that the degree of slipping is very much greater when the conformity is high.

Since equation 8.21 contains two unknowns, namely c and ϕ, the complete

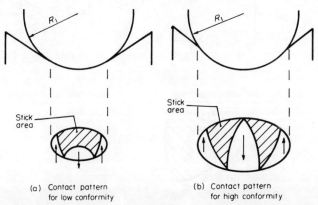

(a) Contact pattern   (b) Contact pattern
for low conformity   for high conformity

Figure 8.15   Stick–slip patterns in conforming contacts

195

solution of this problem requires that the value of the tractions for each particular elemental cylinder must be inserted and the equilibrium of the system must then be considered. The value of the traction for any cylindrical element is of course given by equations of the type 8.18 and 8.19 and the equilibrium conditions (figure 8.14) are

resolving horizontally

$$F = \int T_{SL} + \int T_{ST}$$

taking moments about O

$$FR_1 = \int T_{SL}\, z + \int T_{ST}\, z$$

where the integrals are taken over the limits of the appropriate sticking and slipping regions, taking due account of their directions.

It is not possible to obtain simple analytical solutions to this problem, but it is worth noting that at very high conformity the stick areas become very small and slip must occur throughout most of the contact zone. For the limiting conditions where slip occurs everywhere in the contact zone, it can easily be shown that

$$\frac{F}{N} = \frac{0.08\mu b^2}{R_1^2}$$

This method of solution may also be adopted to deal with a ball rolling around a conforming circular track such as occurs in ball thrust bearings, although such solutions necessitate the use of a digital computer. The resultant stick–slip patterns of the type shown in figure 8.16 occur.

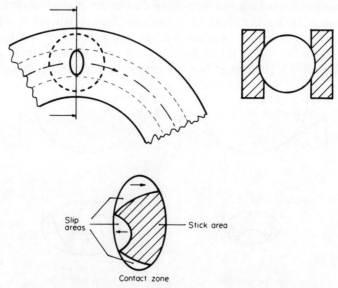

Figure 8.16   Stick–slip patterns in ball thrust bearings

196

We are now in a position to make a qualitative analysis of the stick–slip patterns which could occur due to an element rolling along a conforming track and also subjected to either a driving or braking traction, (figure 8.17).

From figure 8.17 we may surmise that if a body is rolling along a conforming track while subjected to a braking traction, the two microslip effects, (figures 8.17a and 8.17c), will tend to increase the slip at the extremities of the contact ellipse and reduce the slip near to the centre of the contact, that is, slip is reduced in the region of high local contact pressure. This has the effect of reducing wear due to the microslip, and has an optimum condition when the slip in the high pressure region is eliminated. A detailed presentation of these arguments may be found in references 8 and 9.

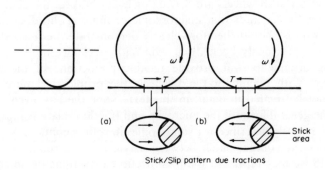

Stick/Slip pattern due tractions

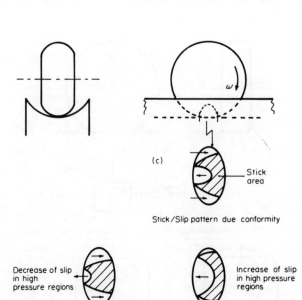

Stick/Slip pattern due conformity

Figure 8.17   Combinations of slip patterns due to different effects

197

The method of solution proposed in the foregoing is approximate in that it neglects any interaction between adjacent elemental cylinders. It does, however, lead to reasonable predictions of the microslip effects for a wide range of rolling contact situations, and should therefore enable designers to ascertain the probable effects which would occur in such cases.

8.4 TYRE–ROAD CONTACTS

A very common example of rolling wheels subjected to tangential tractions where T is generally less than μN, occurs in the wheels of cars and locomotives. Whether such wheels are driving or being braked, the control of the vehicle usually demands that macroscopic or limiting slip does not occur. All such wheels should therefore clearly demonstrate the microslip effects which have previously been discussed. With steel locomotive-wheels the preceding arguments will provide reliable predictions of the microslip behaviour. With pneumatic-tyred wheels the problem is more complex and cannot be treated in detail in this text. None the less, even with such wheels the general form of the microslip and traction relationship is similar to that previously described for solid bodies in rolling contact.

Consider the contact patch between a tyre and the road. As shown in figure 8.18 we may represent the total frictional tractions by an equivalent force $F \leqslant \mu N$ acting through some point such as X. This system may be

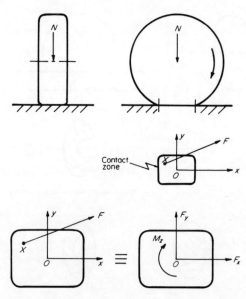

Figure 8.18   Resolution of the frictional effects in the tyre–road contact

198

resolved into the three components acting at some origin of coordinates O where

(a) F_x is the driving or braking traction which will produce the type of microslip effects already described.

(b) F_y is the transverse tangential traction which will also produce transverse microslip. This force is the 'cornering force', which as the name implies is the force present when cornering and producing the change of direction of the vehicle.

(c) M_z is a torque called the 'self-aligning torque' which is endeavouring to straighten the course of the vehicle. It is self-evident to the driver through the feel of the steering wheel, and will only exist when the vehicle is cornering. Clearly when the vehicle travels along a straight course both F_y and M_z are zero.

Since F_x, F_y and M_z all arise from the frictional tractions the resultant microslip velocities V_x, V_y and ω_z must be in the opposite sense to the frictional tractions and clearly for a straight course $V_y = \omega_z = 0$. The form of the relationship is similar to that discussed in earlier sections of this chapter. Figure 8.19 shows the relationship between V_x and F_x and the similarity to figure 8.11 is apparent. As $F_x \rightarrow \mu N$ macroslip, that is, skidding, will ensue.

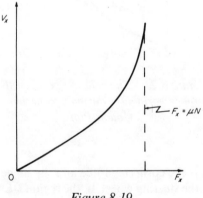

Figure 8.19

Returning to the more general case, consider a car being driven around a left-hand turn. The force system and the slip velocity components are as shown in figure 8.20 and the path of motion lags behind the centre-line of the tyre by the slip angle θ. This slip angle may be interpreted as the angular microslip arising due to the self-aligning torque M_z. The relationships between F_y, M_z and θ are then most vividly seen in the form shown in figure 8.21[10]. Here we see that in the region ABC, as we get increased cornering force F_y, the self-aligning torque increases and the slip angle increases—our common driving experience. In the region CD there is very little change in the

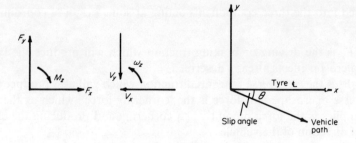

Figure 8.20

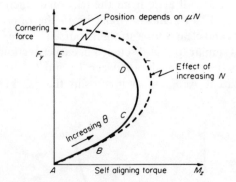

Figure 8.21   The relation between the
major   variables   during   cornering—
Gough Plot

self-aligning torque as the cornering force and slip angle increase, that is, we
have lost the 'feel' at the steering wheel. In the region *DE*, which is close to the
point where full skidding occurs, there is a reversal in the feel at the wheel
as the car drifts around the corner. It is also worth noting that the effect of the
friction coefficient μ and the normal load N are fairly modest in the region
ABC but become critical as the point *E* is approached.

The nature of the road surface and the tread pattern on the tyre become
increasingly important in wet conditions. In such cases it is possible for
continuous films of fluid to form between the tyre and the road and in the
limit leads to loss of control due to 'aquaplaning'. This phenomenon is then
more a problem in lubrication than dry friction, and is briefly mentioned in
chapter 11. It is in effect an example of elastohydrodynamic lubrication
within the regime where film thickness is defined by equation 11.5.

200

REFERENCES

1. F. P. Bowden and D. Tabor. *The Friction and Lubrication of Solids* Part II. Oxford University Press, (1964), 288.
2. J. Poritsky. *J. appl. Mech., Trans. Am. Soc. mech. Engrs,* **72**, (1950), 191.
3. J. A. Jefferis and K. L. Johnson. *Proc. Instn mech. Engrs,* **182**, (1968), 281.
4. J. Halling and S. K. Sen Gupta. *Wear,* **24**, (1973), 127.
5. K. L. Johnson. *J. appl. Mech., Trans. Am. Soc. mech. Engrs,* **80**, (1958), 339.
6. J. Halling. *Wear,* **7**, (1964), 516.
7. D. J. Haines and E. Ollerton. *Proc. Instn. mech. Engrs,* **177**, (1963), 95.
8. J. Halling. *J. mech. Engng Sci.,* **6**, (1964), 64.
9. K. L. Johnson (Ed. J. B. Bidwell). *Rolling Contact Phenomena,* Elsevier, New York, (1962), 6.
10. V. E. Gough. *Auto. Engr,* (April, 1954).

9

Lubricant Properties
and Testing

9.1 INTRODUCTION

The term lubricant generally suggests oil or grease simply because they are
the most common lubricants in use; but they are not exclusive and, in fact,
any fluid can be used as a lubricant in the right circumstances. In modern
applications a very wide range of fluids are used as lubricants. Air or gas
bearings are now quite common, there are numerous examples of the use
of water as a lubricant and there is an increasing use of process fluids, one
rather exotic example being the use of liquid sodium as a lubricant in nuclear
reactors. In some situations solid lubricants are used but the main emphasis
in this chapter is on fluid lubricants.

The most important single property of a lubricant is its viscosity, and a
substantial portion of this chapter is devoted to viscosity measurement and
viscous behaviour in general. In the limited space available it is impossible
to consider all fluids in detail and, because of their importance, the remainder
of the chapter concentrates on oils and greases for which there are standard
tests laid down for the evaluation of many properties, including viscosity.

9.2 VISCOSITY

The viscosity of a fluid may be defined qualitatively as its resistance to flow,
and liquids are often described as 'thick' or 'thin' which is really an arbitrary,
visual assessment of their viscosity.

The foundations of modern viscous theory were laid down in the seventeenth century by Newton who proposed a method of quantifying the viscosity of a fluid. Newton based his postulations on a model of fluid flow which considered the flow to be equivalent to a large number of thin layers of fluid sliding over each other rather like a pack of playing cards, as shown in figure 9.1.

The internal friction or viscosity of the fluid gives rise to shear stresses, τ between the relative sliding layers which act in such a way that they tend to retard the faster moving layer and accelerate the slower layer. Newton

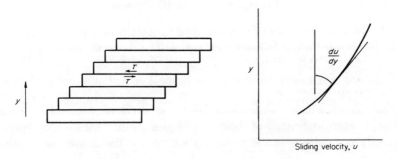

Figure 9.1   Newtonian model of fluid flow

postulated that the viscous shear stresses were directly proportional to the shear strain rate. It can be shown that the rate of shear is equal to the velocity gradient, du/dy and hence

$$\tau \propto \frac{du}{dy}$$

or

$$\tau = \eta \frac{du}{dy} \tag{9.1}$$

The constant of proportionality, η, is sometimes referred to as the viscosity, coefficient of viscosity or absolute viscosity but the title *dynamic viscosity* has been adopted in modern practice and it is recommended that this name be used to avoid confusion.

Fluids whose viscous behaviour is described by equation 9.1 are referred to as newtonian fluids, and most common fluids, particularly pure fluids and fluids which have a simple molecular structure, fall into this category. Fluids which do not behave in this manner are referred to as non-newtonian fluids. Examples of non-newtonian fluids are greases, multiphase fluids and simple fluids under extreme conditions, and their behaviour is discussed in more detail later in the chapter. In the SI system the unit of dynamic viscosity is 1 Ns/m^2 but in practice the commonly used unit is the *poise* (P) which is a CGS unit.

$$1 \text{ poise} = 1 \text{ dyne s/cm}^2 = 1 \text{ g/cm s}$$

and

$$1 \text{ poise} = 0.1 \text{ Ns/m}^2$$

The poise is quite a large unit and it is even more common to use its submultiple, the *centipoise* (cP) where

$$1 \text{ centipoise} = 1 \times 10^{-2} \text{ poise}$$

The dynamic viscosity of different fluids covers a wide range of values as may be seen from the following examples. The dynamic viscosity of air is of the order of 0.02 cP while water has a viscosity of the order of 1 cP. Lubricating oils have viscosities in the range 2–400 cP and molten bitumen up to 700 cP.

Instruments which are used to measure the viscosity of a fluid are known as viscometers and many of these do not give a direct measure of dynamic viscosity but determine the value η/ρ where ρ is the density of the fluid. The same ratio occurs frequently in fluid flow problems and is called the *kinematic viscosity* of the fluid for which the symbol v is conventionally used that is

$$v = \eta/\rho \tag{9.2}$$

The SI unit of kinematic viscosity is $1 \text{ m}^2/\text{s}$ but in practice the commonly used unit is the *stoke* (S), which is a CGS unit.

$$1 \text{ stoke} = 1 \text{ cm}^2/\text{s} = 1 \times 10^{-4} \text{ m}^2/\text{s}$$

The stoke being a large unit it is more common to use its submultiple, the centistoke (cS).

The viscosity of a fluid is not a constant, it is a function of temperature and pressure and viscous properties vary from one fluid to another. The newtonian model of fluid flow is adequate for the definition of dynamic viscosity but it does not explain the factors which give rise to the viscous effects. To get a better understanding of this we must consider the fluid in greater detail, in fact, we must look at a fluid on a molecular scale.

All fluids consist of a large number of molecules which behave in a very complex manner, but there are two main features of their behaviour which contribute towards the viscous effects

(a) There is an attractive force between all molecules of the fluid. The force of attraction between any two molecules depends upon the distance separating them and decreases very rapidly as the separation increases.

204

(b) The molecules of a fluid are in continuous random motion regardless of any bulk movement of the fluid, and the mean velocity of the molecules increases with temperature.

If we look again at the newtonian model of fluid flow (figure 9.1) we must accept that a fluid does not actually separate into a number of layers since there is a continuous movement of molecules throughout the fluid. However, the model is still a reasonable approximation for steady flows, since the quantity of fluid and the conditions in each hypothetical layer will not vary with time. If we consider two adjacent layers of fluid it is clear that the inter-molecular attraction forces between the molecules in the two layers will give rise to an overall force which will oppose any relative sliding motion of the two layers. This effect is equivalent to a shear stress acting at the inter-face.

If two adjacent layers of fluid are moving at different velocities the momentum of the fluid in each layer is not the same, and consequently the mass transfer between layers is accompanied by momentum transfer. When molecules from a slow moving layer pass into an adjacent layer moving at a higher velocity they tend to reduce the momentum of that layer, which is equivalent to applying a shear force opposing the motion. Similarly, molecules from the fast layer tend to increase the momentum of the slow layer, which is equivalent to applying a shear force in the direction of motion.

The viscous properties of a fluid are the result of the combined effects of the intermolecular forces and the momentum transfer. The relative contribution from each of these two sources depends upon the type of fluid concerned. In liquids the molecules are close together and quite slow moving, with the result that the intermolecular forces dominate their viscous behaviour. On the other hand gas molecules are widely spaced and have higher velocities than those in liquids and it is the momentum transfer which is predominant.

Viscosity data are available from many sources but a useful collection of data are presented in the *Engineering Sciences Data Sheets*[1-3] covering common liquids, gases and petroleum products. These also include extensive references to other sources of information.

9.2.1 Effect of Temperature of Viscosity

The viscosity of a liquid is due almost entirely to intermolecular forces. As the temperature is increased the liquid expands, the molecules move farther apart and the intermolecular forces decrease with the result that the viscosity decreases. Examples of dynamic viscosity–temperature curves for common liquids are shown in figure 9.2, and for typical lubricating oils in figure 9.3. Since the variation in density of a liquid with temperature is small the curves of figures 9.2 and 9.3 also represent the variation of kinematic viscosity.

It is convenient to have a mathematical equation relating viscosity and

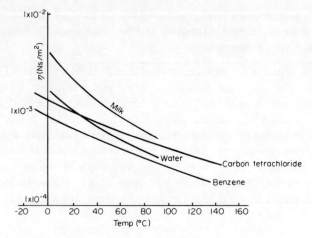

Figure 9.2 Examples of viscosity–temperature relationships for liquids at atmospheric pressure

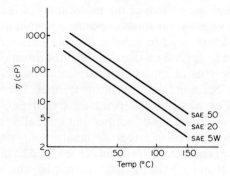

Figure 9.3 Typical viscosity–temperature relationships for mineral oils at atmospheric pressure

temperature, and several empirical formulae have been proposed for this purpose. None of these empirical equations are universally applicable but one such equation which is simple and gives reasonable accuracy for liquid lubricants is

$$\log \eta = A + B/T \qquad (9.3)$$

where A and B are constants and T is the absolute temperature.

With gases, momentum transfer is the dominant contribution to their viscosity. As the temperature of a gas is raised, the velocity of the molecules is increased. This gives an increase in momentum transfer and consequently the dynamic viscosity. Therefore, the effect of temperature on the viscosity of a gas is opposite to that for a liquid. The dynamic viscosity of air as a function of temperature is shown in figure 9.4. At constant pressure the density of a gas is inversely proportional to the absolute temperature and therefore the kinematic viscosity of a gas is more sensitive to temperature than the dynamic viscosity. Useful information and further references on viscosity–temperature characteristics are to be found in the *Engineering Sciences Data Sheets*[1,2,3].

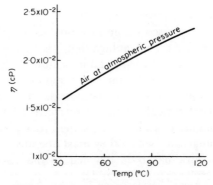

Figure 9.4 Viscosity–temperature curve for air

The variation of viscosity with temperature is a very important feature of a lubricant. For lubricating oils the *viscosity index* (V.I.) has come into common use and gives an approximate indication of the effect of temperature on kinematic viscosity. The viscosity index classification was devised by Dean and Davies[4] in 1929 and was based on two standard oil series. The standard oils chosen were Pennsylvanian oils and Gulf Coast oils. The Gulf Coast oils were thought to have the greatest variation of viscosity with temperature and were assigned a viscosity index of 0, while the Pennsylvanian oils were thought to have the least variation and were assigned a viscosity index of 100. Following this, any other oil should have a V.I. between 0 and 100.

The method of determining the V.I. of an oil sample is shown diagramatically in figure 9.5. The kinematic viscosity of the oil sample is

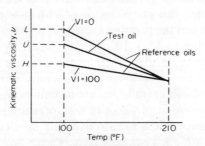

Figure 9.5 Method of viscosity index classification

207

determined at 100 °F and 210 °F. Two reference oils are chosen, one from each standard series such that they both have the same kinematic viscosity as the sample at 210 °F. The viscosity index of the sample is then determined from the formula

$$\text{V.I.} = \frac{L - U}{L - H} \times 100 \qquad (9.4)$$

where L, U and H are as shown in figure 9.5.

The viscosity index is based on an arbitrary scale and many anomalies can arise in its use. It has the advantage that it is simple, but because of its deficiencies it can only be regarded as an indication of the rate of change of viscosity with temperature. With improved refining, the use of additives and synthetic lubricants, V.I. values well outside the range 0–100 are now encountered. Most commerical lubricating oils have V.I. in the region of 100, while automobile engine oils have V.I. of the order of 160 because of the wide temperature range over which they must operate.

A great deal of work has been devoted to the viscosity–temperature characteristics of petroleum products and it is now possible to determine with reasonable accuracy the kinematic viscosity of a product at any temperature knowing only its viscosity at two different temperatures. Two graphical methods have been devised for this purpose known as the ASTM (American Society for Testing and Materials)[5] method and the 'Refutas' method[6]. Both these methods are based on the Walther equation

$$\log_{10} |\log_{10}(v + a)| = b + c \log_{10} T \qquad (9.5)$$

where a, b and c are constants which are dependent on the constitution of the oil.

9.2.2 Effect of Pressure on Viscosity

When the pressure of a liquid or gas is increased the molecules are forced closer together. This increases the intermolecular forces and the viscosity increases. Typical viscosity–pressure characteristics are shown in figure 9.6. With mineral oils the changes of viscosity with pressure in general only become significant when the pressure exceeds approximately 2×10^7 N/m². When the pressure rises to 3.5×10^7 N/m² the viscosity of a typical mineral oil is approximately twice that at atmospheric pressure. As the pressure is further increased, the rate of change of viscosity increases, until at very high pressure mineral oils cease to behave as liquids and become more like a waxy solid. The variation of viscosity with pressure is very important in situations which involve elastohydrodynamic lubrication, that is, the lubrication of heavily loaded contacts such as gears and rolling bearings.

It is convenient to have a mathematical relationship between viscosity and pressure and many empirical equations have been suggested for this

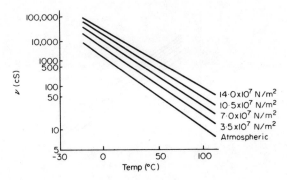

Figure 9.6 Effect of pressure on viscosity of
SAE 40 oil

purpose. None of these are universally applicable but one such equation which gives reasonable accuracy for liquids and is convenient for mathematical manipulation is

$$\eta = \eta_0 e^{\alpha p} \qquad (9.6)$$

where η is the dynamic viscosity at pressure p, η_0 is the dynamic viscosity at atmospheric pressure and α is the pressure exponent of viscosity.

Useful data and further references on viscosity–pressure characteristics are given in *Engineering Sciences Data Sheets*.

9.2.3 SAE Classification of Lubricating Oils

In the automobile industry lubricating oils are commonly classified by their SAE (Society of Automotive Engineers) number. In general, the higher the SAE number, the greater is the viscosity of the oil, but the number does not specify the actual viscosity, it simply denotes that the viscosity–temperature curve lies within a band which is specified at temperature of 0 °F or 210 °F. Multigrade oils are given a double number. For example, an oil designated SAE 5w/30 means that at 0 °F the oil falls in the SAE 5 grade and at 210 °F it falls in the SAE 30 grade. In the classification lubricants are divided into two types, crankcase oils and transmission oils. Typical viscosity–temperature curves for a range of SAE oils are shown in figure 9.7 and further details of the method of classification may be obtained from the Society of Automotive Engineers Handbook.

9.2.4 Non-newtonian Behaviour

For a newtonian fluid a graph of shear stress against rate of shear is a straight line passing through the origin as shown in figure 9.8. For such a fluid the viscosity is given by the slope of the line and therefore its viscosity can be determined by a single measurement to obtain one point on the graph. Any

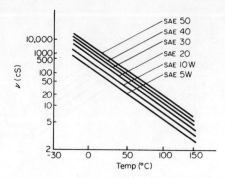

Figure 9.7 Typical viscosity–temperature graphs for mineral oils at atmospheric pressure

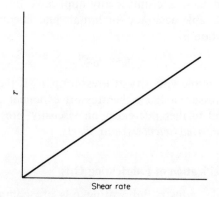

Figure 9.8 Shear stress–shear rate relationship for a newtonian fluid

fluid which does not behave in the manner described above is classified as a non-newtonian fluid and it is clear that such fluids do not have a single value of viscosity. In connection with non-newtonian fluids it is common to use the term *apparent viscosity*, which is simply the ratio of shear stress to shear rate at a specified shear rate. The departure from newtonian behaviour can take many forms. It may simply be a departure from the linear shear stress–shear rate curve or the stress may vary with the length of time that shearing has taken place or the fluid may behave in a pseudo-solid manner. Before discussing these different forms of behaviour it is useful to consider the factors which give rise to these non-newtonian effects.

Non-newtonian behaviour is in general a function of the structural complexity of a fluid. Liquids such as water, dilute solutions, benzene and light oils are newtonian. Such liquids are considered to have a loose molecular structure which is not affected by shearing action. Highly dispersed suspensions of solid or liquid particles may also behave as newtonian fluids if the structure is loose, allowing the shear to take place in the fluid separating the particles.

The most common non-newtonian fluids are suspensions in which the suspended particles form a structure which interferes with the shearing of the suspension medium in a manner dependent upon the shear rate. Examples of non-newtonian fluids are water–oil emulsions and polymer-thickened oils, and an extreme example is greases which are a concentrated suspension of particles in an oil base. Non-newtonian behaviour can be broadly classified under the following headings.

Plastic or Bingham Fluids

These are fluids which require a significant shear stress before flow or deformation commences, but once this shear stress has been exceeded the shear stress is proportional to the shear rate as shown in figure 9.9.

After being deformed a Bingham plastic retains its shape. This may or may not be a desirable property. In the case of modelling clay such a property is essential if the model is not to collapse, but in the case of, for example, paint, it would be very undesirable since it would retain all the brush marks.

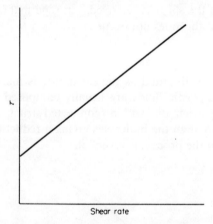

Shear rate

Figure 9.9   Shear stress–shear rate relationship for a plastic or Bingham fluid

Thixotropy

Thixotropy is a time-dependent loss of consistency when the fluid is sheared. Ideally, the process is reversible in the sense that the viscosity returns to its original value when the shear is removed and sufficient recovery time is allowed, as shown in figure 9.10. This is referred to as a temporary viscosity loss but in some instances the viscosity never returns to its original value and

211

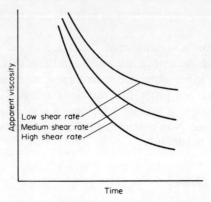

Figure 9.10 Thixotropic behaviour

this is a permanent viscosity loss. It is thought that thixotropic fluids have a structure which is being continuously broken down under the shearing action and which, at the same time, rebuilds itself. The breakdown of the structure progresses with time, giving a reduction in apparent viscosity until a stage is reached where the structure is being rebuilt at the same rate and the apparent viscosity attains a steady value. Ideally the structure returns to its original form when the shear is removed, but a permanent loss of viscosity is experienced when this does not occur.

Pseudo-Plastic Flow

This is a 'thinning' of the fluid as the shear rate is increased as shown in figure 9.11. Pseudo-plastic fluids are usually composed of long molecules which are randomly orientated with no connected structure. Application of a shear stress tends to align the molecules giving a reduction in the apparent viscosity. In general the process is reversible.

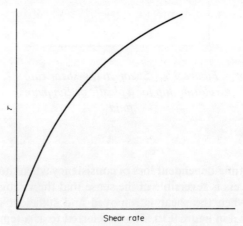

Figure 9.11 Pseudo-plastic behaviour

212

Dilitancy

This is a 'thickening' of the fluid when the shear rate is increased, as shown in figure 9.12. Fluids which exhibit dilitancy are usually suspensions having a high solid content and their behaviour can be related to the arrangement of the particles. In the unstressed state the particles adopt a close-packed formation which gives a minimum volume of voids, but when the stress is applied the particles move to an open-packed formation dilating the voids. This leads to a situation where there is insufficient fluid to fill them and gives an increased resistance to flow. The same phenomenon can be observed when walking on wet sand where the pressure of one's foot produces a dry patch.

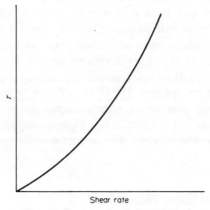

Shear rate

Figure 9.12   Non-newtonian behaviour referred to as dilitancy

Elasticity

Elasticity is where the application of a constant stress simply produces a constant deformation without initiating flow. When the stress is removed the fluid recovers completely and returns to its undeformed shape. There is a limiting value of shear stress which can be applied, that is, an elastic limit; once this is exceeded flow takes place and the elastic effects may be swamped by the viscous effects. Fluids which exhibit significant elastic behaviour are referred to as viscoelastic fluids.

9.3 MEASUREMENT OF VISCOSITY

The importance of viscosity is reflected in the number of instruments which have been devised to measure this property, and a thorough description of all the different instruments would require a book in itself. However, the instruments can be conveniently classified into three main groups, according to the physical measurements employed

(i) Instruments in which the viscosity is determined from the flow of the fluid.

(ii) Instruments in which the viscosity is determined from the motion of a solid object through the fluid.

(iii) Rotational viscometers, in which the viscosity is determined by shearing the fluid between two relatively rotating surfaces.

The instruments vary widely in their complexity and applicability. In general, the simple instruments are only suitable for measuring the viscosity of fluids in which the non-newtonian effects are small or negligible. When non-newtonian effects are significant, more complex instruments must be used and rotational viscometers are best suited for this purpose. The main types of viscometer and their uses are described briefly in the following sections. Some of these instruments have facilities built-in for controlling and measuring the temperature of the test fluid. Where these facilities are not provided it is usual to locate the viscometer in a temperature-controlled liquid bath. The majority of viscometers require visual monitoring and the bath liquid must be transparent. The choice of bath liquid depends upon the range of temperature over which measurements are to be made. Water is suitable for the range 0 °C to 90 °C. For higher temperatures clear mineral oils are usually used and ethyl alcohol or acetone allow measurements to be made down to −54 °C.

9.3.1 Capillary and Efflux Viscometers

Instruments in which the viscosity is determined from the flow of the fluid can be subdivided into two types, capillary viscometers and efflux viscometers.

Capillary Viscometers

In capillary viscometers the viscosity is determined from the time taken for a given volume of fluid to flow through a capillary tube. For satisfactory operation the flow in the capillary must be laminar and the measurements are based on Poiseuille's Law for steady viscous flow in a pipe which takes the form

$$\eta = \frac{\pi d^4 \, \Delta p \, t}{128 l q} \tag{9.7}$$

where d and l are the diameter and length of the capillary respectively, Δp is the pressure difference across the capillary and q is the volume of fluid flowing through the capillary in time t.

Equation 9.7 is valid only for fully developed laminar flow and in some cases it is necessary to make corrections to allow for 'end' effects. There are two main corrections which may be made. First there is the kinetic energy

214

correction, which makes allowance for the fact that some of the available pressure difference is taken up in accelerating the fluid at entry to the capillary, thus reducing the pressure difference available to overcome the viscous stresses. Second, the rapid changes in velocity at inlet and exit also influence the flow and act in such a way that they effectively increase the length of the pipe. The magnitude of this effect is a function of capillary diameter. When the two corrections are incorporated equation 9.7 takes the modified form

$$\eta = \frac{\pi d^4 \Delta p \, t}{128(l + nd)q} - \frac{m\rho q}{128\pi(l + nd)t} \tag{9.8}$$

where m and n are empirical constants. m normally lies in the range 1 to 1.5 and the *British Standard Specification* 188[7] recommends a value of $n = 3$.

To minimise end effects the capillary tube should have a length–diameter ratio of 300 or greater and the Reynolds number (Re) for the flow should not exceed 1000 where

$$Re = \frac{\rho d v}{\eta} \tag{9.9}$$

v being the mean velocity of flow, given by

$$v = \frac{q/t}{\pi d^2/4} \tag{9.10}$$

Capillary viscometers may be subdivided into two types, absolute viscometers and relative viscometers. With absolute viscometers the viscosity is calculated from the fundamental equation 9.7. These viscometers are designed to minimise end effects and their accuracy is mainly dependent on the accuracy with which the capillary dimensions can be measured. The best known absolute viscometer is the Bingham viscometer[8]. The majority of capillary viscometers in common use are relative viscometers. These are calibrated with liquids of known viscosity in order to obtain the viscometer constant and, in general, give greater accuracy than absolute viscometers since dimensional errors are eliminated. Relative viscometers are usually gravity operated, in that the pressure difference is provided by the hydrostatic head of the test fluid which is dependent on the density of the fluid. If Δh is the hydrostatic head, $\Delta p = \rho g \, \Delta h$ and equation 9.7 becomes

$$\eta = \frac{\pi d^4 \rho g \, \Delta h \, t}{128 l q}$$

or

$$v = \frac{\pi d^4 g \, \Delta h \, t}{128 l q} = Bt \tag{9.11}$$

where B is a constant for a particular viscometer. For low viscosity fluids where the flow time is short it may be necessary to correct for end effects and then

$$v = Bt + C/t \qquad (9.12)$$

where C is of a similar form to the coefficient of $1/t$ in equation 9.8 and its numerical value is normally supplied with the instrument.

The Ostwald U-type viscometer and its modified forms are by far the most commonly used relative viscometers. The design shown in figure 9.13 is that adopted by the British Standards Institution and is known as the BS/U.

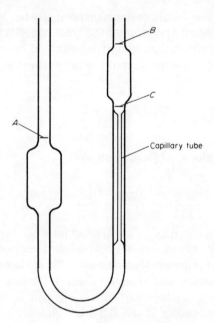

Figure 9.13   Ostwald U-tube viscometer

In the right-hand limb the reservoir is situated above a vertical capillary tube, which is connected to the receiving bulb at a lower level in the left-hand limb by a U-tube. In use the fluid level is first adjusted so that the meniscus is at mark A. The fluid is then drawn into the right-hand limb until the meniscus is about 0.5 cm above B and then released. The time taken for the meniscus to fall from B to C is noted and the viscosity calculated from equations 9.11 or 9.12.

There are eight recommended sizes of BS/U viscometers which cover a range from 1–1500 centistokes. Detailed instructions and information are given in BS 188[7] which also lays down the method of calibration using freshly distilled water.

Efflux Viscometers

With efflux viscometers the viscosity is determined from the time taken for a given volume of liquid to discharge under gravity through a short tube orifice in the base of the instrument. There are three types of efflux viscometer in common use, the Redwood, Saybolt and Engler viscometers. The three instruments all have the form shown in figure 9.14, and differ only in the quantity of fluid discharged and the dimensions, particularly the dimensions of the orifice. The Redwood viscometer[9] is commonly used in the United

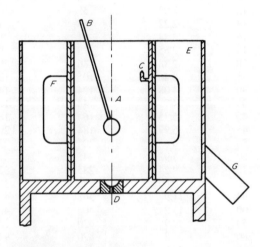

A – Container for test liquid
B – Plug
C – Test liquid level setting
D – Discharge orifice
E – Water jacket
F – Water agitator
G – Heater

*Figure 9.14   Typical arrangement of an efflux
viscometer*

Kingdom while the Saybolt viscometer is used mainly in Europe and the USSR. The three types of efflux viscometer are all empirical in that they use arbitrary scales, the viscosities being quoted in terms of the efflux time, for example, Redwood seconds or Saybolt seconds. In the case of the Engler instrument the viscosity is quoted in Engler degrees, this being the time of efflux divided by that for water at the same temperature. The viscosity measurements from these instruments can be converted to kinematic viscosity units at the same temperature and a conversion table is given in table 9.1.

217

TABLE 9.1   VISCOSITY CONVERSION TABLE

Kinematic Viscosity (centistokes)	Engler (degrees)	Redwood No. 1 (seconds)	Saybolt Universal (seconds)
2	1.14	31	32
4	1.31	36	39
6	1.48	41	46
8	1.66	46	52
10	1.84	52	59
12	2.02	58	66
14	2.22	65	74
16	2.44	71	81
18	2.65	78	90
20	2.88	86	98
25	3.46	105	120
30	4.08	124	142
35	4.71	144	164
40	5.35	164	187
45	5.99	184	209
50	6.65	204	232
60	7.92	245	279
70	9.24	285	325
80	10.60	326	371
90	11.9	368	419
100	13.2	406	463
150	19.9	620	700
200	26.8	820	940
250	33.0	1010	1160
300	40.0	1230	1410
400	53.0	1640	1870
500	66.0	2040	2320
600	79.0	2430	2800
700	93.0	2820	3250
800	105	3250	3700
900	118	3650	4200
1000	133	4100	4750
1500	199	6100	7000
2000	260	8100	9200
2500	325	10100	11600
3000	400	12300	14000
4000	530	16100	18500
5000	660	20000	23000

This Table may be used for the *approximate* conversion of viscosity units *at the same temperature*.

9.3.2 Falling-Body Viscometers

The most common type of falling-body viscometer is the falling-sphere viscometer shown in figure 9.15. In this instrument the viscosity is determined from the terminal velocity of a steel sphere falling through the test fluid. When the sphere is falling at its terminal velocity v through the fluid it is in equilibrium under three forces, its weight, the buoyancy force and the drag force, that is

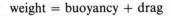

$$\text{weight} = \text{buoyancy} + \text{drag}$$

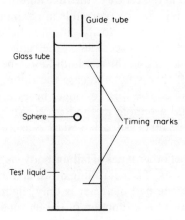

Figure 9.15   Falling-sphere vis-cometer

The weight and buoyancy forces are readily determined from the size of sphere and the densities of the sphere and the liquid. For a sphere of radius r moving at velocity v through an infinite fluid, Stokes derived the expression

$$\text{drag} = 6\pi\eta r v$$

Therefore, for equilibrium

$$\tfrac{4}{3}\pi r^3 \rho_s g = \tfrac{4}{3}\pi r^3 \rho g + 6\pi\eta r v$$

or

$$\eta = \frac{2}{9}\frac{r^2(\rho_s - \rho)g}{v} \tag{9.13}$$

where ρ_s and ρ are the densities of the sphere and the liquid respectively.

Stokes formula only applies to a sphere moving through an infinite fluid and it is sometimes necessary to apply a correction factor, F, to equation 9.13 to allow for the finite tube diameter, such that

$$\eta = \frac{2}{9}\frac{r^2(\rho_s - \rho)g}{v} \times F$$

Numerous correction factors have been proposed but a useful one is the Faxen factor given by

$$F = 1 - 2.104(d/D) + 2.09(d/D)^3 - 0.9(d/D)^5$$

where d is the diameter of the sphere and D is the internal diameter of the tube.

It is normal practice to use a steel sphere and the terminal velocity is determined by timing the descent over a known distance. It is important that the ball is released in the centre of the tube and that it does not have any attached air bubbles. BS 188[7] recommends a maximum ratio of ball–tube diameter of 0.1 and that the ball velocity should not exceed 1 cm/s. Working within the recommendations, viscosities in the range 10 to 2000 poise can be measured. For visual timing the test liquid must be clear enough for the sphere to be visible. If this is not the case the timing may be carried out electrically, and then of course it is no longer necessary to use a glass tube. Using a metal tube and a remote method of releasing the ball, the falling-sphere viscometer can be very easily adapted to measure viscosity at pressures greater than ambient.

There are a number of other types of falling-body viscometers. The rolling-sphere viscometer consists of a ball rolling down an inclined tube filled with test fluid. In this type, the ball diameter is only slightly less than the tube diameter and the viscosity is a function of the terminal velocity. With very small clearances the rolling-sphere viscometer has been used to measure the viscosity of gases and can also be used for high-pressure measurements. For reliable results these instruments must be calibrated.

The falling coaxial-cylinder viscometer consists of two coaxial cylinders with their axes vertical. The outer cylinder is clamped and the clearance space between the two cylinders is filled with the test fluid. In operation the inner cylinder is released and falls under the action of gravity. The viscosity is calculated from the expression

$$\eta = \frac{W}{2\pi l v} \log_e\left(\frac{D_o}{D_i}\right) \tag{9.14}$$

where W is the weight of the inner cylinder plus any additional weights, l is the length of the outer cylinder, D_o and D_i are the diameters of the outer and inner cylinders respectively and v is the speed of descent. This type of viscometer is particularly useful for measuring high viscosities and has the facility for varying the shear rate which is useful for non-newtonian fluids.

The band viscometer is similar in operation to the falling-cylinder viscometer. A thin band is located in a narrow gap between two parallel shear blocks as shown in figure 9.16. The gap is filled with the test fluid and when the band is released it falls under the action of gravity, its motion being opposed by the shear stresses induced in the test fluid.

220

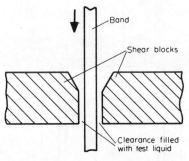

Figure 9.16   Band viscometer

The viscosity is determined from the speed of descent using the following equation

$$\eta = \frac{Wv}{2bhd} \qquad (9.15)$$

where W is the applied load, v is the velocity of descent, b is the width of the shear blocks, d is the length of the shear blocks in the direction of sliding and h is the thickness of one of the shear films. This type of instrument is useful for non-newtonian measurements and has been used for a wide range of viscosities, but low viscosity fluids are difficult to retain in the shear gap.

9.3.3   Rotational Viscometers

Rotational viscometers consist essentially of two elements, one of which is stationary while the other is rotated by an external drive, the space between the two members being filled with test fluid. The viscosity measurements are made either by applying a fixed torque and measuring the speed of rotation, or by driving the rotating element at a constant speed and measuring the torque required or the reaction torque on the stationary member.

There are two main types of rotational viscometer commonly known as the rotating-cylinder viscometer and the cone and plate viscometer.

The rotating-cylinder viscometer consists of two concentric cylinders with the annular clearance filled with test fluid. One of the cylinders, usually the inner one, is driven while the other is stationary as shown in figure 9.17. There are a number of commercial rotating-cylinder viscometers available and these are usually supplied with cylinders of different diameters giving different radial clearances. These can be driven at varying speeds allowing a range of viscosities and shear rates to be accommodated. The main shearing takes place in the annular clearance, but in some cases the shear forces on the ends of the rotating cylinder may be significant and they must then be allowed for or taken into account by calibration with a known liquid. The cone and plate viscometer is shown in figure 9.18 and consists of a plane surface and a conical surface, either of which may be rotated. The clearance

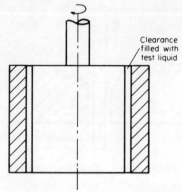

*Figure 9.17   Basic arrange-
ment of rotating-cylinder viscometer*

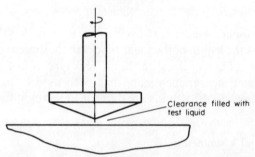

*Figure 9.18   Basic arrangement of cone and
plate viscometer*

space is filled with the test fluid and the included angle of the cone is generally large to ensure a substantially constant shear rate in the clearance space.

The cone and plate arrangement has been employed to study properties other than viscosity. For example, the Weissenberg rheogoniometer[10] which is basically a cone and plate viscometer can be used to measure elastic effects in viscoelastic fluids. When a viscoelastic fluid is sheared in the instrument, normal stresses are generated and the resulting forces tend to separate the two surfaces. These forces are a measure of the elastic properties of the fluid.

When using rotational viscometers problems can arise due to shear heating. The energy dissipated in shearing goes to heating the test fluid and if the temperature rise is significant, ambiguous results may be obtained.

9.3.4   High-Pressure Viscometers

The general description of viscometers presented in the preceding sections covers only the types in common use and is by no means a comprehensive treatment of the subject. Numerous other instruments have also been devised, mainly for investigations which are outside the scope of those described. Many of these have been devised to measure extremely high or low viscosities which fall outside the range of normal instruments, others have been specially designed to investigate non-newtonian phenomena and some instruments have been developed to operate with very small quantities of test fluid for situations where supplies are limited.

The choice of instrument to be used for measuring the viscosity of a particular fluid is to some extent based on experience, but for newtonian fluids the simpler types of instrument such as the capillary, efflux or falling-sphere viscometers are usually adequate. For non-newtonian fluids the most common instruments used are the rotational viscometers.

In heavily loaded situations such as occur in gear teeth and rolling bearings the pressures in the loaded region can be extremely high and a thorough knowledge of the high-pressure properties of a lubricant are essential. Viscosity measurement at high-pressure presents many problems and is usually regarded as a specialist field. The vast majority of viscometers have been designed for operation at ambient pressure and cannot be conveniently adapted for use at high pressures. Of the common instruments the falling-sphere viscometer can be modified for this purpose as described in section 9.3.2. For high-pressure use, very careful design is required to ensure safety and this usually results in poor accessibility. The necessity for an external pressurisation system can also make such an instrument very expensive.

Another type of instrument which offers many advantages for high-pressure viscometry is the disc machine shown diagramatically in figure 9.19. This type of machine does not require an external pressurisation system but instead the high pressures are generated internally in an elastohydro-dynamic film developed between two relatively moving surfaces. In its

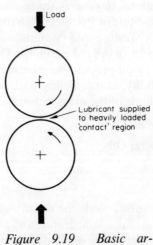

Figure 9.19 Basic arrangement of a disc machine

223

simplest form the disc machine consists essentially of two discs with independent drives. In operation the discs are set in motion, the test fluid is introduced between the discs in the region of minimum clearance and the discs are then loaded together and separated only by a thin elastohydrodynamic film. By driving the discs at different speeds and measuring the film thickness and the viscous torque on the discs, the viscosity can be determined. Sophisticated measuring techniques are required for the disc machine and its use as a viscometer is still being perfected, but it is a very versatile instrument which can be used for non-newtonian measurements.

9.4   LUBRICATING OILS

Originally, lubricating oils were derived solely from animal and vegetable sources, but this changed with the discovery of petroleum oils and now the great majority of lubricating oils come from petroleum, that is, mineral or hydrocarbon oils. One of the reasons for this change is that animal and vegetable oils are more chemically reactive than mineral oils and deteriorate more rapidly in use. However, animal and vegetable oils are very 'oily' and have good emulsifying properties and are still used in certain applications where these properties are desirable, for example textile oils, compressor oils and metal-cutting oils.

   Mineral oils are complex substances containing a wide range of mineral hydrocarbons and it is impossible to assess their performance simply from chemical tests. Consequently, oils must be submitted to many tests to measure their physical characteristics. For comparison purposes the tests must produce reproducible results and therefore it is essential that they are carried out under specified conditions. In general, the test methods applied are those laid down either by the Institute of Petroleum[9] in the United Kingdom or by the American Society for Testing and Materials in the USA[11]. These are usually referred to as the IP or ASTM methods. The tests laid down cover a wide range of lubricating oil properties including such features as rate of deterioration, inflammability, compatibility and suitability for use under extreme conditions and some of the main features are described below.

9.4.1   Properties of Mineral Oils

Relative Density

The relative density of a substance is the density of that substance divided by the density of water at the same temperature and pressure. In the petroleum industry the relative density of an oil is determined at 60 °F or may be corrected to that temperature using standard tables or formulae; most mineral oils have relative densities in the range 0.85 to 0.95. The density of a fluid is required for flow rate calculations and for the conversion of kinematic viscosity to dynamic viscosity and the density of an oil in g/cm^3 is very nearly numerically equal to the relative density.

Specific Heat and Thermal Conductivity

The specific heat and thermal conductivity of an oil are important parameters when the oil acts as a coolant or heat transfer medium. Most mineral oils have specific heats in the range 0.44 to 0.48 and thermal conductivity of approximately 3×10^{-4} cal/cm s °C. Neither of these quantities is very sensitive to temperature changes.

Acidity and Alkalinity

Even well-refined oils contain traces of weak organic acids, but in general these are harmless. More important are the acids which are formed in use by combustion or oxidation and acids introduced by contamination. The acidity of mineral oils is expressed as a neutralisation number, this being the weight of potassium hydroxide in milligrams required to neutralise one gram of oil.

In some lubricants alkalinity is introduced to give them special properties. Most modern engine oils are alkaline to allow them to neutralise fuel combustion products. In some soluble cutting oils alkalinity is introduced to protect ferrous metals from corrosion and to stabilise the emulsion. Alkalinity is expressed as a base number which is the number of milligrams of potassium hydroxide which are equivalent to the alkali present in one gram of oil. (Tests IP 1 and IP 139)

Oxidation Stability

Mineral oils are not very reactive in the chemical sense, but they can oxidise when exposed to oxygen or air at elevated temperatures and this has a strong influence on the life of an oil. The rate of oxidation is a function of temperature and the operating conditions. The products of oxidation vary but in general they consist of acidic compounds, sludge and lacquers, which may cause the oil to become corrosive, to increase in viscosity and to deposit insoluble products on working surfaces. (Tests IP 48, 56, 114, 148, 157)

Flashpoint

Flashpoint may be defined as the lowest temperature at which the vapour given off by an oil, when heated in a standard piece of apparatus, ignites momentarily on the application of a flame. The flashpoint is a measure of the fire hazard and is also useful in determining whether an oil has been contaminated. (Tests IP 34, 35 and BS 2839)

225

Foaming

Excessive foaming of a lubricant due to churning can lead to inadequate lubrication and other problems. This is particularly true for crank-case oils and hydraulic oils since the pump cannot deal efficiently with the frothy mixture. (Test IP 146)

Pour Point

The pour point is the temperature at which oil ceases to flow freely under specified conditions and this imposes a lower limit on the oil working temperature. (Test IP 15)

Demulsibility

Mixtures of pure mineral oil and pure water separate out quite quickly but if the mixture is contaminated separation takes much longer and a stable emulsion may be formed. In many industrial applications oils come in contact with steam and water and it is essential that the oil should separate rapidly, allowing it to be recirculated. The most common test is to pass saturated steam into a measured volume of oil under standard conditions and to observe the time taken for the resulting emulsion to separate into oil and water layers.

Some industrial oils are deliberately manufactured so that they will form stable emulsion. The so-called soluble oils used in metal cutting are an example. (Test IP 19)

Extreme-Pressure Properties

A number of machines are used for evaluating and comparing lubricants under extreme-pressure conditions. All these machines involve heavily loaded regions where sliding or combined sliding and rolling take place and the parameters measured include friction, wear and seizure load. Due to the differences in the various machines, correlation of results is poor and lubricants can really only be compared using the same machine. A commonly used machine is the four-ball tester shown in figure 9.20.

9.4.2 Additives and Solid Lubricants

Most modern lubricating oils have chemical compounds added to them to improve the characteristics of the straight mineral oil. Additives are included for many purposes and some of the main ones are described below.

226

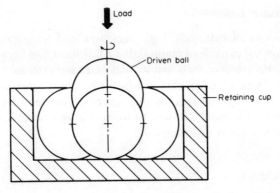

Figure 9.20   Basic arrangement of four-ball testing machine

Viscosity-index Improvers

These are added to an oil to reduce the rate of change of viscosity with temperature and are usually high molecular weight polymers. There is a tendency for the polymers to be broken down when subjected to high rates of shear which will give a general reduction in both viscosity index and viscosity. Some viscosity-index improvers also act as pour-point depressants.

Pour–point Depressants

When mineral oils are cooled, waxy crystals are precipitated, eventually combining to form a rigid structure and this process determines the pour point. Pour-point depressants delay the formation of a rigid structure by forming a coating on the waxy crystals. Complex polymers and other types of products are used as pour-point depressants.

Oxidation Inhibitors

Oxidation inhibitors prevent or reduce the rate of formation of oxidation products such as acidic products and insoluble compounds. By implication, oxidation inhibitors also act as corrosion inhibitors. Different types of oxidation inhibitors are used, depending on whether they are required for high or low temperature applications.

Detergents or Dispersants

These are used mainly in engine oils to keep the engine clean by holding insoluble material in suspension and preventing the formation of sludge or deposits. For high temperature conditions organo-metallic salts are the usual type of additive, while polymer compounds are mostly used for low temperature conditions.

227

Extreme Pressure Additives

Under conditions of extremely high pressures and temperatures straight mineral oils are not capable of maintaining a lubricant film between surfaces. These conditions are frequently met with in gears and extreme pressure additives allow a mineral oil to operate satisfactorily. The most common extreme pressure additives are compounds of sulphur, chlorine and phosphorus and these react with metal surfaces to form coatings which exhibit plastic flow and have low shear resistance.

Corrosion Inhibitors

These are alcohols, esters, amines, organic acids or soaps which are adsorbed on, or react with, metal surfaces to form protective films.

Anti-foam Additives

These are usually silicone based compounds which minimise foaming in situations where this might present problems.

Emulsifiers

These are used to stabilise oil–water emulsions and are particularly useful for metal-cutting oils, fire resistant hydraulic fluids and similar fluids. Typical emulsifiers are petroleum sulphonates and metallic soaps.

Solid Lubricants

These materials are lubricants in their own right and are used as dispersions in mineral oils for special applications. At high temperatures they are advantageous in that they are non-volatile and can provide a lubricant film when the base oil has disappeared. Graphite, molybdenum disulphide and PTFE are examples of solid lubricants (see chapter 6).

9.5 GREASES

A grease is a stabilised mixture of a liquid lubricant and a thickening agent and may include additives to improve or impart particular properties. In the majority of greases the liquid lubricant is a mineral oil but synthetic lubricants such as silicone are sometimes used, particularly in greases which have to operate over a wide temperature range such as in aircraft applications. An indication of the typical working temperature range for different liquid bases is given in table 9.2, but the range may be extended by the use of additives.

228

TABLE 9.2   USEFUL TEMPERATURE
RANGE OF DIFFERENT GREASES

Liquid base	Useful temperature range
Mineral oils	$-30\,°C–100\,°C$
Silicone	$-40\,°C–150\,°C$
Diester and polyester	$-60\,°C–\ 90\,°C$

The thickening agents include metallic soaps, clays and silica. The main metallic soap thickeners are calcium, lithium and sodium soaps in the form of fibres. In general there is no lower temperature limit for the soaps, but at higher temperatures structural changes may occur, and since this temperature is usually below the upper limit for the liquid component, it imposes an upper temperature limit on the grease. With clay and silica thickeners there appears to be no practical lower or upper limit to their use and they can certainly be used up to several hundred degrees celcius.

The structure of greases is such that they are normally self-supporting in the static state and this property allows greases to act both as a lubricant and a seal against contamination. When sheared, greases behave as a viscous fluid but are very non-newtonian and their properties at any one time are strongly dependent on their previous usage. Many standard tests for greases are laid down by both the Institute of Petroleum and the American Society for Testing and Materials and some of the main ones are described below.

Consistency

In this test a standard cone is allowed to penetrate a sample of grease maintained at 25 °C. The consistency is measured in terms of the depth of penetration (in tenths of a millimetre) after a period of five seconds. Usually, two tests are made. The first test is made on unworked grease, that is, grease straight from a container, but these results tend to be unreliable, since some working occurs when the sample is removed from the container. A more reliable second test is made on the grease after it has been subjected to a period of working under standard conditions. (Tests IP 50, 167, ASTM. D.217)

The National Lubricating Grease Institute (NLGI) of America has classified greases in grades depending on their consistency and these gradings, shown in table 9.3, are widely used. Consistency gives only a broad indication of performance. Clearly a soft grease, say grade 1, will set up a lower resistance-to-motion in a bearing than a hard grade 5 grease, but two greases of the same grade may give completely different bearing performance.

229

TABLE 9.3   NLGI CLASSIFICATION OF
GREASES

Grade number	Worked penetration range. (Tenths of a millimetre at 25 °C)
000	445–475
00	400–430
0	355–385
1	310–340
2	265–295
3	220–250
4	175–205
5	130–160
6	85–115

Oxidation Stability

Resistance to oxidation is assessed by oxidising thin layers of grease in a bomb filled with oxygen under pressure and heated. The degree of oxidation is determined from the fall in pressure over a given time. The test has some correlation with the storage life of greases but is not reliable for predicting service life. (Tests IP 142, ASTM. D.942)

Drop Point

The drop point is the temperature at which a drop falls from the orifice of a standard cup and this gives an indication of the transition from solid to fluid state. In practice, drop points are usually considerably higher than normal maximum service temperatures. (Tests IP 31, 132, ASTM. D.2265, BS 894)

Oil Separation

There is a natural tendency for oil to separate out from greases and this is often referred to as 'bleeding'. When grease is stored in a container it is quite common to see some free oil on the surface or at the bottom, but the basic structure of the grease is not changed if the amount of free oil is small. Oil separation tests are carried out by measuring the loss of oil from a sample of grease, supported on a gauze or filter paper, when acted on by gravitational or centrifugal forces. Such tests give an indication of the stability of a grease but seem to have little relevance to the service of grease in bearings. (Tests IP 121, ASTM. D.1742)

Mechanical Stability

When greases are subjected to severe mechanical working, such as in rolling bearings, their consistency may change considerably. To evaluate this characteristic prolonged worked penetration tests may be carried out.

Extreme-Pressure Properties

For heavy duty service a grease must have good EP properties to protect against frictional damage to the bearing surfaces. The EP properties of greases are normally evaluated using the Timken Wear and Lubricants Tester or a four-ball machine (see figure 9.20). (Tests ASTM. D.2266)

Bearing Performance Tests

The foregoing simple laboratory tests do not provide reliable information on the performance of a grease in service, and for this purpose it is necessary to build test rigs in which test bearings are run under suitable conditions. Leading rolling-bearing manufacturers have developed their own performance tests for evaluating greases and these involve a wide variety of test bearings and conditions. A standard form of bearing rig test which has more general application is also used. (Test IP. 168)

Grease Additives

Additives are often included in greases to impart special qualities. Examples are corrosion inhibitors, oxidation inhibitors and EP additives. Solid lubricants such as graphite or molybdenum disulphide are also used to reduce wear and prevent seizure.

REFERENCES

1. Introductory memorandum on the viscosity of liquids. *Engineering Sciences Data* Item No. 65001, Instn mech. Engrs.
2. Approximate data on the viscosity of some common liquids. *Engineering Sciences Data* Item No. 66024, Instn mech. Engrs.
3. A guide to the viscosity of liquid petroleum products. *Engineering Sciences Data* Item No. 67015, Instn mech. Engrs.
4. E. W. Dean and G. H. B. Davis. Viscosity variations of oils with temperature. *Chem. metall. Engng*, **36**, (1929), 618–19.
5. Charts Published by American Society for Testing and Materials.
6. Refutas Chart available from Baird and Tatlock, London.
7. Determination of the viscosity of liquids in CGS units. *BS 188*, (1957), British Standards Institution.

8. E. C. Bingham. *Fluidity and Plasticity*. McGraw-Hill, New York, (1922).
9. *I.P. Standards for Petroleum and its Products*. Institute of Petroleum, London. (Annually).
10. K. Weissenberg. *Principles of Rheological Measurement*. Nelson, London, (1949).
11. *ASTM. Standards on Petroleum Products and Lubricants*. American Society for Testing and Materials. Philadelphia. (Annually).

10

Hydrodynamic Lubrication

10.1 INTRODUCTION

The introduction of a film of fluid between components with relative motion forms the solution of a vast number of tribological problems in engineering. In chapter 9 we have seen how lubricants may be supplied to the contact at high pressure. However, in many cases the viscosity of the fluid and the geometry and relative motion of the surfaces, may be used to generate sufficient pressure to prevent solid contact without any external pumping agency. If the bearing is of a convergent shape in the direction of motion, the fluid adhering to the moving surface will be dragged into the narrowing clearance space, thus building-up a pressure sufficient to carry the load. This is the principle of hydrodynamic lubrication, a mechanism which is essential to the efficient functioning of the whole of modern industry. Motor vehicles, locomotives, machine tools, engines of all types, domestic appliances, aircraft, surface and underwater vessels, gearboxes, pumps and spacecraft are only a small part of an almost endless list of equipment and machines which rely heavily on hydrodynamic films for their operation. Although usually so beneficial, hydrodynamic films sometimes occur in situations where they are undesirable or even dangerous. For example, care has to be taken to prevent the formation of such a film of water between the pantograph of an electric locomotive and the conductor in wet weather, and the tread pattern of a motor tyre is an attempt to prevent 'aquaplaning', which is the build-up of a hydrodynamic film between the tyre and the road, resulting in a loss of grip.

10.1.1 Notation

B	Width of bearing
C	Radial clearance of journal bearing
e	Eccentricity
F	Friction force per unit width
f	Coefficient of friction
H_1	h_i/h_o
h	Film thickness
h_i	Inlet film thickness
h_m	Reference film thickness ($= h_o$ for inclinded slider, $= h_i - h_o$ for Rayleigh step, $= C$ for journal bearing)
h_o	Outlet film thickness for sliders
	Minimum film thickness for rollers
$\bar{h}$	h/h_m
h^*	Film thickness at point of maximum pressure
K	Load capacity factor for inclined slider

$$= \frac{\log_e(1 + n)}{n^2} - \frac{2}{n(2 + n)}$$

L	Length of bearing
$L_1 \, L_2$	Length of inlet and outlet sections of Rayleigh step bearing
n	$\dfrac{h_1}{h_o} - 1$
N	Rotational speed
p	Pressure (gauge for liquid bearings; absolute for gas bearings)
p_a	Atmospheric pressure
p_s	Pressure at step of Rayleigh bearing
$\bar{p}$	p/p_a
p'	$\dfrac{ph^2}{6\eta UL}$
q	Reduced pressure $\left(= \dfrac{1}{\alpha}(1 - e^{-\alpha p}) \right)$
R	Radius
S	Sommerfeld number $\left(= \dfrac{\eta N}{W} LD\left(\dfrac{R}{C}\right)^2 \right)$
t_i	Inlet temperature
t_o	Outlet temperature
U	Surface velocity in tangential direction
u	Fluid velocity in tangential direction
V	Surface velocity in normal direction
W	Load
x	Coordinate along bearing
$\bar{x}$	x/L
y	Coordinate in axial direction
z	Coordinate across film

α Viscosity–pressure coefficient

α_t Viscosity–temperature coefficient

ε Eccentricity ratio $\left(= \dfrac{e}{C} \right)$

η Absolute viscosity

θ Angle

Λ Bearing number $\left(= 12\eta \dfrac{(U_1 + U_2)}{2} \dfrac{L}{h_m^2 p_a} \right)$

ρ Density

ρ_i Density at inlet

ρ_o Density at outlet

ρ' $= \dfrac{\rho_o}{\rho_i}$

τ Shear stress

ϕ Attitude angle

10.2 THEORY

The variation of lubricant pressure in the bearing is described by the Reynolds equation, which is derived and fully presented in the appendix. The solution of the full equation involves great difficulties, but we will use a much simplified version in the first instance, by making the following assumptions:

(1) The fluid is incompressible.
(2) The fluid is newtonian (the shear stress is directly proportional to the shear strain rate).
(3) The fluid properties remain constant; effects due to variation in temperature and pressure being neglected.
(4) Inertia and turbulence effects are negligible.
(5) The solid bodies remain rigid.
(6) The film is of sufficiently small thickness that the fluid pressure can be considered constant through the thickness of the film, (but not of course along its length).
(7) The bearing is infinitely wide.

It will be shown later how some of these assumptions may be relaxed to adequately describe certain situations, although there are many cases where the errors induced by these assumptions are negligible.

10.2.1 Longitudinal Motion (Sliding)

In the situation where one surface slides over another (in the x direction) with no normal motion and a fluid between them, if the above assumptions are made the Reynolds equation reduces to

$$\frac{dp}{dx} = 12\eta \left(\frac{U_1 + U_2}{2} \right) \left(\frac{h - h^*}{h^3} \right) \tag{10.1}$$

235

where h^* is the film thickness at the position of maximum pressure where $dp/dx = 0$. U_1 and U_2 are the velocities in the x direction of the lower and upper surfaces, respectively, and must be arranged such that the film thickness h is invariant with time.

Consider two flat plates having velocities U_a and U_b as shown in figure 10.1a. If the system is resolved into its component parts a and b, we can see that a is simply a rigid body motion of both surfaces and the fluid at a uniform

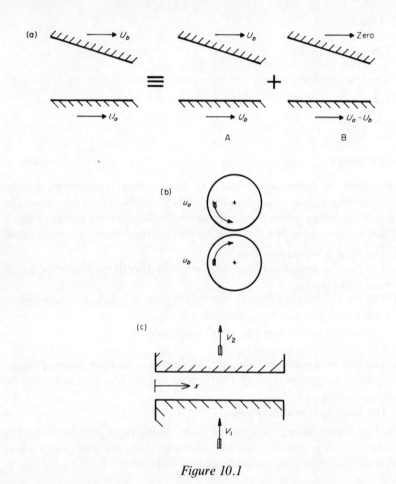

Figure 10.1

velocity U_b, while B represents a system *where h is invariant with time*, to which equation 10.1 may be applied putting $U_1 = U_a - U_b$ and $U_2 = 0$. Thus the Reynolds equation becomes for two sliding surfaces

$$\frac{dp}{dx} = 12\eta\left(\frac{U_a - U_b}{2}\right)\left(\frac{h - h^*}{h^3}\right)$$

(10.2)

236

Any situation with longitudinal velocities may be dealt with in the same way. For example, two discs rolling on fixed axes with circumferential velocities U_a and U_b, as in figure 10.1b, produce a Reynolds equation

$$\frac{dp}{dx} = 12\eta\left(\frac{U_a + U_b}{2}\right)\left(\frac{h - h^*}{h^3}\right) \tag{10.3}$$

since in this case *h is clearly invariant with time* and equation 10.1 applies.

10.2.2 Normal Motion (Squeeze)

In this case there 'is no relative sliding of the surfaces ($U_1 = U_2 = 0$), but there is movement normal to the surfaces. Consider the two parallel flat plates in figure 10.1c with respective normal velocities V_1 and V_2. Common-sense tells us that a pressure will be developed in the fluid if $V_1 - V_2$ is positive, and that the fluid will flow outwards from the point of maximum pressure. The Reynolds equation confirms this, since making the same assumptions as before, it becomes

$$\frac{dp}{dx} = 12\eta(V_2 - V_1)\left(\frac{x - x^*}{h^3}\right) \tag{10.4}$$

where x^* is the coordinate of the position of maximum pressure ($dp/dx = 0$). This situation is often called 'squeeze film lubrication'.

10.2.3 Combined Longitudinal and Normal Motion

The build-up of pressure in a bearing where both types of relative motion are present can be found by a simple summation of the two effects, thus

$$\frac{dp}{dx} = 12\eta\left(\frac{U_1 + U_2}{2}\right)\left(\frac{h - h^*}{h^3}\right) + 12\eta(V_2 - V_1)\left(\frac{x - x^*}{h^3}\right) \tag{10.5}$$

To find the actual pressure distribution it is necessary to integrate the equation. Two unknown quantities will then be present, the integration constant and the value of x^*. These are determined by the incorporation of two relevant boundary conditions. The above equation may be applied to any pair of surfaces, such as those shown in figure 10.2, provided that the appropriate velocity components are resolved to obtain the appropriate values of U_1, U_2, V_1 and V_2.

237

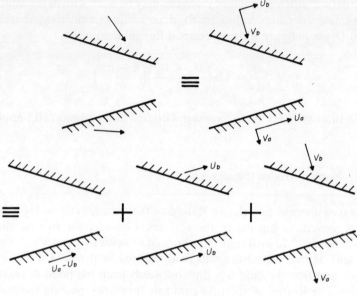

Figure 10.2

10.3 APPLICATION OF REYNOLDS EQUATION TO SLIDING BEARINGS

We have discussed hydrodynamic action on the basis of a convergent clearance space through the length of the bearing, figure 10.3 illustrating some of the shapes which satisfy this condition for successful operation. All of these forms and many others occur in practice, either because they are manufactured in that way, or due to subsequent wear or deformation. However, fortunately for the bearing designer, in his analysis of many different slider profiles published in 1918, Lord Rayleigh concluded that there is little to

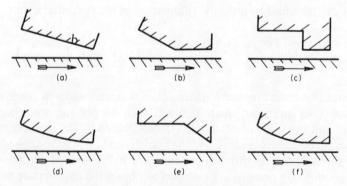

Figure 10.3 Slider bearing shapes

238

choose between the various forms of bearing, given the same inlet and outlet film thicknesses. The slider of figure 10.3c, often called the Rayleigh step, actually produces the highest peak pressure[1].

10.3.1 Plane-Inclined Slider

By far the most common form of lubricated slider bearing is the plane-inclined pad illustrated in figure 10.3a. As an example of the application of the Reynolds equation to slider bearings, we will determine the pressure distribution and load capacity for such a configuration. The bearing and notation are illustrated in figure 10.4.

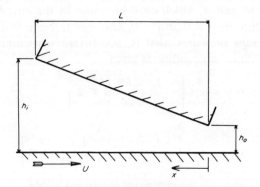

Figure 10.4   Slider bearing notation

We can use equation 10.1 since there is no normal motion present, putting $U_1 = -U$, $U_2 = 0$.

$$\frac{\mathrm{d}p}{\mathrm{d}x} = -6\eta U\left(\frac{h - h^*}{h^3}\right)$$

This equation must be integrated with respect to x to yield the pressure distribution

$$p = -6\eta U \int \left(\frac{h - h^*}{h^3}\right) \mathrm{d}x$$

The film thickness can be expressed at any point as

$$h = h_o\left(1 + \frac{nx}{L}\right)$$

239

where

$$n = \frac{h_i}{h_o} - 1$$

$$p = -6\eta U \left[\int \frac{dx}{h_o^2 \left(1 + \dfrac{nx}{L}\right)^2}\, dx - \int \frac{h^*}{h_o^3 \left(1 + \dfrac{nx}{L}\right)^3}\, dx \right]$$

$$= -\frac{6\eta U}{h_o^2} \left[-\frac{L/n}{1 + \dfrac{nx}{L}} + \frac{h^*}{2h_o} \frac{L/n}{\left(1 + \dfrac{nx}{L}\right)^2} + A \right]$$

where A is a constant of integration. It will be noted that h^* is the value of h where $dp/dx = 0$, that is, where the pressure has a maximum value. We have two unknowns h^* and A, which must be found by the introduction of two boundary conditions: $p = 0$ at $x = 0$, and $x = L$. Note that pressures are expressed as gauge pressures, that is, $p = 0$ represents ambient pressure. Substitution of these two conditions gives

$$0 = -6\eta U \left[-\frac{L}{n} + \frac{h^*}{2h_o} \frac{L}{n} + A \right]$$

and

$$0 = -6\eta U \left[\frac{-L/n}{1 + n} + \frac{h^*}{2h_o} \frac{L/n}{(1 + n)^2} + A \right]$$

The solution of these two simultaneous equations yields

$$h^* = 2h_o \frac{(1 + n)}{2 + n}$$

and

$$A = \frac{L}{n(2 + n)}$$

These can now be substituted in the pressure distribution equation to give

$$p = \frac{6\eta U L}{h_o^2} \left[\frac{n\dfrac{x}{L}\left(1 - \dfrac{x}{L}\right)}{(2 + n)\left(1 + \dfrac{nx}{L}\right)^2} \right] \tag{10.6}$$

A further integration of the pressure gives the normal load capacity of the bearing per unit width

$$W = \int_{x=0}^{x=L} p\, dx \tag{10.7}$$

240

For the inclined slider the load capacity is given by

$$W = \frac{6\eta U L^2}{h_o^2}\left[\frac{\log_e(1 + n)}{n^2} - \frac{2}{n(2 + n)}\right]$$ (10.8)

or

$$W = \frac{6\eta U L^2}{h_o^2} K$$

where

$$K = \frac{\log_e(1 + n)}{n^2} - \frac{2}{n(2 + n)}$$

The maximum load capacity depends on the value of K, which in turn depends on the inlet–outlet film thickness ratio. By putting $dK/dn = 0$ we can find the value of n for optimum load capacity. This occurs when n is approximately 1.2, or the inlet film thickness is 2.2 times that at the outlet. In this condition $K = 0.0267$. The load capacity is, however, very insensitive to n—if n is reduced to 0.6 the value of K only falls to 0.0235, while if n is increased to 2.0, K is only reduced to 0.0246.

Similar values of this optimum ratio apply to the other slider shapes. When we remember that in the great majority of engineering applications, the minimum film thickness will be in the order of 0.02 mm to 1 mm, it will be appreciated that the angles, curvatures, steps, etc., which transform a rigid surface into an efficient bearing are extremely small and often present manufacturing problems. This accounts for the popularity of the pivoted pad bearing, which is free to take up its own angle to the other surface. This bearing type and some of its derivatives are illustrated in figure 10.5.

If the pressure distribution for a plane-inclined slider is plotted, a curve

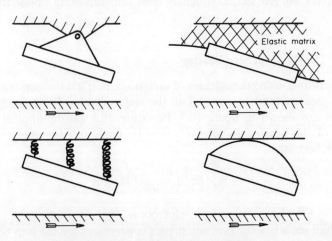

Figure 10.5   Self-stabilising sliders

241

such as that of figure 10.6 is obtained. Clearly the centre of pressure lies to the rear of the centre of the pad. Indeed, we know that

$$h^* = \frac{2h_o(1 + n)}{2 + n}$$

If we substitute $h^* = h_o[1 + (nx^*/L)]$ we obtain an expression for the position of maximum pressure

$$\frac{x^*}{L} = \frac{1}{2 + n}$$

x^*/L varies from $\frac{1}{2}$ when $n = 0$ to 0 as $n \to \infty$. In other words, x^* always lies in the rear half of the pad. For equilibrium of the pivoted pad, the centre of pressure must coincide with the pivot position. It is possible, for a given set of conditions, to calculate exactly the correct position of the pivot, but this is usually a futile exercise, since we do not have complete control over the

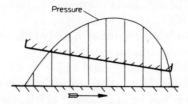

Figure 10.6

conditions in which the bearing operates. Moreover, a pad pivoted centrally to allow motion in either direction functions quite satisfactorily, although theory indicates that it should never generate a film. The explanation for this is that the increases in pressure and temperature in the film distort the pad to a convex curved shape, which is then self-stabilising about the centre pivot.

10.3.2 The Rayleigh Step Bearing

It can be shown, using the calculus of variations, that it is the step bearing that has the greatest load capacity of all the slider shapes. The bearing and its notation are shown in figure 10.7. Equation 10.1 can be applied to each section of the bearing in turn, $U_1 = -U$, $U_2 = 0$. After integration the equation becomes

$$p = -6\eta U\left(\frac{h - h^*}{h^3}\right)x + A$$

since $(h - h^*)/h^3$ is constant for each section.

We shall see when the pressure profile is produced for the step bearing in figure 10.7 that, since the curve is discontinuous at the location of the maxi-

242

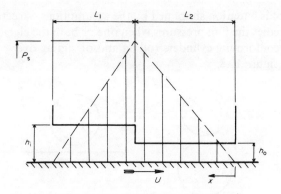

Figure 10.7 Step bearing notation

mum pressure, h^* must be simply regarded as a constant of integration, having no physical significance, that is, $h^* \neq h_i$, $h^* \neq h_o$.

The boundary conditions in this case are $p = 0$ at $x = 0$ and $p = p_s$ at $x = L_2$ for the region where $h = h_o$; $p = p_s$ at $x = L_2$ and $p = 0$ at $x = L$ for the region where $h = h_i$. The resulting pressure distribution is shown in figure 10.7, the maximum pressure occurring at the step.

The load capacity is given by

$$W = p_s \times \frac{L}{2}$$

which results in the following expression when the value of p_s, obtained by substituting the boundary conditions into the expression for p, is inserted

$$W = \frac{3\eta U L L_2 (L - L_2)(a - 1)}{(L_2 a^3 + L - L_2)h_o^2} \tag{10.9}$$

where $a = h_i/h_o$. For optimum performance $a = 1.866$ and $L_1/L_2 = 2.549$. With these optimum parameters

$$W = \frac{6\eta U L^2}{h_o^2} = 0.0342$$

which is an improvement on the load capacity of the inclined slider (0.0342 as compared with 0.0267).

10.4 CONTACTS IN THE FORM OF NON-CONFORMING DISCS

Many engineering situations may be represented with some accuracy by two cylinders or a cylinder and a plane. We immediately think of rolling element bearings, the deformed pivoted pad and many others, including gear teeth contacts which operate by a combined rolling and sliding motion.

243

The Reynolds equation simplified by the assumptions of section 10.2 may be used to predict the film pressures when one or both the elements have the form of non-conforming cylinders rolling and/or sliding over each other, as illustrated in figure 10.8.

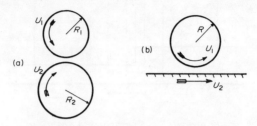

Figure 10.8   Equivalent cylinder

10.4.1   The Equivalent Cylinder

The cylinder and plane shown in figure 10.8b are more easily analysed than the two-cylinder situation of figure 10.8a. However, the two cylinders of radius R_1 and R_2 may be represented by an equivalent cylinder of radius R and a plane, where

$$R = \frac{R_1 R_2}{R_1 + R_2}$$

and the surface velocities of the two components and the film thickness variation in case a are preserved in case b. The two situations may be considered identical for the purposes of calcuation, except for the calculation of horizontal components of pressure, as we shall see later in this chapter.

10.4.2   Cavitation

The Reynolds equation 10.1 may be used to predict the performance of the bearing by the insertion of values for U_1 and U_2 and the film thickness in terms of the distance through the bearing

$$h = h_o + R(1 - \cos \theta) \qquad (10.10)$$

(See figure 10.9 for illustration and notation.)

It will be immediately obvious from both the figure and equation 10.10 that in this case, unlike the sliders previously dealt with, the convergent portion of the bearing is followed by an equally long divergent portion before ambient pressure is once again achieved. If we apply our boundary conditions as for the sliders, namely $p = 0$ at $\theta = -\pi/2$ and $+\pi/2$, then a pressure distribution results which is antisymmetrical about the horizontal axis, indicating zero load capacity, (curve a of figure 10.9). This implies that

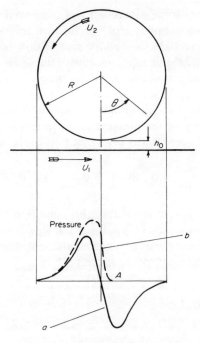

Figure 10.9

there exist in the liquid, negative pressures of the same magnitude as the positive pressures. In the vast majority of cases the liquid is unable to sustain these negative pressures and hence, in this region, the film will rupture and the air, which is almost inevitably present dissolved in the oil, will appear as bubbles. This phenomenon, known as cavitation, and well-known to anyone with experience of bearing operation, results in the majority of the clearance space being composed of alternate fingers of liquid and air across the width of the bearing. This phenomenon has received considerable attention, especially by Floberg[2] and Dowson[3], and for the student of hydrodynamic design has two consequences. If the pressure falls to the vapour pressure of the lubricant and low pressure air bubbles are allowed to form due to boiling of the liquid, their subsequent collapse against solid surfaces can create such enormous local pressures that damage in the form of pitting (cavitation erosion) can result. This form of cavitation, although all too familiar to the hydraulics engineer, does not often occur in bearing practice. The other consequence is that it is no longer adequate to use the simple boundary condition that $p = 0$ when $\theta = \pi/2$ at the outlet from the film. Experimental measurement indicates the pressure in the fluid has the form illustrated in figure 10.9 curve b, falling to the saturation pressure of the liquid at some point A in the divergent zone. As the liquid is usually drawn from an ambient pressure source, the saturation pressure may be regarded as ambient ($p = 0$).

The representation in analytical terms of what happens at this boundary has been the subject of some discussion and controversy. However, there is now almost universal acceptance of one form of this boundary condition. To the left of A (figure 10.9) the volume rate of flow of fluid has two components: $[(U_1 + U_2)/2]h$ due to the motion of the surfaces and an additional flow due to the pressure gradient dp/dx. To the right of A the flow is simply $[(U_1 + U_2)/2]h$. Therefore for continuity at A $dp/dx = 0$ at A. This leads us to the Reynolds cavitation boundary condition, which agrees well with experience and states that the pressure curve terminates with zero gradient at some unknown position in the divergent part of the film.

$$p = \frac{dp}{dx} = 0 \quad \text{at} \quad \theta = \theta_2 \qquad 0 < \theta_2 < \frac{\pi}{2} \qquad (10.11)$$

It has been shown by Dowson[3] that this condition can in some cases be complicated by the existence of small subambient negative pressures for a small distance before the pressure becomes ambient. The effect of such areas is, however, very small and may be neglected.

10.4.3 Load Capacity

Due mainly to the more complex boundary condition at the outlet when the liquid cavitates, the solution of even our simplified Reynolds equation to give the force components on the cylinders is no longer as simple as in the previous analyses. Martin[4] and Purday[5] performed the analysis for a cylinder of parabolic shape, having a film thickness given by

$$h = h_o + \frac{x^2}{2R}$$

The film shape for such a cylinder differs significantly from the circular shape only in regions of large film thickness which contribute little to the performance. Floberg in 1961 presented a solution for a circular cylinder, showing that the normal load capacity is given by

$$W = \alpha(U_1 + U_2)\eta \frac{R}{h_o} \qquad (10.12)$$

The difference between Floberg's result and that for a parabolic cylinder with the same h_o and R is negligible. The value of α depends upon the ratio R/h_o as shown in table 10.1. Also shown in the table are values of θ_2, the position of the cavitation boundary. The values of α given are from calculations of the normal force on the plane surface, the value of α for the disc being different only at very low values of R/h_o and then by only a small amount. From the table it can be seen that, for most practical cases, α may be given the approximate value of 2.45. We can also see that as R/h_o increases, the cavitation boundary approaches the position of minimum film thickness.

246

TABLE 10.1

R/h_o	α	θ_2^0
10	1.399	11.31
10^2	2.215	3.82
10^3	2.411	1.22
10^4	2.442	0.385
10^5	2.447	0.122
10^6	2.451	0.039

10.5 THE JOURNAL BEARING

We come now to the most familiar and widely used form of hydrodynamic bearing—the journal bearing. A rotating shaft is supported in a bush which completely or partially surrounds it with a small clearance as in figure 10.10a. If a load is applied to the journal it will be displaced from the centre, thus forming, as it rotates, a convergent clearance space which is conducive to the building-up of a lubricating film to support the load.

As the pressurised film is created, the journal moves round the bearing in the same sense as the rotation until it reaches an equilibrium position as shown in figure 10.10b. Note that the line of centres of the journal and the bush does not coincide with the line of action of the load, but is displaced by an angle ϕ. The distance between the centres, the eccentricity e, divided by the radial clearance of the bearing C, is called the eccentricity ratio ε. Obviously $\varepsilon = 0$ represents concentricity, and $\varepsilon = 1$ represents contact between the two surfaces. If the bearing is unwrapped, the form of the clearance can be clearly seen, as in figure 10.10c.

The film thickness at any point may be written as

$$h = C(1 + \varepsilon \cos \theta) \tag{10.13}$$

providing $C/R \ll 1$. This is almost invariably the case, typical values of C/R ranging from 0.0004 to 0.004. The film thickness may be substituted into our Reynolds equation 10.1, together with $U_1 = U$, $U_2 = 0$ and $x = R\theta$

$$\frac{\mathrm{d}p}{\mathrm{d}\theta} = \frac{6\eta U R}{C^2} \left[\frac{1}{(1 + \varepsilon \cos \theta)^2} + \frac{h^*}{C(1 + \varepsilon \cos \theta)^3} \right] \tag{10.14}$$

The solution of this equation was first achieved by Sommerfeld[6] by means of the Sommerfeld transformation

$$1 + \varepsilon \cos \theta = \frac{1 - \varepsilon^2}{1 - n \cos \gamma}$$

but the resulting equation for p does, of course, contain two unknown constants, which require evaluation by the substitution of two boundary conditions.

247

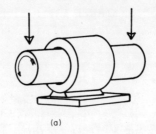

(a)

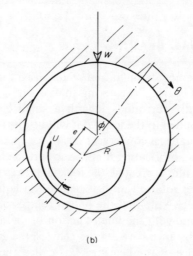

(b)

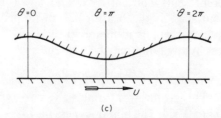

(c)

Figure 10.10 (a) Journal bearing, (b) journal bearing notation, (c) development of journal bearing clearance

10.5.1 Boundary Conditions

Unlike the slider bearings we considered previously, where we usually know that the pressures at the inlet and outlet of the film are ambient, the journal bearing, being a cyclic device, has a continuous film of fluid round its circumference. Of course in the bearing of finite length, the lubricant can flow out

248

and in from the bearing sides, but for the infinitely long journal bearing presently being considered this is precluded. Therefore the condition which must be satisfied is that

$$p_\theta = p_{\theta+2\pi}$$

This is not much help for analysis of the bearing and we need to know the pressure at some point. In practice this datum pressure is often the pressure of the lubricant supply where it is introduced into the clearance of the bearing. This supply may or may not be close to ambient pressure and, although always located in the lower pressure area of the bearing, it need not coincide with $\theta = 0$.

Sommerfeld assumed a pressure p_o at $\theta = 0$ and $\theta = 2\pi$. This produces the so-called 'full Sommerfeld solution', with a pressure distribution given by equation 10.15 and illustrated in figure 10.11.

$$p - p_o = \frac{6\eta U R \varepsilon \sin \theta(2 + \varepsilon \cos \theta)}{C^2(2 + \varepsilon^2)(1 + \varepsilon \cos \theta)^2} \tag{10.15}$$

As with the discs of section 10.4.2, because there is a divergent clearance present, impossibly large negative pressures can be predicted by these boundary conditions. In a few cases it is possible that these negative pressures are eliminated by a sufficiently high p_o, but in general the predicted distribution has the form shown in figure 10.11, with the negative shaded portion.

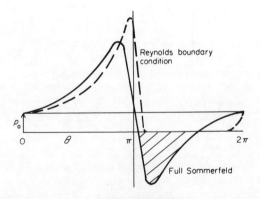

Figure 10.11

Unlike the case of the discs, the inclusion of the negative pressure, even with $p_o = 0$, does not imply zero load capacity, since the positive and negative pressure areas occur in opposite halves of the bearing, and therefore both tend to move the journal in the same direction. Since the negative pressures would assist the bearing to support the load, the full Sommerfeld solution will overestimate the load capacity.

We have seen in section 10.4.2 that it is closer to practical experience to

249

dismiss these large negative pressures, and to use the Reynolds cavitation boundary condition that

$$p = \frac{\mathrm{d}p}{\mathrm{d}\theta} = 0 \quad \text{at} \quad \theta = \theta_2 \qquad \pi < \theta < 2\pi$$

The inclusion of this condition increases the difficulty in solving the Reynolds equation 10.14, but the pressure distribution that is finally achieved is shown in figure 10.11.

10.5.2 Partial Journal Bearings

Since almost half the clearance of the full journal bearing is occupied by low pressure or cavitated fluid, there are many cases where, provided that the load is in an approximately constant direction, this part of the bearing may be dispensed with altogether. Indeed, this is often a distinct advantage because, although the low pressure region contributes little to the load capacity, it does add to the viscous drag, as we shall see later. The result is the partial journal bearing, as illustrated in figure 10.12.

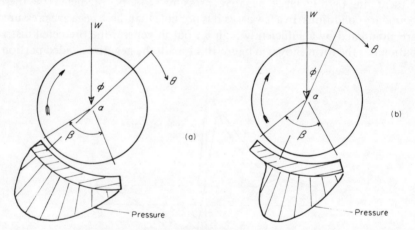

Figure 10.12   Partial journal bearings

The partial journal bearing has the same form of film thickness as the full journal bearing, and may be analysed using the Reynolds equation 10.14.

The boundary conditions, however, are different for partial bearings. The cyclic form is no longer present and usually the inlet boundary condition is known to be that the pressure is ambient at the beginning of the bearing, ($p = 0$ at $\theta = \alpha$). At the outlet it will either be sufficient to put the pressure equal to ambient at the end of the arc ($\theta = \alpha + \beta$), as in figure 10.12a, or the Reynolds boundary condition will have to be adopted, as in figure 10.12b. Which condition is applicable will depend on the condition of operation and the length of the divergent clearance space.

250

10.5.3 Load Capacity of Journal Bearings

As the load on the journal is increased from zero, the centre of the journal will move into a more and more eccentric position, with decreasing minimum film thickness. If the Reynolds equation is applied using the full Sommerfeld solution, assuming no cavitation, theory indicates that the journal moves at right angles to the load line; as shown in figure 10.13. Experience, however, tells us that this is not a very realistic solution. In practice the locus of the journal centre is very different from the full Sommerfeld prediction, and the Reynolds boundary condition must be used adequately to describe what happens.

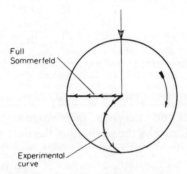

Full
Sommerfeld

Experimental
curve

Figure 10.13   Locus of shaft centre within clearance circle with increasing load

It is often useful to express the load capacity of a journal in terms of the Sommerfeld number, S, a dimensionless parameter which is given by

$$S = \frac{\eta N}{W} LD \left(\frac{R}{C} \right)^2 \qquad (10.16)$$

and which depends upon the eccentricity of the bearing. Rearranging equation 10.16

$$\frac{W}{L} = \left(\frac{1}{S} \right) \eta ND \left(\frac{R}{C} \right)^2 \qquad (10.17)$$

This gives the load per unit width of the journal in terms of the Sommerfeld number and the properties of the fluid and the bearing. Values of the Sommerfeld number for different eccentricity ratios, assuming the Reynolds boundary condition and ambient pressure at $\theta = 0$, are presented by Pinkus and Sternlicht[7]. These are shown in table 10.2.

It will be appreciated that, since the angle ϕ between the line of centres and load line varies with the operating conditions, even if the bearing is designed

251

TABLE 10.2

ε	S	ϕ	$\dfrac{F_j R}{WC}$
0	∞	71	—
0.1	0.247	69	—
0.2	0.123	67	2.57
0.3	0.0823	64	1.90
0.4	0.0628	62	1.53
0.5	0.0483	58	1.32
0.6	0.0389	54	1.20
0.7	0.0297	49	1.10
0.8	0.0211	42	0.962
0.9	0.0114	32	0.721
0.95	0.00605	23	0.568
1.0	0	—	—

so that the supply is located at $\theta = 0$ for one particular case, it will not be so for other loads, speeds and viscosities. In practice we find that the performance is virtually unaffected by the position of the inlet within $\pm 30°$ of $\theta = .0$. Indeed, there is remarkably little difference between the performance of the full journal bearing and the 180°, or even 150°, partial bearing. However, if the inlet should move a considerable distance into the convergent zone, the load capacity will be reduced.

From the table we can see that the Sommerfeld number falls, and hence the load capacity rises rapidly as the eccentricity ratio approaches unity. Indeed, the load capacity virtually doubles as ε increases from 0.8 to 0.9. However, in practice we cannot approach $\varepsilon = 1$ too closely, since in this region the films are becoming so small that surface roughness and other irregularities of manufacture begin to have a significant effect on the performance.

10.6 VARIABLE VISCOSITY—THE REDUCED PRESSURE CONCEPT

At this stage we may usefully relax one of the assumptions stated in section 10.2. It is well known that the viscosity of most fluids is markedly dependent on the pressure, and in many engineering situations extremely high pressures can occur, for example, in the restricted contacts between gear teeth and between rolling elements and their tracks. For such situations we must include this change of viscosity in our theory. A widely used relationship for the effect of pressure on viscosity, especially for lubricating oils, is

$$\eta = \eta_0 \, e^{\alpha p} \tag{10.18}$$

where η_0 is some reference viscosity and α is known as the pressure exponent of viscosity. Remembering that in our Reynolds equation the pressure gradient, and therefore the instantaneous pressure, is a function of viscosity and that viscosity itself increases with pressure, we can see that this effect rapidly escalates the pressure gradient, and therefore the pressure, to very large values. Because the viscosity is now a function of pressure the integration of Reynolds equation as it stands is no longer a simple matter. We therefore introduce a parameter q, known as the reduced pressure, which is given by

$$q = \frac{1}{\alpha}(1 - e^{-\alpha p}) \qquad (10.19)$$

It should be noted that q has a maximum value of $1/\alpha$, and if we substitute for p from this relationship into equation (10.5), we obtain

$$\frac{dq}{dx} = 12\eta_0 \frac{U_1 + U_2}{2}\left(\frac{h - h^*}{h^3}\right) + 12\eta_0(V_2 - V_1)\left(\frac{x - x^*}{h^3}\right) \qquad (10.20)$$

We can see that equation 10.20 is of exactly the same form as the original Reynolds equation 10.5, except that p is now replaced by q. Since η_0 is a constant, equation 10.20 can be readily integrated. Thus, any pressure distribution found for the constant viscosity case can be used for a fluid where the viscosity is defined by equation 10.18, by the substitution

$$p = \frac{1}{\alpha}\log_e(1 - \alpha q) \qquad (10.21)$$

As an example of this procedure, we may consider the case of two discs rolling with some degree of sliding. For a constant viscosity we obtain the pressure distribution shown in figure 10.14.

If the fluid has a viscosity which depends exponentially on pressure, the

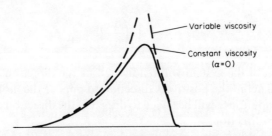

Figure 10.14 *Pressure distribution with variable viscosity*

253

solution is transformed using equation 10.21 to give the new pressure distribution. It will be noted that in the latter case the pressure is tending to an infinite value, although the load remains finite, but this does not, of course, occur. What happens in practice is that the very large pressures produce deformation of the bodies, which redistributes the pressure over a finite area. This situation is dealt with in the next chapter under the title 'elastohydrodynamic lubrication.'

10.7 SHEAR STRESSES AND TRACTION IN HYDRODYNAMIC FILMS

Although one of the functions of lubrication is the reduction of friction forces, however thick the lubricant film may be there will still remain tangential forces opposing the motion. Considerable reduction of the coefficient of friction can be effected by the provision of a fluid film, typically from 1 to 0.001, but even such relatively small frictional forces will result in the dissipation of energy and consequent loss of efficiency of the machine. It is usually, therefore, the objective of the designer to reduce these forces to as low·a value as possible.

To evaluate the drag on the solid boundaries of a lubricated contact such as that illustrated in figure 10.15, we use the definition of newtonian viscosity

$$\tau = \eta \frac{du}{dz} \tag{10.22}$$

where τ is the shear stress at the surface and du/dz is the rate of shear. The drag per unit width is then given by integrating the shear stress along the length of the bearing at the top and bottom surfaces

$$F_1 = \int_0^L \eta \left(\frac{du}{dz}\right)_{z=0} dx$$

$$F_2 = -\int_0^L \eta \left(\frac{du}{dz}\right)_{z=h} dx \tag{10.23}$$

Consideration of the system will readily tell us the direction of the friction forces. F_1 will act in the positive x direction to oppose the motion, while the force on the top surface will be trying to drag it in the direction of the velocity, the negative x direction.

For the typical velocity distribution shown in figure 10.15 clearly du/dz is not necessarily the same for each boundary and, therefore, the frictional drag on the two surfaces will be different. The value of du/dz will depend on the

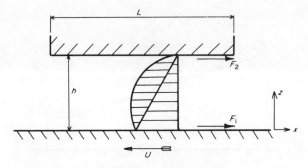

Figure 10.15

physical properties of the fluid and the type, geometry and velocity of the bearing surfaces. Some particular bearing types are analysed in the following sections.

10.7.1 Friction in Sliding Bearings

The velocity distribution across the film in a bearing consists of two parts, as shown in figure 10.15. Fluid theory tells us that there is a linear distribution due simply to the parallel velocity of the surfaces, and also a parabolic distribution due to the pressure gradient. The velocity of the flow at a transverse coordinate z is

$$u = -U\left(1 - \frac{z}{h}\right) - \frac{z}{2\eta}(h - z)\frac{\mathrm{d}p}{\mathrm{d}x}$$

To get the velocity gradient we differentiate with respect to z, remembering that we have assumed that the pressure does not vary across the thickness of the film

$$\frac{\mathrm{d}u}{\mathrm{d}z} = \frac{U}{h} - \frac{1}{2n}(h - 2z)\frac{\mathrm{d}p}{\mathrm{d}x}$$

On the lower surface $z = 0$

$$\left(\frac{\mathrm{d}u}{\mathrm{d}z}\right)_{z=0} = \frac{U}{h} - \frac{h}{2\eta}\frac{\mathrm{d}p}{\mathrm{d}x} \tag{10.24}$$

and on the upper surface $z = h$

$$\left(\frac{\mathrm{d}u}{\mathrm{d}z}\right)_{z=h} = \frac{U}{h} + \frac{h}{2\eta}\frac{\mathrm{d}p}{\mathrm{d}x} \tag{10.25}$$

These velocity gradients are evaluated by substitution of the film shape and the pressure distribution. For example, the inclined slider shown in figure 10.16 has a film thickness

$$h = h_\mathrm{o}\left(1 + \frac{nx}{L}\right)$$

255

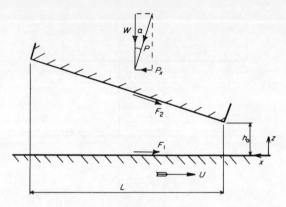

Figure 10.16

So that for the lower surface, substituting equation 10.24 in equation 10.23, we get an expression for the friction force per unit width of the bearing

$$F_1 = \int_0^L \left(\frac{\eta U}{h} - \frac{h}{2} \frac{dp}{dx} \right) dx$$

$$= \left[\frac{\eta U \frac{L}{n} \log_e \left(1 + \frac{nx}{L} \right)}{h_o} \right]_0^L - \frac{1}{2} \int_0^L h \frac{dp}{dx} dx \qquad (10.26)$$

$$= \frac{\eta U L}{n h_o} \log_e (1 + n) - \frac{1}{2} \left[(hp)_0^L - \int_0^L p \frac{dh}{dx} dx \right]$$

Note that since $p = 0$ at $x = 0$ and $x = L$

$$(hp)_0^L = 0$$

For small inclinations the term $p \, dh/dx$ is the component of the normal pressure on the inclined surface resolved parallel to the other surface. If the total resultant force on the inclined surface in this direction is denoted by P_x then

$$F_1 = \frac{P_x}{2} + \frac{nUL}{nh_o} \log_e (1 + n) \qquad (10.27)$$

Similarly

$$F_2 = \frac{P_x}{2} - \frac{\eta U L}{nh_o} \log_e (1 + n)$$

The value of P_x can be found by the triangle of forces P, W and Px. Since α is a very small angle $P = W$ and $\alpha = P_x/W = nh_0/L$. Therefore, the surface

256

drags per unit width, since we have already evaluated W (equation 10.8) are given by

$$F_1 = \frac{\eta UL}{h_o}\left[\frac{4\log_e(1+n)}{n} - \frac{6}{2+n}\right]$$

$$F_2 = -\frac{\eta UL}{h_o}\left[\frac{2\log_e(1+n)}{n} - \frac{6}{2+n}\right]$$

(10.28)

Similar expressions can be derived for other forms of sliding bearings. Notice that the drag on the moving lower surface is greater than that on the upper surface. Care must be taken, therefore, when experimental measurement of friction are made, that the value obtained is the one relevant to any subsequent operations.

10.7.2 Friction in Rolling/Sliding Discs

For rolling and sliding discs, such as those represented in figure 10.17 by an equivalent cylinder and a plane, the same equation (10.26) for the drag on the plane will apply, with slight modification of the coordinates

$$F_1 = \int_{\theta_1}^{\pi/2} \left(\frac{h}{2R}\frac{dp}{d\theta} - \frac{\eta}{h}(U_1 - U_2)\right)R\,d\theta$$

(10.29)

where θ_1 is the inlet and $\pi/2$ is the outlet. Note that for $\theta > \theta_2$ (the cavitation boundary)

$$p = 0 \quad \text{and} \quad \frac{dp}{d\theta} = 0$$

$$F_1 = [hp]_{\theta_1}^{\theta_2} - \int_{\theta_1}^{\theta_2} p\frac{dh}{d\theta}\,d\theta - \eta\int_{\theta}^{\pi/2}\frac{(U_1 - U_2)}{h}R\,d\theta$$

(10.30)

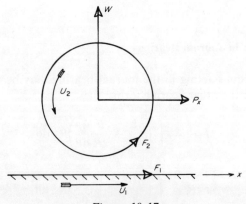

Figure 10.17

257

The first term in this equation is zero, since the pressure is zero at the two boundary positions. The second term represents the resultant pressure force on the cylinder in the x direction, P_x. The last term requires the substitution of the film shape and subsequent integration. If this is undertaken we find that

$$F_1 = -\frac{P_x}{2} - \eta(U_1 - U_2)A$$

$$F_2 = -\frac{P_x}{2} + \eta(U_1 - U_2)A$$

(10.31)

where A is a function of R/h_o.

Once again the forces on the two individual elements are different, except in the case of rolling without slip ($U_1 = U_2$), when the rolling friction is given by $P_x/2$ on each component. Purday's analysis[5] gives a value for A of 2.84, using $\theta_1 = -\pi/2$ as the inlet boundary, but this neglects the viscous drag from fluid in the cavitated region. If the maximum possible contribution from this region is included, then A has a value of 3.48 for cases where R/h_o is large.

The value of P_x is given by

$$P_x = 4.5\eta(U_1 + U_2)\left(\frac{R}{h_o}\right)^{1/2}$$

(10.32)

for the usual large values of R/h_o. For the plane of course $P_x = 0$. If the real situation consists of two discs, clearly there will be a horizontal force on each member. If the two discs are of radius R_a and R_b respectively and R is the equivalent cylinder radius, then it can be shown that

$$P_{x_a} = \frac{R}{R_a}P_x$$

$$P_{x_b} = \frac{R}{R_b}P_x$$

10.7.3   Friction in Journal Bearings

The friction torques acting in the journal bearing may be evaluated in the same way as for the other bearing forms.

$$F_j R = \int_0^{2\pi} \left(\frac{\eta U}{h} + \frac{h}{2R}\frac{dp}{d\theta}\right)R^2 \, d\theta$$

$$F_b R = \int_0^{2\pi} \left(\frac{\eta U}{h} - \frac{h}{2R}\frac{dp}{d\theta}\right)R^2 \, d\theta$$

258

where the subscripts j and b denote the journal and bush respectively. If the full Sommerfeld condition (see section 10.5.1), including negative pressures, is used then these equations, like those for load capacity dealt with earlier, are relatively simple to solve, giving the following expressions for the surface friction torques per unit axial length

$$F_j R = -\frac{4\eta U\pi R^2}{C} \frac{(1 + 2\varepsilon^2)}{(2 + \varepsilon^2)(1 - \varepsilon^2)^{1/2}}$$

$$F_b R = \frac{4\eta U\pi R^2}{C} \frac{(1 - \varepsilon^2)^{1/2}}{2 + \varepsilon^2} \tag{10.33}$$

A solution using the more relevant Reynolds boundary condition (see section 10.5.1) is more difficult to achieve. Values of the friction on the journal using this boundary condition are presented by Pinkus and Sternlicht[7] in the form

$$\frac{F_j}{W} \frac{R}{C} = -\frac{\varepsilon \sin \phi}{2} - \frac{2\pi^2 S}{(1 - \varepsilon^2)^{1/2}} \tag{10.34}$$

where S is the Sommerfeld number

$$S = \frac{\eta N}{W} LD\left(\frac{R}{C}\right)^2$$

as in equation 10.16. Values of $F_j R/WC$ taken from this reference are found in table 10.2. The value of this parameter falls with increasing eccentricity ratio and hence, since R/C is a constant. F_j/W, which may be regarded as a form of coefficient of friction, also decreases with increasing eccentricity. Figure 10.18a illustrates the forces on the bearing components from the lubricant film, while figure 10.18b shows these forces reversed, that is, the force on the film. If we consider the equilibrium of the oil film by considering the forces acting on it, the moments due to the friction are $F_j R$ and $F_b R$ in the directions shown, assuming $R = R_j \simeq R_b$. There is an additional

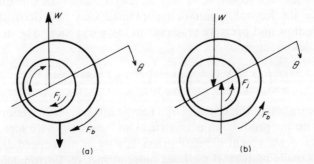

Figure 10.18 (a) Friction forces on bearing components, (b) forces on lubricant film

259

couple due to the normal loads W on the journal and bush being displaced from each other by a perpendicular distance $e \sin \phi$. For equilibrium

$$F_b R + F_j R + W \varepsilon C \sin \phi = 0 \tag{10.35}$$

This gives an equation for F_b which, when written with that for F_j, yields a pair of expressions having a pattern which by now is becoming familiar (compare with equations 10.27 and 10.31).

$$F_j R = -\frac{W C \varepsilon \sin \phi}{2} - \frac{2\pi^2 S W C}{(1 - \varepsilon^2)^{1/2}}$$

$$F_b R = -\frac{W C \varepsilon \sin \phi}{2} + \frac{2\pi^2 S W C}{(1 - \varepsilon^2)^{1/2}} \tag{10.36}$$

Note that the friction on the journal, which is of course the value used for calculation of power loss, is always greater than the friction on the bush, except for the concentric case. We must remember this if friction is being measured experimentally in the usual way, namely by the reaction torque on the stationary bush.

10.8 FINITE LENGTH BEARINGS

In the analysis so far we have considered the bearing to be infinitely wide and the forces calculated have been for unit width of such a bearing. In practice the lubricant will flow across the width of the bearing and out of the ends, thus modifying such parameters as the pressure distribution, load capacity and friction forces. This does not mean that the earlier results have no value, but that care must be taken to apply them only in situations where the side flow does not make them too inaccurate. Also the performance of a finite bearing is often expressed in terms of the infinite bearing and side-leakage factors. Thus, for example, the load capacity

$$W_{\text{finite}} = (\text{side-leakage factor}) \times W_{\text{infinite}}$$

The analysis developed so far will be referred to as the one-dimensional case, since the Reynolds equation contained only one derivative $\mathrm{d}p/\mathrm{d}x$, precluding flow and pressure gradients in the y direction. The new form of the Reynolds equation is

$$\frac{\partial}{\partial x}\left(h^3 \frac{\partial p}{\partial x}\right) + \frac{\partial}{\partial y}\left(h^3 \frac{\partial p}{\partial y}\right) = 12\eta\left(\frac{U_1 + U_2}{2}\right)\frac{\partial h}{\partial x} \tag{10.37}$$

The integration of this equation is a formidable task. Much effort has been expended on the production of analytical and analogue solutions, although in many cases the recent increase in availability of high-speed digital computers has made numerical methods more attractive. Fortunately we need not concern ourselves with the mechanics of solution, but will investigate the significance of the results for particular bearing forms.

260

10.8.1 The Finite Slider Bearing

If we consider the slider bearing illustrated in figure 10.19a, we can see that the moving surface will drag the lubricant into the bearing, but that unlike the infinitely wide case some of the lubricant will flow out of the bearing sides, as shown in figures 10.19b and c, and therefore the pressure generated in the narrowing clearance will be reduced. Obviously, the magnitude of this effect will depend upon the ratio of the width of the bearing to its length, the narrower the bearing the more its pressure will be reduced and vice versa.

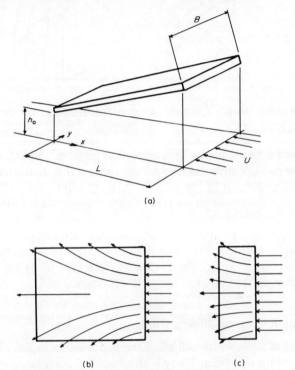

(a)

(b)   (c)

Figure 10.19   Side leakage in slider bearing

Fortunately, the conclusion drawn by Reynolds that the actual shape of the slider surface is relatively unimportant is still valid, and so if we investigate the case of the plane inclined slider, this will give a good indication of what happens for other shapes.

The notation and the bearing are shown in Figure 10.19a. As before

$$h = h_o\left(1 + \frac{nx}{L}\right)$$

and the Reynolds equation is

$$\frac{\partial}{\partial x}\left(h^3\frac{\partial p}{\partial x}\right) + \frac{\partial}{\partial y}\left(h^3\frac{\partial p}{\partial y}\right) = 6\eta U\frac{\mathrm{d}p}{\mathrm{d}x} \qquad (10.38)$$

261

The pressure distribution resulting from the solution of equation 10.38 is a function of both x and y and will have the form illustrated in figure 10.20a, as compared with 10.20b, which is the one-dimensional case. Along the edges of the bearing, $y = 0$ and $y = B$, the pressure is of course ambient.

The load capacity of the bearing is naturally reduced by side leakage. The extent to which this is so can be seen from figure 10.21a, which is based on

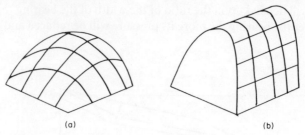

(a)   (b)

Figure 10.20   Pressure distribution in finite and in-
finite slider bearing

results derived numerically by Jakobsson and Floberg[8], Hays[9] and Muskat et al[10]. For a pad of conventional dimensions with $B \simeq L$, the load capacity is only about a half that of the infinite slider, and until we make the width about five times the length or more, the one-dimensional solution is plainly inadequate.

When we consider the friction forces we must remember that, since the load capacity is reduced, the surfaces must come closer together to carry the same load and therefore the friction coefficient is likely to rise. This is indeed the case, as the curves of figure 10.21b show.

10.8.2 Finite Rolling and Sliding Discs

The inclusion of finite width will affect the performance of discs in the same manner as that for the slider. There is an additional complication in that the cavitation boundary will no longer fall at a constant coordinate θ_2, but will be curved. This can be seen in figure 10.22a, which represents a typical pressure distribution between a finite cylinder and a plane as calculated by Dowson and Whomes[11]. The cavitation boundary occurs at a decreasing value of θ as we depart from the centre line, until at the edge it coincides with the position of minimum film thickness ($\theta = 0$).

Once again the load capacity will reduce as the bearing becomes narrower. However, because the pressure is very slow to build-up through the bearing, and the effective load carrying region is a narrow strip near the position of minimum film thickness as shown in figure 10.22b, the load capacity of the discs is relatively unaffected by side leakage for all but very narrow discs and very low values of R/h_o. In fact, it is as though the performance of the bearing were defined by the dimensions of the load carrying region. Side leakage

262

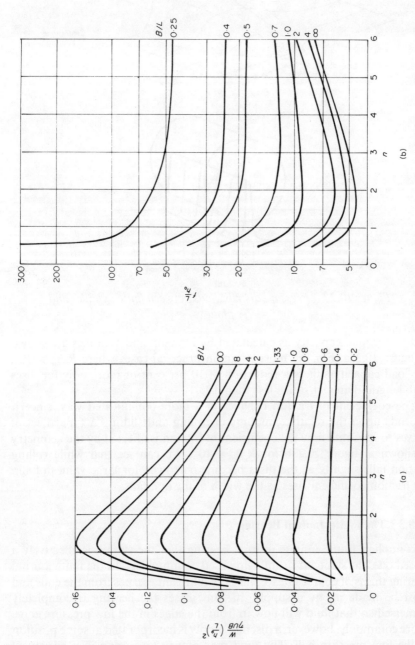

Figure 10.21 (a) Load capacity of finite slider bearing[7], (b) friction in finite slider bearing

263

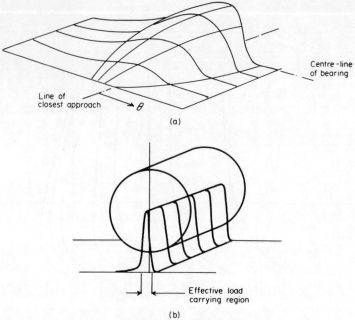

Line of
closest approach

θ

Centre-line
of bearing

(a)

Effective load
carrying region

(b)

*Figure 10.22   Pressure distributions between finite cylinder and
plane*

factors for load capacity as calculated by Dowson and Whomes are shown
in figure 10.23a. For the most common practical cases where $R/h_o > 10^4$,
the load capacity suffers a reduction of 10 per cent or more only for discs
with length–diameter ratios less than 0.2.

The coefficient of friction behaves in a more complicated way, since it
depends upon the relative amounts of rolling and sliding. A typical set of
curves to illustrate how the coefficient of friction is affected by the geometry
is shown in figure 10.23b for $R/h_o = 10^3$. We can see that while rolling
friction falls slightly as the discs become narrower, for large values of slip
the friction coefficient rises as the width of the disc falls.

10.8.3   The Finite Journal Bearing

It is implicit in the infinite journal bearing that the system is effectively a
closed one, the same lubricant being used continuously. In the finite journal
bearing this is of course not so, since the lubricant escapes from the ends and
must be made up by a supply. In some cases the bearing is completely
immersed so that fluid will flow in from the edges in the low pressure zone.
More commonly, however, a discrete supply is incorporated at some position
in the low pressure half. The fluid may enter the bearing in a variety of
ways—a single hole, multiple holes, a slot—each of which will give a slightly
different performance.

264

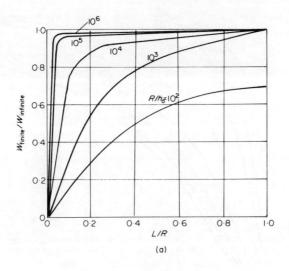

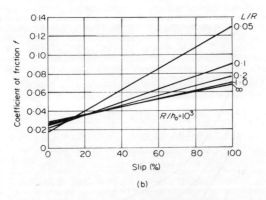

Figure 10.23   Load capacity and friction for a
finite cylinder and plane[11]

Because the fluid is permitted to flow sideways, we have the familiar reduction in load capacity compared with that of the infinite journal bearing. Of course, the greater the length–diameter ratio, the greater the load capacity. Also as the eccentricity becomes larger, the load capacity approaches that of the infinite bearing. Typical curves based on data from Fuller[12] are shown in figure 10.24a. Because of difficulties in manufacture and alignment, journal bearings seldom have a length–diameter ratio of more than one. From the curves we can see that for practical bearings a considerable reduction in load capacity will be experienced, except when the eccentricity ratio is very large. Because of the reduced load capacity and hence the smaller film thickness, the coefficient of friction is much higher for the finite bearings, especially at low eccentricity ratios. The variation of coefficient of friction with length–diameter ratio and eccentricity ratio is shown in figure 10.24b.

265

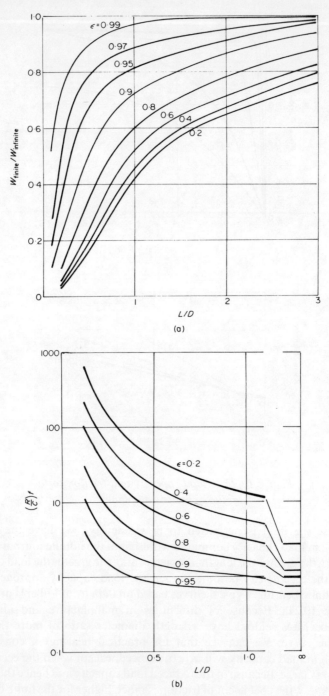

Figure 10.24 (a) Load capacity of finite journal
bearing[12], (b) friction in finite journal bearing

266

The Short Bearing Theory

We have seen that the Reynolds equation 10.39 for the finite journal is difficult to solve analytically

$$\frac{\partial}{R^2 \partial\theta}\left(h^3 \frac{\partial p}{\partial\theta}\right) + \frac{\partial}{\partial y}\left(h^3 \frac{\partial p}{\partial y}\right) = \frac{6\eta U}{R}\frac{\partial h}{\partial\theta} \qquad (10.39)$$

In an attempt to circumvent these difficulties, Ocvirk and Dubois[13] considered a bearing where pressure gradients around the circumference were very small compared with those along the length. This has the effect of reducing equation 10.39 to

$$\frac{\partial}{\partial y}\left(h^3 \frac{\partial p}{\partial y}\right) = \frac{6\eta U}{R}\frac{\partial h}{\partial\theta} \qquad (10.40)$$

Just as we initially considered an infinitely long bearing, this represents an infinitely short bearing. The results are only applicable to situations where the pressure gradients in the direction of motion are not large. For instance the short bearing theory is most unsuitable to describe the performance of discs where very high pressure gradients are experienced, and is not recommended for the majority of slider bearings. For short journal bearings running at low eccentricities, however, very useful results can be obtained very simply.

Integrating equation 10.40 twice with respect to y gives

$$p = \frac{6\eta U}{R}\frac{dh}{d\theta}\frac{y^2}{2h^3} + \frac{Ay}{h^3} + B$$

The boundary conditions are that $p = 0$ at $y = \pm L/2$ where y is measured from the centre line of the bearing. This yields

$$p = \frac{3\eta U}{Rh^3}\frac{dh}{d\theta}\left(\frac{4y^2 - L^2}{4}\right)$$

Of course this equation gives a circumferential pressure distribution at any value of y which is proportional to $dh/d\theta$ and $dh/d\theta$ is negative for half of the circumference, as shown in figure 10.25. For practical purposes we only

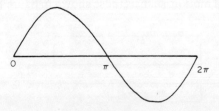

Figure 10.25

consider positive pressures by simply deleting the contribution from the range $\theta = \pi$ to 2π. The load capacity then becomes

$$W = \frac{\eta U L^3}{4C^2} \frac{\varepsilon}{(1 - \varepsilon^2)^2} [\pi^2(1 - \varepsilon^2) + 16\varepsilon^2]^{1/2} \tag{10.41}$$

and the friction coefficient

$$f = \frac{2\pi^2 C S}{R(1 - \varepsilon^2)^{1/2}} \tag{10.42}$$

where S is the Sommerfeld number as before. To get some appreciation of the range of applicability of the short bearing theory, reference should be made to figure 10.26. We can see that for bearings of the more common proportions, $L/D \leq 1$, the short bearing theory gives good results for moderate eccentricity ratios ($\varepsilon \leq 0.6$).

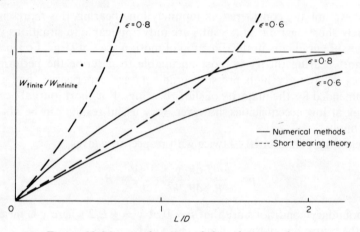

Figure 10.26   Application of short bearing theory

10.9   THERMAL EFFECTS

Throughout our discussion of hydrodynamic lubrication there has been one principle underlying all our arguments: namely, that pressure is generated by the lubricant being drawn into a narrowing clearance space. It is at first somewhat surprising, therefore, to find that two parallel flat surfaces with relative longitudinal motion, such as those shown in figure 10.27, do generate

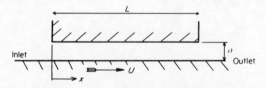

Figure 10.27   Parallel surface bearing

268

a pressure within the film, in spite of the absence of any convergence. This means that such a device, which is called a parallel surface bearing, will carry a load. Why should this be so? Clearly if we look at our usual Reynolds equation for such a slider

$$\frac{dp}{dx} = 6\eta U\left(\frac{h - h^*}{h^3}\right)$$

since h has the same value everywhere

$$\frac{dp}{dx} = 0$$

resulting in no pressure build-up. The discrepancy between theory and practice comes from two of our initial assumptions which are no longer applicable: that the fluid is incompressible and that the fluid properties remain constant.

As the lubricant is drawn into the bearing by the moving surface, it will be subjected to shear due to the velocity gradient across the film. The energy involved in this shearing will appear as heat, raising the temperature more and more as it passes through the gap. This has two effects. The first is to reduce the viscosity of the fluid, a typical viscosity–temperature curve having been shown in chapter 9. The second is to try to increase the volume of the fluid. It is this latter effect which generates the pressure. The fluid is unable to expand, due to the restricted clearance and therefore the pressure rises. This effect is sometimes called the thermal wedge.

In order to analyse this system we must return to the Reynolds equation in a fuller form than we have used so far. Neglecting side flow

$$\frac{d}{dx}\left(\frac{\rho h^3}{12\eta}\frac{dp}{dx}\right) = \frac{d}{dx}\left(\frac{\rho U h}{2}\right) \tag{10.43}$$

Integrating this with respect to x gives

$$\frac{\rho h^3}{12\eta}\frac{dp}{dx} = \frac{\rho U h}{2} + A$$

or

$$\frac{dp}{dx} = \frac{6\eta U}{h^2} + \frac{12 A\eta}{\rho h^3}$$

and integrating again, remembering that h is a constant

$$p = \frac{6U}{h^2}\int \eta\, dx + \frac{12A}{h^3}\int \frac{\eta}{\rho}\, dx + B \tag{10.44}$$

where A and B are constants.

In order to solve this equation it is necessary to know the viscosity and density at any point along the bearing, which in turn requires a knowledge of the temperature variation through the contact. To assess this with complete accuracy is obviously a very difficult task, but a good approximation is to assume that the energy dissipation is constant along the bearing and that all this energy goes to raising the temperature of the lubricant, giving a linear temperature distribution.

The density–temperature relationship for most fluids may be written

$$\rho = \rho_i + \alpha_t(t - t_i) + \beta_t(t - t_i)^2 + \cdots$$

where ρ_i is the inlet density. In fact the third and subsequent terms account for a very small percentage of the density change and so

$$\rho = \rho_i + \alpha_t(t - t_i)$$

Since the relationship between density and temperature is now linear and the temperature variation along the bearing is also linear, the density variation can be expressed as

$$\rho = \rho_i + \frac{x}{L}(\rho_o - \rho_i)$$

where ρ_o is the density at the outlet.

We find that the contribution of viscosity variation is much smaller than that of density change. This is partly due to the fact that the decrease in viscosity with a temperature rise is to some extent offset by the increase in viscosity with pressure rise. Therefore in our approximate analysis we shall consider the viscosity to be constant.

If $\rho' = \rho_o/\rho_i$ equation 10.44 becomes

$$p = \frac{6\eta U x}{h^2} + \frac{12 A \eta}{\rho_i h^3} \int \frac{dx}{\left[1 + \dfrac{x}{L}(\rho' - 1)\right]} + B$$

$$= \frac{6\eta U x}{h^2} + \frac{12 A \eta L}{\rho_i h^3 (\rho' - 1)} \log_e\left[1 + \frac{x}{L}(\rho' - 1)\right] + B$$

The boundary conditions are that $p = 0$ at inlet and outlet and so

$$p = \frac{6\eta U L}{h^2}\left[\frac{x}{L} - \frac{\log_e\left[1 + \dfrac{x}{L}(\rho' - 1)\right]}{\log_e \rho'}\right] \qquad (10.45)$$

where

$$\rho' = 1 + \frac{\alpha_t}{\rho_i}(t_o - t_i)$$

We will choose some typical values so that we can appreciate the magnitude of this effect

$$\alpha_t \text{ for mineral oils} \simeq -0.00065/^\circ C$$
$$\rho_i \text{ for mineral oils} \simeq 0.9 \text{ g/cm}^3$$

The value of ρ' will depend upon the temperature difference between the inlet and the outlet as shown in figure 10.28a. Since many lubricants boil at about 100 °C, the maximum temperature difference which can occur is about the same value. If we use $t_o - t_i = 100$, the pressure distribution has the form of figure 10.28b, where p' is a dimensionless pressure. The maximum value of p' is about 0.011. For a plane-inclined slider $p'_{max} \simeq 0.042$.

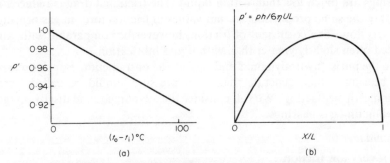

Figure 10.28   Thermal effects in parallel surface bearing

Since the shapes of the pressure distributions in the two cases are very similar, the relative load capacities can be assessed from the values of maximum pressure. We can see therefore that the parallel surface bearing has a load capacity approximately 1/3.5 that of the corresponding inclined slider. Of course it is very seldom that the temperature rise through a bearing is 100 °C. For most practical bearings it is more of the order of 2–20 °C with a correspondingly lower load capacity. Therefore for bearings where the power dissipated in viscous friction is only sufficient to raise the temperature of the lubricant by a few degrees, we are justified in neglecting thermal effects.

10.10   GAS-LUBRICATED BEARINGS

Throughout this chapter on hydrodynamic lubrication we have used the words 'lubricant' and 'fluid' without indicating their nature, and although we probably think automatically of water, oil, grease, and similar substances, there is no reason why the fluid should not be a gas, such as air. Our physical feel for the viscosity of an oil between our fingers, but not for the same effect with air, is due to our fingers lack of sensitivity at viscosities below about 5 cP, rather than to the absence of this property.

As long ago as 1854 Hirn[14] proposed air as having great advantages as a lubricant for bearings, although it was not until 1897 that Kingsbury[15] demonstrated the viability of such a proposal. Today gas bearings are very much a part of our industrial life, doing useful service in machine-tool spindles, inspection and measuring instruments, turbomachinery, gyroscopes, dental drills and many other more specialised applications.

The hydrodynamic gas bearing operates on exactly the same principle

271

as the liquid counterpart, namely the drawing of the fluid into a convergent clearance with the consequent rise in pressure. However, if we look at any of the equations we have derived for pressure distribution in hydrodynamic bearings, we find that, at least in the first approximation, the pressure is proportional to the lubricant viscosity. Since the viscosity of a gas is much lower than that of a liquid, the pressures and hence the load capacities of gas bearings are much less than with a liquid. The frictional drag is reduced in roughly the same proportion, so that values of friction force in gas bearings are very low. The coefficient of friction, however, is comparable with, and indeed often slightly greater than with liquid lubrication.

To the purist 'hydrodynamic' and 'gas' are a contradiction. For this reason gas bearings which generate their own pressure should be called 'aerodynamic' or 'self-acting'. We will consider the advantages and disadvantages of self-acting gas bearings.

Advantages

(1) Very low friction.
(2) If air is used the lubricant is readily available and free.
(3) The lubricant will not contaminate the bearing surfaces.
(4) Unlike a liquid which boils with quite modest temperature rises and freezes solid at low temperatures, a gas can be used at both very high and very low temperatures.

Disadvantages

(1) Low load capacity. Pressures in self-acting gas bearing are typically 10^5 N/m^2 as compared with 10^7 N/m^2 in a liquid bearing.
(2) Susceptibility to instability. All fluid bearings exhibit a tendency to instability in the form of self-indiced vibrations under certain conditions, but gas bearings are rather more prone to this. There are, however, design methods to overcome this difficulty. Hydrodynamic instability is dealt with in more detail in section 10.11.
(3) Surface finish and machining accuracy of the bearing components must be rather better than for liquid bearings. This is because lubricant films and clearances are typically, but not always, thinner than those experienced in liquid lubricated bearings.

In addition to the foregoing specific advantages and disadvantages we must bear in mind other differences which are introduced by the change of lubricant from liquid to gas.

Since the energy dissipated by the friction forces will be very low, there will be little temperature rise through the bearing. We can therefore in most cases consider the process to be isothermal without much loss in accuracy. The effects of compressibility will become important if the pressures generated are at all high.

There is no cavitation in gas bearings. All films are continuous even though the pressure falls below ambient.

10.10.1 Gas Bearing Theory

We employ a form of the Reynolds equation to describe the situation

$$\frac{\partial}{\partial x}\left(h^3\rho\frac{\partial p}{\partial x}\right) + \frac{\partial}{\partial y}\left(h^3\rho\frac{\partial p}{\partial y}\right) = 12\eta\frac{(U_1 + U_2)}{2}\frac{\partial}{\partial x}(\rho h) \qquad (10.46)$$

Notice that we have to retain the density ρ as a variable. The relationship between pressure and density will be in general polytropic, namely

$$\frac{p}{\rho^n} = \text{constant} \qquad (10.47)$$

In our consideration of gas bearings we must use

$$p \equiv absolute \text{ pressure}$$

since clearly if we regarded $p = 0$ as ambient pressure, equation 10.47 would imply zero density in this condition, which is not true. Substituting for ρ from 10.47 into equation 10.46 and cancelling the constant gives

$$\frac{\partial}{\partial x}\left(h^3 p^{1/n}\frac{\partial p}{\partial x}\right) + \frac{\partial}{\partial y}\left(h^3 p^{1/n}\frac{\partial p}{\partial y}\right) = 12\eta\frac{(U_1 + U_2)}{2}\frac{\partial}{\partial x}(p^{1/n}h) \qquad (10.48)$$

If we consider the case of the infinitely wide bearing, implying no variation in the y direction, and we assume that isothermal conditions prevail that is, $n = 1$, then equation 10.48 reduces to

$$\frac{\mathrm{d}}{\mathrm{d}x}\left(ph^3\frac{\mathrm{d}p}{\mathrm{d}x}\right) = 12\eta\frac{(U_1 + U_2)}{2}\frac{\mathrm{d}}{\mathrm{d}x}(ph)$$

which on integration becomes

$$ph^3\frac{\mathrm{d}p}{\mathrm{d}x} = 12\eta\frac{(U_1 + U_2)}{2}ph + C$$

or

$$\frac{\mathrm{d}p}{\mathrm{d}x} = 12\eta\frac{(U_1 + U_2)}{2}\left[\frac{ph + C}{ph^3}\right]$$

If as usual we define the coordinate of the position where the pressure is a maximum ($\mathrm{d}p/\mathrm{d}x = 0$) by x^*, then

$$\frac{\mathrm{d}p}{\mathrm{d}x} = 12\eta\frac{(U_1 + U_2)}{2}\left[\frac{ph - (ph)^*}{ph^3}\right]$$

and for isothermal conditions when $p/\rho = $ constant, this may be written

$$\frac{\mathrm{d}p}{\mathrm{d}x} = 12\eta\frac{(U_1 + U_2)}{2}\left[\frac{h - (\rho^*/\rho)h^*}{h^3}\right] \qquad (10.49)$$

273

This equation is identical to equation 10.3 for incompressible fluids, except for the inclusion of the density ratio. We can conclude therefore that for bearings where the pressures generated are not high and there is little consequent compression of the gas ($\rho \simeq \rho^*$), the bearing will behave exactly as the liquid bearing, the pressure and load capacity being in the ratio of the viscosities, that is

$$\frac{W_{\text{gas}}}{W_{\text{liquid}}} = \frac{\eta_{\text{gas}}}{\eta_{\text{liquid}}} \simeq \frac{1}{1000}$$

Therefore for gas bearings which do not produce large pressures, we may use the solution developed for incompressible lubrication.

It is usual in gas bearing analysis to consider the equation in terms of dimensionless variables, putting

$$\bar{p} = \frac{p}{p_a} \qquad \bar{x} = \frac{x}{L} \qquad \bar{h} = \frac{h}{h_m}$$

where p_a is ambient pressure and h_m is some reference film thickness. This transforms equation 10.49 into

$$\frac{d\bar{p}}{d\bar{x}} = 12\eta \frac{(U_1 + U_2)}{2} \frac{L}{h_m^2 p_a} \left[\frac{\bar{h} - (\rho^*/\rho)\bar{h}^*}{\bar{h}^3} \right] \qquad (10.50)$$

The term

$$12\eta \frac{(U_1 + U_2)}{2} \frac{L}{h_m^2 p_a}$$

is often referred to as the bearing number or compressibility number and is denoted by Λ. This is a very significant parameter to which we will refer repeatedly in our consideration of the various gas bearing forms which follows.

10.10.2 Gas-Lubricated Slider Bearings

Inclined Slider

If we consider the plane-inclined slider illustrated in figure 10.29a, we may solve equation 10.50 by integrating and inserting the boundary condition that $\bar{p} = 1$ ($p = p_a$) at inlet and outlet. The pressure profile produced is shown in figure 10.29b. We can see that the maximum pressure for a gas bearing occurs at a smaller film thickness than for the incompressible lubricant, so that the centre of pressure is always further towards the trailing edge of the bearing.

If we investigate the effect of bearing number on the pressure profile, we get a family of curves such as those derived by Gross[16] and presented in

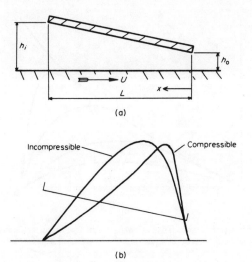

Figure 10.29

figure 10.30a. Notice that for $\Lambda \to 0$ the pressure distribution approximates to that for an incompressible fluid. At $\Lambda \equiv \infty$ the peak pressure occurs at the end of the bearing and in this case

$$\frac{p_{\max}}{p_a} = \frac{h_i}{h_o}$$

The variation of load capacity of the bearing is shown in figure 10.30b. For low bearing numbers we can see that the maximum load capacity occurs when h_i/h_o is between 2 and 3, but for higher bearing numbers the optimum ratio increases.

Algebraic expressions can be produced for the load capacity of an inclined slider for the extreme cases $\Lambda \to 0$ and $\Lambda \to \infty$, but no general formula is available to cover the whole range of bearing numbers. The magnitude of the load capacity may be seen from figure 10.30b, or by reference to table 10.3.

The friction force on each component is given by

$$2F \frac{h_m}{\eta BL(U_1 + U_2)} = \frac{\log_e \bar{h}_i}{\bar{h}_i - 1} \pm \frac{3(\bar{h}_i - 1)}{\Lambda} \frac{W}{LBp_a}$$

where the addition sign is to be used for the faster moving surface.

Step Slider Bearing

We have seen that for incompressible fluids the step bearing is theoretically the most efficient geometrical shape for a slider. It is reasonable therefore to investigate the performance of such a bearing when lubricated with a gas. The analytical procedure is exactly the same as that for the liquid lubricated bearing, namely the application of Reynolds equation to each section of the

275

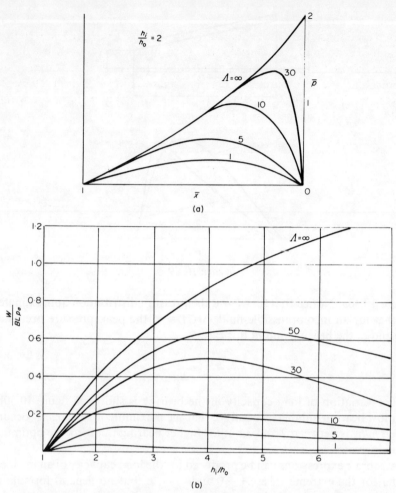

Figure 10.30 (a) Pressure distribution in a slider lubricated with a compressible fluid[16]; (b) load capacity of gas lubricated slider[16]

TABLE 10.3 LOAD CAPACITY OF PLANE-INCLINED GAS
LUBRICATED SLIDER (neglecting side leakage)
(After Gross[16])

Λ \ H_1	1.5	2.0	3.0	4.0	6.0
0.5	0.01091	0.01323	0.01232	0.01034	0.007241
1.0	0.020172	0.02640	0.02063	0.02063	0.01448
5.0	0.0957	0.1234	0.1201	0.1023	0.07243
10.0	0.1486	0.2124	0.2252	0.1980	0.1435
50.0	0.1942	0.3367	0.5618	0.6424	0.5943

bearing, having a pressure p_{step} at the change of clearance. The mass flow in each section may then be equated to give a value for p_{step}. The resulting pressure distribution has typically the form of the curves in figure 10.31. We can see that for low bearing numbers the profile is almost triangular, approaching the incompressible case (see figure 10.7). As Λ increases the pressure in the inlet region is slow to build up, compared with the triangular profile while that in the outlet is even slower to decay. The net effect is to

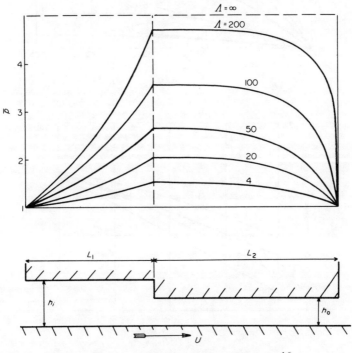

Figure 10.31   Gas lubricated step bearing[16]

increase the area under the curve and therefore more load is supported. Thus for the same step pressure the gas film will usually carry more load than the incompressible film. The load capacity of the bearing can be calculated and typical results are illustrated in figure 10.32a.

The optimum values of step height and ratio L_1/L_2 can be deduced from the curves presented in figure 10.32b.

10.10.3   Gas-Lubricated Journal Bearings

During the past decade the gas journal bearing has become an indispensable part of our industrial life. The low friction energy dissipation, together with its other advantages, has led to its incorporation in many different forms in widely differing applications.

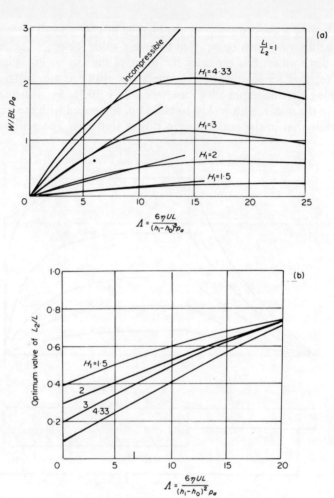

$$\Lambda = \frac{6\eta UL}{(h_1 - h_0)^2 p_a}$$

$$\Lambda = \frac{6\eta UL}{(h_1 - h_0)^2 p_a}$$

Figure 10.32   Gas lubricated step bearing[16]

The gas journal bearing operates in the same way as its liquid counterpart except for the important difference that there is no cavitation zone, the film being continuous. A typical pressure distribution is shown in figure 10.33 (cf. figure 10.11) for a 360° bearing. The subambient portion of the pressure curve will of course add to the load capacity since it applies a suction to the top of the shaft which opposes the applied load.

As with all self-acting gas bearings the load capacity of the bearing is well below that of the comparable liquid bearing, but in the case of the full gas journal bearing, the theoretical load capacity is seldom the practical limit of operation. This is more often dictated by the onset of instability, the most prevalent form of which is known as 'half-speed whirl' and which will be discussed in more detail in section 10.11.

278

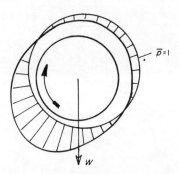

Figure 10.33   Pressure distribution in gas lubricated journal bearing

An investigation into the load capacity of the gas journal bearing will reveal that once again we have the problem of a Reynolds equation to which there is no direct analytical solution. Many methods have been adopted in an attempt to solve this problem, analytical and semi-analytical solutions being produced for very high and very low bearing numbers, but for the general case numerical methods are essential. The most significant contribution in this field is the work of Raimondi[17] who used relaxation methods to produce sets of load–bearing number and attitude angle–bearing number curves for various length–diameter ratios. Typical of these are figure 10.34a and 10.34b, which show load and attitude angle, the angle between the load load and the line joining the shaft and bush centres, for a full journal bearing having a length–diameter ratio of unity.

Partial Journal Bearings

In liquid lubrication we have seen that the use of a partial journal bearing can reduce the frictional torque on the shaft without much affecting the load capacity. In gas lubrication, however, the load capacity is reduced significantly if the low pressure section of the circumference is removed because the suction of the subambient pressure does contribute to the load capacity. The friction will also be reduced. The most common reason for the use of the partial gas journal bearing is the suppression of half-speed whirl. This will be discussed in section 10.11.

10.10.4   Other Gas Bearing Types

We have dealt with the gas bearing slider and journal bearing, but many other types of gas bearing are frequently used in industry. A few of the more important ones are described below.

279

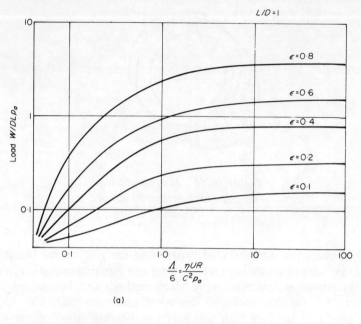

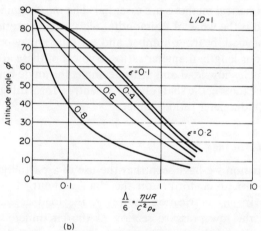

Figure 10.34 Gas lubricated journal bearing: (a) load capacity; (b) attitude angle

The Tilting-pad Journal Bearing

The tilting-pad journal bearing is illustrated in figure 10.35a. The pads, of which there may be any number, are mounted on pivots so that the pads may adopt their own angle to the shaft surface. Although it is obviously more complicated to manufacture and assemble, the load capacity is somewhat reduced and design data is not so readily available, it does have some

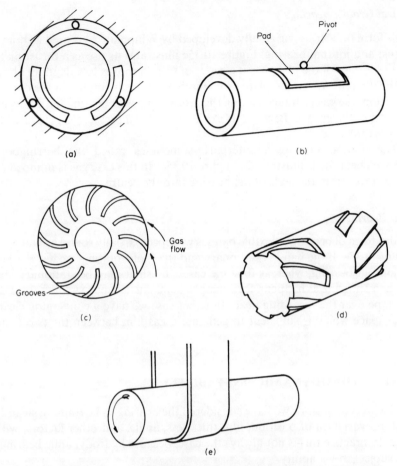

Figure 10.35 (a) Tilting-pad journal bearing; (b) pivoted sector bearing; (c) spiral groove thrust bearings; (d) spiral groove journal; (e) foil bearing

distinct advantages over the full 360° journal bearing. It is almost entirely free from half-speed whirl and therefore may be used at high speeds, where this instability precludes the use of full journal bearings. If, as is usual, the pads are permitted to pivot axially as well as circumferentially, they are capable of absorbing shaft misalignment. Any foreign matter or debris may escape from the bearing clearance by way of the gaps between the pads.

This type of bearing has been applied with great success to rotating turbomachinery.

A derivative of the tilting-pad bearing is the 'window pad' bearing or 'nutcracker' bearing, illustrated in figure 10.35b, which combines the load capacity of the full journal bearing with the whirl resistance introduced by the tilting pad.

281

Spiral Groove Bearings

This form of bearing, originally developed by Whipple[18,19] is used as both a thrust and journal bearing. Figure 10.35c illustrates its use as a thrust plate. In the surface of the plate is cut a series of spiral grooves to a depth of about 0.01–0.05 mm. When the plate or its mating surface is rotated in the correct direction, the gas is dragged along the grooves into the bearing until it meets the end of the groove. Because its exit is restricted a pressure is generated to support the load.

The journal bearing counterpart, sometimes called the herringbone grooved bearing, is illustrated in figure 10.35d. In this case gas is pumped by the grooves from the ends of the bearing into the centre.

Foil Bearings

In this form of bearing a flexible band is wrapped partially round a shaft as in figure 10.35e. If the band is stationary and the shaft rotating, we have a type of journal bearing, whereas in some cases the shaft may be stationary and the band moving as happens, for example, in magnetic tape machines to guide the tape over the recording head. In either case we have a converging clearance space which is sufficient to generate a gas film between the two components.

10.11 HYDRODYNAMIC INSTABILITY

In the preceding work we have considered the bearing to be running steadily with no variation of position, film thickness, angle, and other factors, with time. In practice this is not always the case, because hydrodynamic bearings are subject to instability.

In general we can differentiate between two forms of instability. The first type, which is often called synchronous whirl when found in journal bearings, is caused by a periodic disturbance outside the bearing such that the bearing system is excited into resonance. Since the lubricant film can be considered an elastic system, even though the stiffness is not constant, it will be capable of such a resonance. For example, if the speed of a rotating shaft in a journal bearing is progressively increased, at some stage an out-of-balance or similar periodic force will produce resonance of the bearing. On further increasing the speed we may pass through the resonant frequency. Both thrust and journal bearings are subject to instability, but we will restrict our discussion to journal bearings, since it is here that the problem is most frequent and troublesome.

The other form of instability is induced in the lubricant film itself and is called 'half-speed whirl'. The reason for this is that the displacement of the shaft centre under a load is not in the direction of the load (see figure 10.10). If, in figure 10.36, the shaft centre is displaced from O to P, the restoring

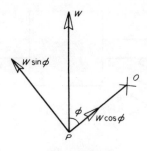

Figure 10.36

force due to the lubricant film pressure does not act along OP, but has a component $W \sin \phi$ which will cause the shaft to move in a circumferential direction and to whirl; that is, at the same time as the shaft rotates about its own centre P, the shaft centre tends to rotate about the bearing centre O. If the whirl takes place at half the rotational speed of the shaft, this will coincide with the mean rotational speed of the lubricant, since the circumferential velocity of the lubricant varies from zero at the bush to the shaft speed at the shaft. Because the film shape and fluid then have no relative circumferential motion, no hydrodynamic film is produced and the lubrication fails, often with disastrous consequences. This phenomenon is encountered in all full journal bearings, but is especially prevalent in gas bearings, because of their low damping capabilities, because the attitude angle ϕ is in general higher than for the comparable liquid bearing and because of their frequent use in high speed applications.

10.11.1 The Prediction and Suppression of Instability

Synchronous Whirl

In order to investigate the synchronous instability of any rotor/bearing system, it is necessary to evaluate the characteristics of that system: shaft inertia and flexibility, stiffness and damping characteristics of the bearing films, etc. A model for such a system is shown in figure 10.37a. The analysis presents problems since the bearing properties themselves change with the rotational speed and whirling speed.

If the rotating part is not perfectly balanced, the locus of the shaft centre will be a circle or some other closed orbit about the steady equilibrium position. The size of the orbit will depend on the bearing stiffness relative to the rotor stiffness in the way indicated in figure 10.37b. If the whirl orbit remains the same in successive rotations, the whirl is stable. This will be the case until a 'critical' speed of the system is reached. At this point the rotational frequency of the rotor unbalance and the natural frequency of the system will be in resonance and the whirl orbit will increase. If the rotational speed is

283

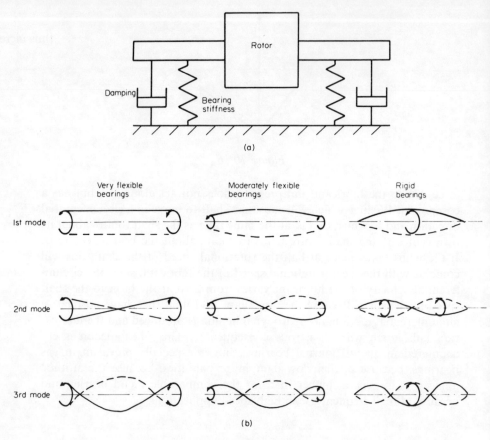

| | Very flexible bearings | Moderately flexible bearings | Rigid bearings |

1st mode

2nd mode

3rd mode

(b)

Figure 10.37   (b) (After MTI/Rensselaer Polytechnic Institute Gas Bearing Design Manual)

kept at approximately this value, the magnitude of the whirl will increase until either it reaches a stable position with a large amplitude limited by the damping capacity, or the orbit increases until failure takes place. However, if these critical speeds are passed through rapidly, the whirl orbit does not have time to grow sufficiently for failure to occur. A typical case is shown in figure 10.38.

It is usual procedure therefore to design the bearing system so that the critical speeds do not coincide with the most commonly used running speeds. This may be done either by increasing the bearing stiffness so that the critical speeds are very high, or reducing the stiffness so that the critical speeds are quickly passed through and normal operation takes place where the attenuation is large. The stiffness may be increased by reducing the bearing clearance, but this often imposes too stringent manufacturing and assembly conditions.

Another means of suppressing or allowing for whirl is the introduction of extra damping into the system. This is often done by flexibly mounting

284

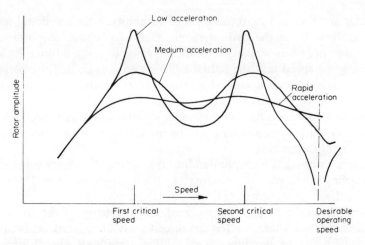

*Figure 10.38 (After MTI/Rensselaer Polytechnic Institute
Gas Bearing Design Manual)*

the bearing housings in rubber 'O' rings or metal diaphragms. This introduces another element into the analysis, which must be allowed for as illustrated in Figure 10.39.

It is clear from the above considerations that rotating parts to be supported in hydrodynamic bearings should be balanced as carefully as possible in an attempt to eliminate or at least minimise synchronous instability.

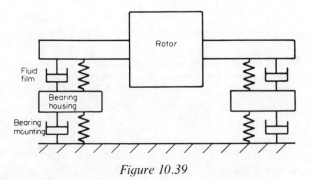

Figure 10.39

Half-Speed Whirl

This form of instability is induced in the hydrodynamic film itself and is independent of rotor balancing. It is always present in hydrodynamic bearings, but its amplitude only becomes large enough to be apparent at a certain threshold rotational frequency. The system will not pass through this danger region with a further increase in rotational speed; on the contrary this will only accelerate failure.

The value of the threshold speed is dealt with by Marsh[20]. Hydrostatic

285

bearings are also subject to this form of instability, if the rotational speed is sufficiently high for the hydrodynamic effects to become important. The rotational speed will once again be in the region of 1.5 to 2 times the lowest critical whirl speed and the actual whirl speed will be half the rotational speed for a liquid bearing and in the range 0.25 to 0.45 the rotational speed for a gas bearing. In this case it is termed fractional-frequency whirl.

The most important ways of preventing or minimising half-speed whirl are by the introduction of damping and by interfering with the circumferential symmetry of the bearing.

If extra damping is incorporated into the system the energy generated by the whirling can be partially destroyed by the damping medium. This is often done by flexibly mounting the bearings as described earlier.

Many devices are employed to break the symmetry of the full journal bearing, although there is often the price to pay of reduced load capacity, especially with gas bearings. Some of these methods are illustrated in figure 10.40. The methods incorporated range from the simple axial groove, figure 10.40a, cut in the low pressure half of the bearing to the tilting-pad arrangement of figure 10.40c. The 'lemon' bearing has deliberate ovality

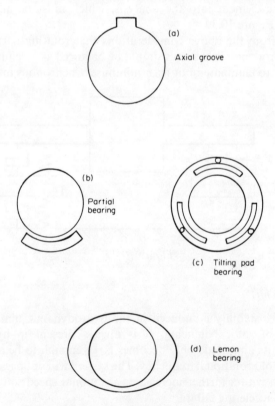

(a) Axial groove

(b) Partial bearing

(c) Tilting pad bearing

(d) Lemon bearing

Figure 10.40   Instability-inhibiting devices

286

of the bore. Bearings can have two, three, four, or any number of lobes and indeed often accidental out-of-roundness is sufficient to delay the onset of whirl. While the partial bearing of figure 10.40b, and tilting-pad bearing of figure 10.40c are practically proof against half-speed whirl, the others will raise the threshold speed, hopefully beyond the operating range.

The more heavily loaded the bearing, the higher the threshold speed, since the higher the eccentricity the more closely the attitude angle approaches zero (see figure 10.13). The problem is therefore acute in vertically mounted spindles, where often there is negligible loading.

REFERENCES

1. Lord Rayleigh. Notes on the theory of lubrication. *Phil. Mag.*, **35**, No. 205, (Jan. 1918), 1–12.
2. L. Floberg. Lubrication of two cylindrical surfaces considering cavitation. Trans. Chalmers University of Technology, No. 234, Institute of Machine Elements, (1961).
3. D. Dowson. Investigation of cavitation in lubricating films supporting small loads. Proc. Conf. on Lubrication and Wear, Paper 49, *Inst. mech. Engrs*, (1957).
4. H. M. Martin. Lubrication of gear teeth. *Engineering*, **102**, (1916), 199.
5. H. F. P. Purday. *Streamline Flow*. Constable, London, (1949).
6. A. Sommerfeld. Zur hydrodynamischen theorie der schmiermittelreibung. *Z. angew. Math. Phys.*, **50**, (1904), 97–155.
7. O. Pinkus and B. Sternlicht. *Theory of Hydrodynamic Lubrication*. McGraw-Hill, New York, (1961).
8. B. Jakobsson and L. Floberg. The rectangular plane pad bearing. Trans. Chalmers University of Technology, No. 203, Institute of Machine Elements, (1958).
9. D. I. Hays. Plane sliders of finite width. *Trans. Am. Soc. Lubric. Engrs*, **1**, No. 2, (1958).
10. M. Muskat, F. Morgan, and M. W. Meres. The Lubrication of plane sliders. *J. appl. Phys.*, **11**, (March, 1940).
11. D. Dowson and T. L. Whomes. Side leakage factors for a rigid cylinder lubricated by an isoviscous fluid. *Proc. Instn mech. Engrs*, **181**, 30, (1967), 165–70.
12. D. D. Fuller. *Theory and Practice of Lubrication for Engineers*. Wiley, New York, (1966).
13. G. B. Dubois and F. W. Ocvirk. Analytical derivation and experimental evaluation of short-bearing approximation for full journal bearings. NACA Rep. 1157, (1953).
14. G. Hirn. 'Sur les principaux phénomènes qui présentent les frottements médiats'. *Bull. Soc. ind. Mulhouse*, **26**, (1854), 188–277.
15. A. Kingsbury. Experiments with an air-lubricated journal. *J. Am. Soc. nav. Engrs*, **9**, (1897), 267–92.
16. W. A. Gross. *Gas Film Lubrication*. Wiley, New York, (1962).
17. A. A. Raimondi. A numerical solution for the gas-lubricated journal bearing of finite length. *Trans. Am. Soc. Lubric. Engrs*, **4**, (1961), 131–55.
18. R. T. P. Whipple. Theory of the spiral grooved thrust bearing with liquid or gas lubricant. Atomic Energy Research Est., Harwell, Berks., T/R 622, (1951).
19. R. T. P. Whipple. Herringbone pattern thrust bearing. Atomic Energy Research Est., Harwell, Berks., T/M 29, (1951).
20. H. Marsh. The stability of aerodynamic gas bearings. Ministry of Aviation Report, (1964).

11

Elastohydrodynamic Lubrication

11.1 HIGHLY LOADED CONTACTS

In the previous chapter we discussed the formation of a hydrodynamic film of lubricant to support a normal load without examining the effects of the size of this load or, more usefully, the value of the load per unit area. We now look more closely at 'highly loaded' contacts, where loads act over relatively small contact areas. Such contacts are to be found in the so-called 'line contacts' of gear teeth and roller bearings and the 'point contact' of ball-bearings. As the contact areas in the latter cases are typically only about one-thousandth of those occurring in such situations as journal bearings, the mean pressures will be about one thousand time greater. We may appreciate that such high pressures will affect the behaviour so that the hydrodynamic solutions which were used to study journal and pad bearings will have to be modified. Indeed we shall find that these high pressures can lead both to changes in the viscosity of the lubricant and elastic deformation of the bodies in contact, with consequent changes in the geometry of the bodies bounding the lubricant film.

If, for example, we apply the hydrodynamic equations of the earlier chapter to the contact between a pair of gear teeth with

$$W = 10^5 \text{ N/m} \qquad \eta = 3.6 \times 10^{-3} \text{ Ns/m}^2$$
$$U_1 + U_2 = 10 \text{ m/s} \qquad R = 0.05 \text{ m}$$

the predicted film thickness according to equation 10.12 is 0.044 μm. Since it is virtually impossible to produce gear teeth smooth or straight to this order of accuracy, this thickness of film will not prevent metallic contact of the surface asperities thus producing disastrous wear of the contacting surfaces. In practice the system operates quite satisfactorily with an adequate lubricant film, which leads us to the conclusion that our simple hydrodynamic theory is no longer applicable, calling for a re-examination of the assumptions made in section 10.2.

The work of Bell[1] indicates that any non-newtonian behaviour of the lubricant will have a detrimental effect on the film thickness so that, while in some cases the lubricant will certainly cease to be newtonian, this can in no way explain the increase in film thickness. Clearly side leakage will also act to reduce the film thickness.

While we have seen in the last chapter that temperature rise through the bearing (the thermal wedge) does introduce additional load capacity, we must also remember that this effect is very small and cannot possibly account for the discrepancy.

11.1.1 Notation

b	Half the hertzian contact width
E	Young's modulus
E'	Given by $\dfrac{1}{E'} = \dfrac{1 - v_1^2}{E_1} + \dfrac{1 - v_2^2}{E_2}$
F_R	Rolling friction per unit width
F_S	Sliding friction per unit width
h	Film thickness
h_m	Minimum film thickness in Martin analysis
$\bar{h}$	h/h_m
K	Thermal conductivity
L	Half length of the cylinder
p	Pressure
p_o	Maximum hertzian pressure $= \sqrt{\dfrac{WE'}{\pi R}}$
q	Reduced pressure
q_m	Maximum pressure in Martin analysis $= \dfrac{0.2W^{3/2}}{(\eta U)^{1/2}R}$
q_α	Pressure at which the viscosity is increased by a factor e (2.718) $= 1/\alpha$
R	Radius
U	Rolling speed $= \dfrac{U_1 + U_2}{2}$
U_1, U_2	Surface velocities
W	Load per unit width
α	Pressure viscosity exponent

289

γ   Temperature viscosity exponent
η   Absolute viscosity
ν   Poisson's ratio
φ   Ellipticality factor

11.1.2 Variable Viscosity

When we introduce the change of viscosity with pressure, we find a significant increase in film thickness. The effect of variable viscosity has been examined in section 10.6 for a lubricant with an exponential viscosity–pressure characteristic, using the reduced pressure concept.

$$p = \frac{1}{\alpha} \log_e(1 - \alpha q) \qquad \text{(10.21 repeated)}$$

Figure 10.14 shows that the pressure in the variable viscosity case rises more quickly and reaches a higher value than for the isoviscous fluid and hence the film thickness will be increased. Of course when q becomes equal to $1/\alpha$ the predicted pressure is infinite, but if we limit the solution to situations which avoid this, we can assess the effect of variable viscosity. Blok[2] performed this analysis using an exponential pressure–viscosity relationship and predicted a film thickness two-and-a-half times that predicted by the isoviscous Martin solution. Other people, using different relationships, have produced similar values.

Although this effect is beneficial, it is insufficient to explain the discrepancy between theory and practice. There must be yet another effect contributing to the generation of larger film thicknesses than those predicted by Martin.

11.1.3 Elastic Deformation

Because the fluid pressure must be very large where high specific loads are encountered, it is no longer reasonable to assume that the bearing surfaces will remain undeformed. If we consider the case of gear teeth or rolling element bearings, we may represent the contact by a plane and equivalent cylinder as in section 10.4.1. Under these conditions the cylinder will flatten locally at the contact so that it has the form illustrated in figure 11.1. The

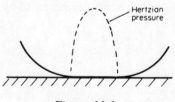

Figure 11.1

290

parallel section will have the effect of increasing the load carrying area and thus increasing the film thickness.

If we use hertzian theory to examine the deformation in the case of dry contact, we find that the distribution of pressure normal to the surface is a half-ellipse, having a maximum value $(WE/\pi R)^{1/2}$, the length of the flattened portion being $8R(W/2\pi)^{1/2}$.

It is this resulting elastic deformation which allows films of practical size in highly loaded contacts. This condition is called elastohydrodynamic lubrication.

11.2 ELASTOHYDRODYNAMIC THEORY

To produce results of use to the designers of highly loaded bearings, we are faced with the simultaneous solution of the Reynolds equation, the elastic deformation equation and the equation relating viscosity and pressure. If isothermal conditions no longer prevail, we must also incorporate the energy equation for the film and the conduction equation for heat passing to and from the solids. The cavitation boundary condition must be employed (see section 10.4.2).

With the resulting complexity, it is hardly surprising that it was not until the introduction of the high-speed digital computer that a full solution became a practical proposition. However, with some inspired assumptions, a very good approximate solution was obtained by Grubin[3] as early as 1949. During the next decade solutions of limited range were produced by Petrusevich[4] and Weber and Saalfeld[5], but it was not until 1959 (when Dowson and Higginson[6] obtained a solution of the inverse hydrodynamic problem involving the solution of the Reynolds equation to give the geometry to produce a specified pressure distribution), that the elastohydrodynamic situation really yielded to analysis. Since that time many solutions of constantly increasing range and accuracy have been achieved, notably by Dowson and Higginson[7,8], Archard, Gair and Hirst[9], and Herrebrugh[10].

Inclusion of temperature variation increases the computation necessary, but solutions have been successfully assompished by Sternlicht, Lewis and Flynn[11], Cheng and Sternlicht[12], and Dowson and Whitaker[13]. A solution which allows for inlet heating due to shearing of the lubricant has been produced by Greenwood and Kauzlarich[14].

11.2.1 Results

Film Shape and Pressure Distribution

Results from some of the foregoing analyses permit us to plot the shape of the distorted cylinder and the pressure distribution for a given set of conditions. Typically these will have the form illustrated in figure 11.2.

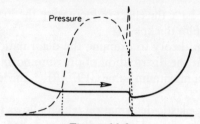

Pressure

Figure 11.2

If we consider first the film shape, we see the convergent inlet zone, followed by a virtually parallel section. At outlet there is a constriction which can amount to a reduction of up to 25 per cent of the parallel film thickness.

The pressure distribution is also illustrated. The curve shown represents a low speed situation, where the pressure is very close to the hertzian 'dry' distribution. There is a slight build-up before the parallel zone and towards the outlet we see the pressure spike and also the cavitation boundary. The spike can be shown theoretically to exist, although because it is extremely narrow, experimental verification of its existence is very difficult. Its presence is important because of the possibility of high subsurface stresses. Of course, the area under the pressure curve must be the same as that under the hertzian semi-ellipse.

If the speed is increased, the pressure distribution will depart more and more from the hertzian as shown in figure 11.3.

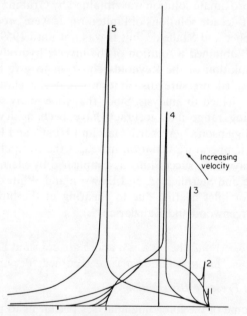

Increasing velocity

Figure 11.3   Effect of velocity on pressure distribution[8]

292

Film Thickness

The most important criterion for determining the success of the lubrication of a contact is the size of the minimum film thickness. Therefore elastohydro-dynamic analyses have usually been aimed at producing a formula which gives the minimum film thickness in terms of the operating conditions. The results are inevitably produced in numerical form, but an empirical fit can be obtained to the results of Dowson and Higginson by a power law of the form

$$h_{o} = 2.6\eta_0^{0.7}U^{0.7}\alpha^{0.54}R^{0.43}W^{-0.13}E'^{0.03} \tag{11.1}$$

where

$$\frac{1}{E'} = \frac{1 - v_1^2}{E_1} + \frac{1 - v_2^2}{E_2}$$

There are two rather surprising aspects of this equation: the film thickness is not very sensitive to changes in load or elasticity. This can be explained if we consider that the build-up of pressure occurs in the convergent inlet zone. Increasing the load or decreasing the elastic modulus leads to greater deformation, a longer parallel zone and hence a larger load carrying area with approximately the same film thickness.

If we substitute the numerical values of our gear example into the equation we now find that the predicted minimum film thickness is 0.5 μm, which is a comfortably large practical film, bearing in mind the magnitude of surface roughness and manufacturing inaccuracies.

11.2.2 Dimensionless Groups

It is often convenient to express the film thickness equation in terms of dimen-sionless groups, but there are so many ways of doing this, each with its own protagonists, that seldom will the reader find two publications which use the same form. Only three such groups are necessary to describe any situation, although Dowson and Whitaker in their original work use four. We will use a system which is based on the work of Greenwood[15] and Johnson[16], since it lends itself to an appreciation of the physical significance of each group. We may express the relationship between the parameters in terms of a power law using three dimensionless groups thus

$$\frac{Wh}{\eta UR} = A\left[\frac{(\eta U)^{1/2}R}{\alpha W^{3/2}}\right]^{a}\left[\frac{(\eta URE')^{1/2}}{W}\right]^{b} \tag{11.2}$$

where the number A and the indices a and b are constant for any particular area of operation. This formulation may be modified, for purposes which will become apparent, to

$$h\left/\frac{4.9\eta UR}{W}\right. = B\left[\frac{1}{\alpha}\bigg/\frac{0.2W^{3/2}}{(\eta U)^{1/2}R}\right]^{a}\left[\left(\frac{WE'}{\pi R}\right)^{1/2}\bigg/\frac{0.2W^{3/2}}{(\eta U)^{1/2}R}\right]^{b} \tag{11.3}$$

where the numerical constant has been adjusted appropriately.

If we examine the terms, we can readily identify four quantities

(1) The minimum film thickness for an isoviscous lubricant and rigid solids. (The Martin solution, section 10.4.3.)

$$h_m = 4.9 \frac{\eta U R}{W}$$

(2) The maximum lubricant pressure in the Martin analysis

$$q_m = \frac{0.2 W^{3/2}}{(\eta U)^{1/2} R}$$

(3) The pressure at which the viscosity is increased by a factor 2.718 (e)

$$q_\alpha = \frac{1}{\alpha}$$

(4) The maximum pressure in dry hertzian contact

$$p_o = \left(\frac{W E'}{\pi R}\right)^{1/2}$$

Equation 11.3 now may be written

$$\frac{\text{e.h.l. film thickness}}{\text{rigid isoviscous film thickness}}$$

$$= B \times \left(\frac{\text{pressure to increase viscosity by e times}}{\text{maximum rigid isoviscous pressure}}\right)^a$$

$$\times \left(\frac{\text{maximum hertzian pressure}}{\text{maximum rigid isoviscous pressure}}\right)^b$$

or in symbols

$$\frac{h}{h_m} = B\left(\frac{q_\alpha}{q_m}\right)^a \left(\frac{p_o}{q_m}\right)^b \tag{11.4}$$

The pressure ratio q_α/q_m may be taken to indicate the effect of variable viscosity on the film thickness, while the pressure ratio p_o/q_m may be identified as the effect of elasticity.

Unfortunately no universal values may be ascribed to B, a and b to cover all operating conditions. We will therefore identify four distinct regions of operation for which the appropriate values must be used.

Isoviscous Lubricant, Rigid Solids

This is of course the Martin solution

$$a = b = 0 \qquad B = 1$$

giving $h/h_m = 1$.

294

Elastic Deformation much more Significant than the Pressure Viscosity Effect

A solution for this case due to Herrebrugh[10] gives

$$a = 0 \qquad b = -0.8 \qquad B = 1.1$$

Thus the film thickness equation becomes

$$\frac{h}{h_m} = 1.1\left(\frac{q_m}{p_o}\right)^{0.8} \tag{11.5}$$

Pressure Viscosity Effect much more Significant than Elastic Deformation

In this case, which is contrary to the previous one, it is the elasticity term which disappears, giving values, according to Blok[17], of

$$a = -\tfrac{2}{3} \qquad b = 0 \qquad B = 0.99$$

Therefore the equation which must be used is

$$\frac{h}{h_m} = 0.99\left(\frac{q_m}{q_\alpha}\right)^{2/3} \tag{11.6}$$

Both Elastic Deformation and Pressure Viscosity Effects are Important

This is the region for which Dowson and Higginson's results as stated in equation 11.1 are appropriate. Rearranging equation 11.1 gives

$$a = -0.54 \qquad b = -0.06 \qquad B = 1.35$$

The film thickness equation therefore becomes

$$\frac{h}{h_m} = 1.35\left(\frac{q_m}{q_\alpha}\right)^{0.54}\left(\frac{q_m}{p_o}\right)^{0.06} \tag{11.7}$$

Clearly it is vital to use the formula appropriate to the operating conditions, by identifying the particular regime in which they lie. This is best achieved by means of a 'Johnson chart'[16], such as figure 11.4, in which lines of constant film thickness are plotted on a graph having axes q_m/q_α and q_m/p_o. Results in each of the four regions described above have been used, together with intelligent interpolation, to produce a map of elastohydrodynamic situations. Approximate boundaries between the regions are indicated. We see that the lines are horizontal in some places, indicating a dependence on q_m/q_α alone, in other places are vertical, indicating that the film thickness depends only on q_m/p_o, and between these two regions the full elastohydrodynamic solution pertains; both q_m/q_α and q_m/p_o affecting the film thickness.

Therefore before attempting to analyse an elastohydrodynamic problem it is essential to compute q_m/q_α and q_m/p_o to determine in which regime the system lies. It is then possible to identify which of the above formulae will lead to the most satisfactory solution.

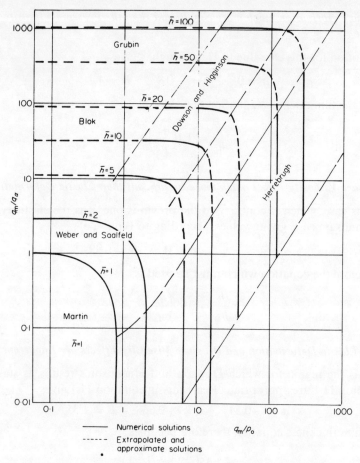

Figure 11.4   Regimes of elastohydrodynamic lubrication[16]

11.3   COMPARISON OF THEORY AND EXPERIMENT

As so much attention has been given to the calculation of the film thickness in elastohydrodynamic lubrication, it is not surprising that many efforts have been made to confirm the theory by direct measurement of the film. This has usually been done, not in the actual gear or bearing contact, but in a disc machine with its simpler geometry and steadier running conditions.

The disc machine consists of usually two, but occasionally four, circular discs which are loaded together, at least one being driven (see figure 11.5). Even in this simple configuration the determination of film thickness is no simple matter, since it is typically of the order of 1 μm or less. Many methods have been employed, but the most successful have been the measurement of the electrical capacitance of the film and the X-ray transmission technique.

296

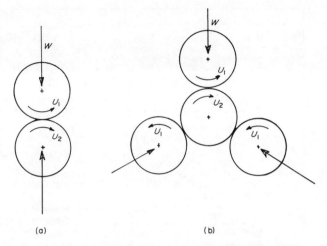

Figure 11.5   (a) 2-disc machine; (b) 4-disc machine

Capacitance measurement, which first yielded results in the work of Lewicki[18], was refined by Crook[19], incorporating a more accurate shape of the deformed cylinders, and was used with great precision by Dyson, Naylor and Wilson[20] in 1965.

The most significant results using X-ray transmission are those by Sibley and Orcutt[21] in 1961.

The general agreement between the theoretical and experimental results is very satisfactory. Figure 11.6 shows the results by Crook and Dyson *et al* for the same oil showing excellent agreement with the theoretical predictions of Dowson and Higginson.

We would expect that the lower limit of film thickness would be given by $\bar{h} = 1.0$, the Martin solution, but values well below this can be seen. This can perhaps be explained by the inadequacy of the inlet boundary condition, which assumes that $p = 0$ at a distance R before the centreline in the Martin theory (see reference 11 of chapter 10).

The general picture can be seen more clearly on the 'Johnson' chart of figure 11.7, in which certain values of film thickness are selected and plotted with coordinates q_m/q_α and q_m/p_o. The order of the results is correct, although the vertical scatter of the points appears to indicate that h is a function of q_m/p_o rather than either q_m/q_α or q_m/q_α and q_m/p_o. Indeed Greenwood[22] points out that a good approximation is given by

$$\frac{h}{h_m} = 2.81 \left(\frac{q_m}{p_o}\right)^{0\cdot8}$$

which, apart from the different numerical constant, agrees well with the prediction of Herrebrugh (equation 11.5). Certainly if we plot the film thickness against q_m/p_o the depedence is no less strong than against q_m/q_α. How-

297

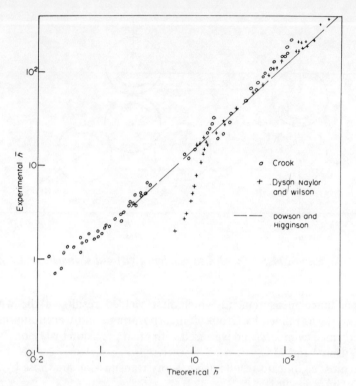

Figure 11.6 Comparison of experimental values of film thickness with theoretical prediction[22]

ever, this can be explained if we write out the three film thickness equations 11.5, 11.6 and 11.7 in terms of the practical quantities

$$h = 2.32(\eta U)^{0 \cdot 6} R^{0 \cdot 6} W^{-0 \cdot 2} E'^{-0 \cdot 4} \alpha^0 \qquad \text{(11.5 rearranged)}$$

$$h = 1.66(\eta U)^{2/3} R^{1/3} W^0 E'^0 \alpha^{2/3} \qquad \text{(11.6 rearranged)}$$

$$h = 2.65(\eta U)^{0 \cdot 7} R^{0 \cdot 43} W^{-0 \cdot 13} E'^{0 \cdot 03} \alpha^{0 \cdot 54} \qquad \text{(11.7 rearranged)}$$

The most appreciable differences between these equations are the indices of E and α. These are just the properties which most experimental workers, using conventional steel and oil, have not varied and it is therefore hardly surprising that over these ranges all equations give similar results.

The shape of the film was investigated by Crook[19] using capacitance methods on a four-disc machine. The central disc was made of glass on which was evaporated a thin chromium electrode. As the electrode passed through the contact the potential was measured and a trace such as that in figure 11.8 produced. The essential features as predicted by theory are clearly shown.

298

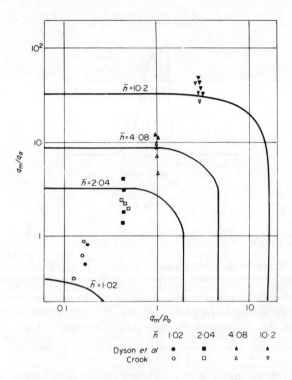

Figure 11.7 Comparison of experimental and
theoretical values of film thickness on the Johnson
chart[22]

The pressure distribution is one of the most difficult measurements to make. Kannel[23] deposited a thin strip of manganin on the surface of the disc and, as this passed through the load carrying region, it produced the records shown in figure 11.9. The local peak in the pressure curve just before the rapid fall at outlet is taken as evidence of the existence of the pressure spike. Even in this experiment the width of the pressure transducer was several times the theoetical width of the spike.

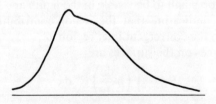

Figure 11.8 Cathode-ray oscillo-
scope trace[19]

299

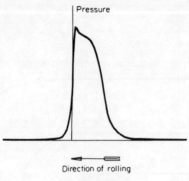

Pressure

Direction of rolling

Figure 11.9

11.4 TRACTION

In elastohydrodynamic lubrication, as in all lubrication mechanisms, surface tractions will be present. If the two surfaces are moving with the same speed (pure rolling in the case of cylinders), energy will be required to compress the lubricant as it enters the contact, the only possible source of this energy being the motion of the surfaces. This will therefore produce a retarding force on each of the surfaces, which we shall call the rolling friction, F_R.

If the surfaces are moving with different speeds then naturally the slower surface will attempt to retard the faster, while the faster surface will exert an equal force on the slower attempting to accelerate it. This force we shall call the sliding friction, F_S. Hence the retarding force on each surface may be written

$$\text{for the slower surface } F_R - F_S$$
$$\text{for the faster surface } F_R + F_S$$

11.4.1 Calculation of Frictional Forces

As in all lubricated contacts the tractions are produced by the velocity gradients at the surfaces on which they act. Typical velocity profiles are shown in figures 11.10a and b.

In pure rolling the velocity gradient in the parallel region, while having a non-zero value, is too small to be visible in the figure and is negligible for all practical purposes, indicating that the major contribution to the rolling friction comes from the convergent inlet region.

The frictional forces on the surfaces are

$$F_1 = \int \left(\eta \frac{du}{dy} \right)_{y=0} dx$$

$$F_2 = - \int \left(\eta \frac{du}{dy} \right)_{y=h} dx$$

300

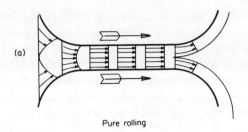

(a)

Pure rolling

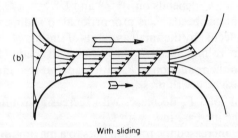

(b)

With sliding

Figure 11.10 Velocity distribution in E.H.L. films: (a) pure rolling; (b) with sliding

where the integrals are taken between limits which embrace the region of significant friction. For isothermal conditions these equations become

$$F_1 = \int \frac{h}{2} \frac{\partial p}{\partial x} \, dx + (U_2 - U_1) \int \frac{\eta}{h} \, dx$$

$$F_2 = \int \frac{h}{2} \frac{\partial p}{\partial x} \, dx - (U_2 - U_1) \int \frac{\eta}{h} \, dx$$

The first term in each can be identified as the rolling friction since it remains when $U_2 = U_1$.

The second term in each equation is the sliding friction. Since the value of this integral is inversely proportional to h, the contribution of the entry zone will be small compared to that of the parallel region and there will be little loss in accuracy if the integral is rewritten

$$F_S = \frac{(U_2 - U_1)}{h_p} \int \eta \, dx$$

integrated over the parallel region where h_p is the parallel film thickness. This may further be simplified

$$F_S = \frac{(U_2 - U_1)}{h_p} \bar{\eta}$$

301

where $\bar{\eta}$ is an effective mean viscosity. An effective viscosity is necessarily introduced, since the actual viscosity at any position will vary with the pressure and the temperature.

A typical graph of friction–sliding speed for the faster surface is shown in figure 11.11. Figure 11.11a showing constant rolling speed with different loads and 11.11b showing constant load with different rolling speeds. There are several points which we should note from the graphs.

(1) At constant rolling speed F_R is virtually unaffected by load, but rises with increasing rolling speed. This confirms the correlation between F_R and h_p since h_p depends on $W^{0\cdot13}$ and $U^{0\cdot43}$.

(2) For low sliding speeds F_S is proportional to sliding speed, indicating that $\bar{\eta}$ is unaffected by the small amounts of sliding.

(3) At constant rolling speed F_S increases with load. This is to a small extent due to the reduction of h_p with load, but to a much larger extent to the increase of $\bar{\eta}$ with pressure.

(4) At constant load F_S decreases with increasing rolling speed. This is because of the increase in h_p with U.

(5) As sliding is increased the traction rises to a maximum and with further increase will fall drastically. This is accounted for largely by the decrease in $\bar{\eta}$ due to the temperature rise within the film resulting from the viscous shearing of the lubricant. It might be expected that the decrease in viscosity would promote a reduction in film thickness leading to a rise in F_S to offset the above effect, but it must be remembered that the film thickness is controlled by the inlet viscosity rather than the actual viscosity within the contact.

Effective Viscosity

The heat generated in the contact is proportional to the product of the shear stress and the rate of strain. Of the two ways in which it is carried away—conduction to the surfaces and convection in the moving lubricant—Crook has shown that, at least for metallic components, the conduction effect is the predominant one[24].

Dowson[25] develops an expression for $\bar{\eta}$ assuming a hertzian pressure distribution, conduction as the only means of heat transfer, and considering the parallel region only

$$\bar{\eta} = \frac{4K}{b\gamma(U_2 - U_1)^2}\left[\frac{\alpha W}{2} + 2b\log_e(U_2 - U_1) + b\log_e\left(\frac{\eta_s\gamma}{2K}\right)\right]$$

where γ is the temperature coefficient of viscosity, K is the thermal conductivity, α is the pressure coefficient of viscosity and η_s is the viscosity of the lubricant at the temperature of the surfaces at ambient pressure.

302

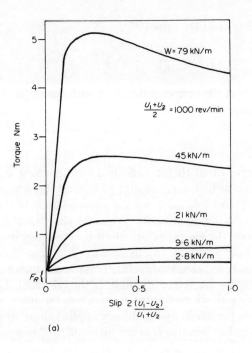

(a)

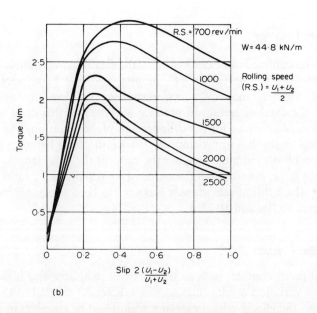

(b)

*Figure 11.11   Effect of load and speed on the traction
in rolling contact*[32]

303

This approximation is valid for values of

$$\eta_x \gamma \frac{(U_2 - U_1)^2}{8K} \gg 1$$

where η_x is the viscosity at the temperature of the surfaces and at a pressure appropriate to the value of x.

Non-newtonian Effects

Since it has been observed that thermal effects do not account entirely for the reduction in traction at high rates of slip, additional rheological factors must be sought. Indeed most lubricants are non-newtonian in that they possess shear elasticity in addition to viscosity. When the time taken to traverse the contact is very small, the elastic deformation becomes significant compared with the viscous deformation. In this case the apparent viscosity will be larger than the viscosity as calculated earlier. This effect is important when the transit time approaches the 'relaxation time' of the lubricant. The 'relaxation time' is defined by η/G, where G is the elastic shear modulus, and is the time taken for the stress produced by a suddenly applied shear strain to fall to $1/e$ of its initial value. For oils this is in the order of 10^{-4} seconds[24].

11.5 THREE-DIMENSIONAL SOLUTIONS

11.5.1 Side Leakage

We have so far confined our examination of the elastohydrodynamic problem to two dimensions. The rollers of our analysis have been operating as sections of infinitely long cylinders, permitting no flow or variation of pressure in the axial direction. In practice we find that this does not result in a great loss of accuracy for the cylinder and plane situation. The effective load carrying region has approximately the width of the hertzian ellipse in the direction of motion and therefore the ratio of the axial length–effective width is very large, even though the nominal length–diameter ratio is small. The way in which this affects the side leakage has been explained in chapter 10 (see figures 10.19a and 10.22b).

11.5.2 Point Contact

In nominal point contact, such as that between a sphere and a plane, not only is there variation of film thickness in a direction parallel to the rolling axis, but also the effective load carrying region will be circular. In this case therefore a new solution which permits flow and variation of the parameters parallel to the rolling axis is necessary.

No complete solution of this case is available but Archard and Cowking[26] present an approximate solution of the form

$$\frac{h}{R} = 1.4\left(\frac{\eta UR}{W}\right)^{0.74}\left(\frac{\alpha W}{R^2}\right)^{0.74}\left(\frac{ER^2}{W}\right)^{0.074}$$

where W is the total load on the sphere.

Experimental verification of this formula is not altogether conclusive. Archard and Kirk's[27] experiments suggest that

$$\frac{h}{R} \propto (\alpha\eta)^{0.57}u^{0.55}R^{-0.38}$$

while Gohar[28], using optical interference to measure the film thickness, presents the empirical formula

$$\frac{h}{R} = 1.28\left(\frac{\eta UR}{W}\right)^{0.7}\left(\frac{\alpha W}{R^2}\right)^{0.49}\left(\frac{ER^2}{W}\right)^{0.1}$$

Interference methods offer an interesting picture of the film shape. Figure 11.12 is taken from the work of Gohar and Cameron[29] and shows the familiar parallel region, which is now approximately circular, and the restriction at exit which extends in a semicircle round the outlet side of the contact.

We have so far considered a sphere and a plane. For other point contact situations, such as a ball rolling in a groove of different radius, the above equations do not pertain. The new model will be in the form of a flat plane and a body with different radii of curvature in the two principal directions, R_x and R_y, the area of contact being elliptical.

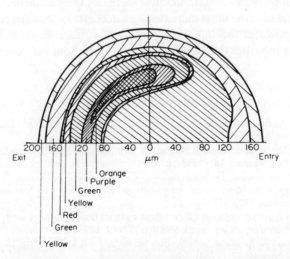

Figure 11.12   Interference pattern in E.H.L. point contact[29]

305

Archard and Cowking define an ellipticity parameter

$$\phi = \left(1 + \frac{2}{3}\frac{R_x}{R_y}\right)^{-1}$$

where R_x is the radius of curvature in the direction of motion and R_y is that measured parallel to the rolling axis. This permits the pressure in the point contact to be related to that in the line contact by

$$q(\text{point}) = \phi q(\text{line})$$

where the effective radius in the line contact is equal to R_x and the same load per unit length is used. Theory suggests that $h \propto \phi^{2/3}$ while experimental results on a crossed cylinders machine indicate that $h \propto \phi^{0.93}$. For the case of the sphere and plane $R_x = R_y$ and $\phi = 0.6$.

11.6 FATIGUE FAILURE

For many years it has been observed that in the rolling lubricated contact of metal surfaces under high load, failure takes place by a characteristic fatigue process known as pitting. A crack starts at or near the surface and propagates inwards at an acute angle to the direction of motion, resulting in the eventual removal of a particle of the metal, leaving a characteristic pit on the surface.

Although the initiation of the cracks remains something of a mystery, it is generally accepted that they propagate by means of a mechanism proposed by Way in 1935[30]. A crack, once formed, is filled with lubricant. As the crack passes through the contact zone, the very high pressures are communicated to the liquid in the crack, producing stresses which propagate the crack. The most important work in this field has been carried out by Dawson[31]. He found that an increase in film thickness increases fatigue life and that for the same film thickness the life is increased by a decrease in the surface roughness. This work indicates the importance of good surface finish in suppressing fatigue and emphasises the need to maintain continuous unbroken lubricated films.

REFERENCES

1. J. C. Bell. Lubrication of rolling surfaces by a Ree–Eyring fluid. *Trans. Am. Soc. mech. Engrs*, **5**, 160.
2. H. Blok. Discussion. Gear Lubrication Symposium. Part I, The Lubrication of Gears. *J. Inst. Petrol.*, **38**, (1952), 673.
3. A. N. Grubin and I. E. Vinogradova. Central Scientific Research Institute for Technology and Mechanical Engineering, Book No. 30, Moscow. (D.S.I.R. Translation No. 337), (1949).
4. A. I. Petrusevich. Fundamental conclusions from the contact-hydrodynamic theory of lubrication. *Izo. Akad.* Nunk SSSR (OTN) **2**, (1951), 209.
5. C. Weber and K. Saalfeld. Schmierfilm bei walzen mit verformung. *Z. angew. Math. Mech.*, **34**, Nos. (1–2). (1954), 54.
6. D. Dowson and G. R. Higginson. A numerical solution to the elastohydrodynamic problem. *J. mech. Engng Sci.*, **1**, No. 1, (1959), 6.

7. D. Dowson and G. R. Higginson. The effect of material properties on the lubrication of elastic rollers. *J. mech. Engng Sci.*, **2**, No. 3, (1960), 188.
8. D. Dowson and G. R. Higginson. New roller bearing lubrication formula. *Engineering*, **192**, (1961), 158.
9. G. D. Archard, F. C. Gair and W. Hirst. The elastohydrodynamic lubrication of rollers. *Proc. R. Soc.*, **262A**, (1961), 51.
10. K. Herrebrugh. Solving the incompressible and isothermal problem in elastohydrodynamic lubrication through an integral equation. *Trans. Am. Soc. mech. Engrs*, **90**, Series F, (1968), 262.
11. B. Sternlicht, P. Lewis and P. Flynn. Theory of lubrication and failure of rolling contact. *Trans. Am. Soc. mech. Engrs*, **83**, Series D, No. 2, (1961), 213.
12. H. S. Cheng and B. Sternlicht. A numerical solution for the pressure, temperature and film thickness between two infinitely long, lubricated rolling and sliding cylinders, under heavy loads. *Trans. Am. Soc. mech. Engrs*, **87**, Series D, (1965), 695.
13. D. Dowson and A. V. Whitaker, A numerical procedure for the solution of the elastohydrodynamic problem of rolling and sliding contacts lubricated by a newtonian fluid. *Proc. Instn mech. Engrs*, **180**, Pt. 3B, (1965–66), 57.
14. J. Greenwood and J. J. Kauzlarich. Inlet shear heating in elastohydrodynamic lubrication. *Trans. Am. Soc. mech. Engrs*, Paper 72-Lub-21, (1972).
15. J. Greenwood. Presentation of elastohydrodynamic film thickness results. *J. mech. Engng Sci.*, **11**, No. 2, (1969), 128.
16. K. L. Johnson. Regimes of elastohydrodynamic lubrication. *J. mech. Engng Sci.*, **12**, No. 1, (1970), 9.
17. H. Blok. Discussion of paper by E. McEwen. *J. Inst. Petrol.*, **38**, (1952), 673.
18. W. Lewicki. Some physical aspects of lubrication in rolling bearings and gears. *Engineer*, **200** (176), (1955), 212.
19. A. W. Crook. Elastohydrodynamic lubrication of rollers. *Nature*, **190**, (1961) 1182.
20. A. Dyson, H. Naylor and A. R. Wilson. The measurement of oil-film thickness in elastohydrodynamic contacts. *Proc. Instn mech. Engrs*, **180**, Pt. 3B, (1965–66), 119.
21. L. B. Sibley and F. K. Orcutt, Elastohydrodynamic lubrication of rolling contact surfaces. *Trans. Am. Soc. mech. Engrs*, A, No. 2, (1961), 234.
22. J. A. Greenwood, A re-examination of elastohydrodynamic film thickness measurements. *Wear*, **15**, (1970), 281.
23. J. W. Kannel. Measurement of pressure in rolling contact. *Proc. Instn mech. Engrs*, **180**, Pt. 3B (1965–66), 135.
24. A. W. Crook. The lubrication of rollers. IV. Measurements of friction and effective viscosity. *Phil. Trans. R. Soc.*, **255A**, (1963), 281.
25. D. Dowson and G. R. Higginson. *Elastohydrodynamic Lubrication*. Pergamon, London, (1966).
26. J. F. Archard and E. W. Cowking. Elastohydrodynamic lubrication at point contacts. *Proc. Instn mech. Engrs*, **180**, Pt. 3B, (1965–66), 47.
27. J. F. Archard and M. T. Kirk. Lubrication at point contacts. *Proc. R. Soc.*, **261A**, (1961), 532.
28. R. Gohar. Oil film thickness and rolling friction in elastohydrodynamic point contact. *Trans. Am. Soc. mech. Engrs, J. Lub. Tech.*, **371**, (July, 1971).
29. R. Gohar and A. Cameron. The mapping of elastohydrodynamic contacts. ASME/ASLE Lubrication Conf., Minneapolis, Preprint No. 66, LC21, (1966).
30. S. Way. Pitting due to rolling contact. *J. appl. Mech.*, **2**, (1935), 49.
31. P. Dawson. The pitting of lubricated gear teeth and rollers. *Power Transmission*, **30**, No. 351, 208.
32. D. Dowson and T. L. Whomes. Effect of surface quality upon the traction characteristics of lubricated cylindrical surfaces. *Proc. Instn mech. Engrs*, **182**, Pt. 1, No. 14, (1967–68).

12

Hydrostatic Lubrication

12.1 EXTERNALLY PRESSURISED BEARINGS

We have seen in chapter 10 that hydrodynamic or self-acting bearings only operate effectively if the two following features exist in their construction:

(a) a film thickness which varies in a preferred manner;
(b) relative tangential movement between the members of the bearing.

Externally pressurised, or hydrostatic, bearings originally attracted attention because of their capability of working in the complete absence of these prerequisites, that is, they can operate with a uniform film thickness and no relative tangential motion. Thus, for hydrostatic bearings both $\partial h/\partial x$, $\partial h/\partial z$ and U and V may be put equal to zero in the Reynolds equation of appendix A1 which then reduces to the following form

$$\frac{\partial^2 p}{\partial^2 x} + \frac{\partial^2 p}{\partial z^2} = 0 \tag{12.1}$$

This is Laplace's equation in two dimensions which must be solved for the particular geometry of the bearing under consideration. This is clearly a much simpler proposition than the original complete form of Reynolds equation. Closed form solutions are easily obtained for a number of simple shapes, such as long rectangular, circular, conical and spherical bearings[1], since these may be described by one coordinate. More complicated shapes may be tackled with series solutions or complex variable methods, but the current widespread availability of digital computers enables numerical solutions for practically any shape of bearing to be obtained in a fairly routine

308

manner. It is also of importance that great confidence can be placed in the solutions obtained, since the boundary conditions in hydrostatic bearings are usually well prescribed; for example problems of cavitation rarely occur.

It must be noted at the outset, however, that a practically useful solution of Laplace's equation is only obtained if the pressure on at least one boundary of a bearing is maintained higher than the pressure at the other boundaries. This immediately implies that the lubricating fluid must be supplied to a hydrostatic bearing at a comparatively high pressure from an external source, whereas hydrodynamic bearings require only that fluid be supplied at a pressure sufficient merely to ensure its presence in the bearing.

Although the matter will be pursued in more detail in the sections which follow, it is obvious that the general level of pressure in an externally pressurised bearing should increase as the load on the bearing increases. This increase in pressure is, in the vast majority of cases, accompanied by a reduction in the film thickness in the bearing so that a finite stiffness is exhibited. It will be shown that it is usually necessary to place a control or compensating device between the bearing and the source of high pressure lubricant, to ensure that a satisfactory performace results. Clearly this means that the behaviour of a given shape of hydrostatic bearing is determined as much by the characteristics of the control elements which are used, as by the details of its internal construction.

12.2 GENERAL DESCRIPTION OF HYDROSTATIC BEARINGS

A section through a typical hydrostatic thrust bearing is shown in figure 12.1. The lower member of the bearing consists of two distinct regions, a number of lands or sills which are separated from the upper member of the bearing by a comparatively thin film of lubricating fluid, and a recess or pocket which has a depth considerably greater than the film thickness over the lands. As fluid is supplied to the recess at a high pressure it is clear that a particular pressure profile exists over the whole area of the bearing, and that the

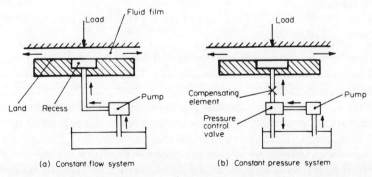

Figure 12.1   Typical arrangements of hydrostatic bearing systems

integral of this pressure distribution must equal the load which is applied to the bearing.

It is obvious that this pressure distribution can only be maintained if fluid is supplied to the recess at a rate equal to the rate at which it escapes over the lands of the bearing. The simplest way of supplying fluid to the recess of a hydrostatic bearing is by means of a 'constant flow' system of the type shown in figure 12.1a. In this arrangement a high pressure pump is assumed to deliver fluid from a reservoir at a constant rate of flow regardless of the pressure which exists in the recess. The general characteristics of this type of system are discussed in some detail in section 12.4, but it is more usual in fact for hydrostatic bearings to be operated in a 'constant pressure' system as indicated in figure 12.1b.

Two additional items are necessary for the successful application of this system, a pressure control valve, and a compensating element. The purpose and characteristics of the various types of compensating elements will not be described until sections 12.5 and 12.6 respectively; it is sufficient to say that they merely provide a resistance to flow. In this system the pressure control valve ensures that fluid is delivered to the compensating element at a constant pressure regardless of the rate at which fluid flows through it.

12.3 VISCOUS FLOW THROUGH RECTANGULAR GAPS AND CIRCULAR CAPILLARY TUBES

It was shown in section 12.1 that the shape of the pressure distribution in a hydrostatic bearing operating with a uniform film thickness in the absence of relative motion may be found by solving Laplace's equation 12.1. However, in order to complete the analysis of any particular bearing it is necessary to relate the pressure distribution in the fluid within the actual bearing itself to either

(a) the flow rate of fluid delivered by the pump in the case of a constant flow system, or
(b) the characteristics of the compensating element in the case of a constant pressure system.

This may be achieved by deducing the rate of flow of fluid through the bearing from its dimensions and the pressure distribution within it by means of elementary viscous flow theory.

Referring to figure 12.2, consider the fluid film of uniform thickness h and constant dynamic viscosity η which extends over an area specified by the x and z coordinates shown. Let us assume that the pressure distribution in this area has been determined so that the pressure profile in the direction of x at a section specified by z_o is as shown, and consider the situation at the point x_o, z_o. Here, the gradient of the pressure profile in the x direction

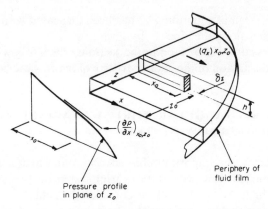

*Figure 12.2   Viscous flow in a thin fluid film
due to pressure gradients*

is $(\partial p/\partial x)_{x_o, z_o}$. The volumetric flow rate of fluid, $(q_x)_{x_o, z_o}$, through the cross-hatched area of height h and width δz in the direction of x at the position x_o, z_o is given by viscous flow theory[2] as

$$(q_x)_{x_o, z_o} = - \frac{h^3 \, \delta z}{12 \eta} \left(\frac{\partial p}{\partial x} \right)_{x_o, z_o} \tag{12.2}$$

The negative sign arises because the flow of fluid is always in the direction in which the pressure decreases, at least for a stationary bearing. Similar reasoning reveals that the volumetric flow rate of fluid $(q_z)_{x_o, z_o}$ in the direction of z at the point x_o, z_o is given by

$$(q_z)_{x_o, z_o} = - \frac{h^3 \, \delta x}{12 \eta} \left(\frac{\partial p}{\partial z} \right)_{x_o, z_o}$$

Although the pressure gradients in the x and z directions are related to each other at every point in the film by Laplace's equation, it is useful to note that the rate of flow of fluid under viscous conditions in, say, the x direction is dependent on the pressure gradient in the x direction but not on the pressure gradient in the z direction.

It is convenient to introduce the properties of one of the most usual types of compensating element in this section, since viscous flow theory[2] also yields a simple expression for the volumetric flow rate of fluid through a straight length of pipe or capillary of circular cross-section. It may be intuitively concluded that the rate of decrease of pressure along the length of the pipe is uniform. The volumetric flow rate q_c of a fluid of dynamic viscosity η through a pipe of diameter d and length l_t when a pressure difference $p_2 - p_1$ is maintained over the length of the pipe is given by

$$q_c = \frac{\pi d^4}{128 \eta} \left(\frac{p_2 - p_1}{l_t} \right) \tag{12.3}$$

311

The negative sign of equation 12.2 has been dispensed with here since the direction of flow along the pipe is obvious.

These expressions will be used in the sections which follow, to pursue the analysis and discussion of the characteristics of hydrostatic bearings.

12.4 LONG RECTANGULAR THRUST BEARINGS IN A CONSTANT FLOW SYSTEM

In order to derive the important characteristics and appreciate the salient features of a constant flow hydrostatic bearing system the very simple arrangement shown in figure 12.3a will be considered in this section; the plane upper member of this bearing has, of course, been omitted in this schematic representation. There are two features which make this system particularly easy to analyse, the absence of a compensating element and the fact that the shape of the bearing leads to an extremely straightforward solution of Laplace's equation, provided that the film thickness remains uniform.

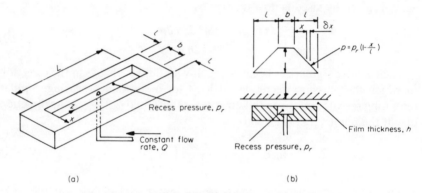

(a)   (b)

Figure 12.3   Long rectangular thrust bearing

It is first assumed that the length of the bearing is much greater than the width of the lands in the direction of the x axis, that is, $L \gg l$. This leads to the simplification that most of the fluid which is supplied to the bearing by the pump leaves by flowing from the recess over the lands in the direction of the x axis, and that comparatively little fluid flows out of the ends of the recess in the z direction. We therefore ignore those parts of the bearing which extend beyond the length of the recess. This is equivalent to assuming that there is infinite resistance to flow in the direction of the z axis at the ends of the recess.

It is also generally postulated in the study of hydrostatic bearings that the pressure of the fluid is uniform over the whole area of the recess. This is justified by the fact that as the depth of the recess in a hydrostatic bearing is about one hundred times greater than the mean film thickness of fluid over its lands, the resistance to the flow of fluid within the recess is much less than that over the lands.

We now assume that a pressure p_r above ambient exists in the recess and set up the x–z axes as shown in order to apply Laplace's equation to this geometry. Following the above introduction it is evident that there can be no change of pressure over the lands in the direction of the z axis, so that equation 12.1 reduces to

$$\frac{\partial^2 p}{\partial x^2} = 0$$

which on integration becomes

$$p = Ax + B$$

where A and B are constants. The boundary conditions for this expression are that $p = p_r$ at $x = 0$ and $p = 0$ at $x = l$; substitution of these in the above gives

$$A = -\frac{p_r}{l} \quad \text{and} \quad B = p_r$$

so that the pressure distribution over both lands is given as

$$p = p_r\left(1 - \frac{x}{l}\right)$$

This is immediately recognised as a linear or uniform pressure gradient, so that

$$\frac{\partial p}{\partial x} = -\frac{p_r}{l}$$

and we are now able to use equation 12.2 to determine the rate of flow of fluid from the bearing. Thus we put

$$\left(\frac{\partial p}{\partial x}\right)_{x_o, z_o} = -\frac{p_r}{l}$$

since this does not vary with x, and determine the volumetric flow rate through an elemental channel of width δz as

$$q_x = \left\{\frac{-h^3}{12\eta}\left(-\frac{p_r}{l}\right)\right\}\delta z$$

Since the expression in parenthesis is not a function of z we may integrate the flow rate q_x between $z = 0$ and L to obtain the total flow rate of fluid out of one side of the recess as

$$\frac{h^3 p_r}{12\eta l}L$$

313

Now we make the important statement of continuity that twice this quantity must equal the total rate of flow of fluid into the bearing from the pump, thus

$$Q = 2\frac{h^3 L p_r}{12\eta l}$$

or

$$p_r = 6\frac{Q\eta l}{Lh^3} \tag{12.4}$$

This is an important expression since it relates the recess pressure to the other parameters of the bearing. Since the pressure distribution in the bearing is now known to be of the form shown in figure 12.3b, it may be easily integrated to yield the total load, W, supported by the bearing. The load supported by the incremental length of land δx is simply $pL\,\delta x$, so that the load supported by the whole bearing is

$$p_r bL + 2\int_0^l p_r\left(1 - \frac{x}{l}\right)L\,dx$$

The first term in the above expression represents the load supported by the uniform pressure in the recess, while the second represents the load carried by the lands. Completion of the integration yields the load capacity of the bearing as

$$W = p_r L(b + l)$$

and eliminating p_r from this expression with the aid of equation 12.4 gives

$$W = \frac{6\eta l}{h^3}\left(\frac{Q}{L}\right)[L(b + l)] \tag{12.5}$$

The physical interpretation of these relationships is that because the bearing is supplied at a constant rate, the recess pressure must rise to overcome the increased resistance to flow which is brought about by a decrease in film thickness.

The rate of change of load capacity with film thickness is known as the stiffness of the bearing and is an important parameter which may be evaluated by differentiating equation 12.5 with respect to h.

$$\frac{\partial W}{\partial h} = -\frac{18\eta l}{h^4}\left(\frac{Q}{L}\right)[L(b + l)] \tag{12.6}$$

The negative sign indicates that the load capacity increases as the film thickness decreases, and that the stiffness increases very rapidly with reduced film thickness because of the presence of the fourth power of the film thickness in the denominator.

314

The expressions obtained here are very simple but in practice greater complication arises because of the inability to operate more than one bearing from a constant flow system and from the need to use bearings of more complicated shape. The first of these complications will be discussed in the following two sections, while the second will be considered in section 12.8.

12.5 THE NEED FOR COMPENSATION IN MULTIBEARING ARRANGEMENTS

It is clear, from studying the bearing geometry and pressure distribution shown in figure 12.3, that a uniform film thickness can only be maintained if the line of action of the applied load is perfectly coincident with the centreline of the bearing. This limitation does not greatly affect the hydrostatic theory which has been presented in section 12.4, but it is obviously of great practical importance, since it implies that a simple hydrostatic thrust bearing of the type so far discussed provides little, if any, resistance to tilting. In figure 12.4a the upper member of the bearing is shown in an inclined attitude following the displacement of the line of action of the load to the right of the bearing's centreline. The film thickness over the left-hand half of the bearing therefore diverges so that the resistance to flow decreases progressively across the width of the land. Without resorting to the analysis involved in the solution of this problem, it may be deduced that the pressure gradient decreases progressively over the left-hand land. By similar reasoning the pressure gradient over the right-hand land must therefore increase with the result that the pressure profile shown in figure 12.4b is produced in the bearing.

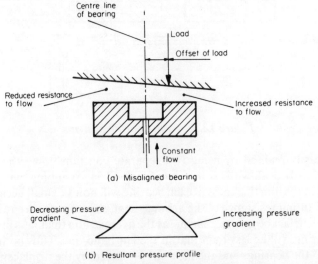

(a) Misaligned bearing

(b) Resultant pressure profile

Figure 12.4   Effect of non-uniform film thickness on pressure distribution in bearing

315

The modified pressure distribution shown in figure 12.4b provides a moment about the central longitudinal axis of the bearing tending to oppose the moment due to the offset load. Unfortunately the tilting capacity which is developed by this means is normally very small in bearings of practical proportions, and it is therefore usual to employ bearings in pairs to support offset loads, as shown in figure 12.5.

For this arrangement to provide a significantly large resistance to tilting it is clearly necessary for the pressures in the recesses of the two bearings to be different, and in the case shown p_2 must be greater than p_1. This condition

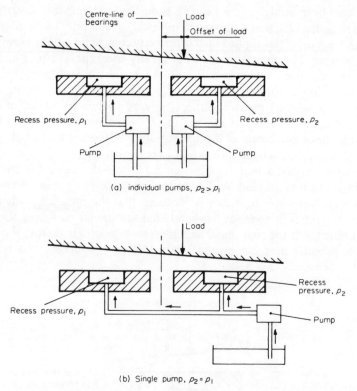

(a) individual pumps, $p_2 > p_1$

(b) Single pump, $p_2 = p_1$

Figure 12.5   Multibearing systems

may be easily realised by using two separate constant flow systems of the type previously dealt with, as shown in figure 12.5a. Assuming that the actual variation of the film thickness in each bearing will now be small because of the enhanced tilting resistance of the arrangement, it follows from equation 12.5 that p_2 will in fact be greater than p_1, as the average film thickness in the right-hand bearing will be less than that in the left-hand one. Thus both the load applied to the bearings and the moment produced by the displacement of its line of action from the centreline of the bearings, will be supported by this arrangement.

316

It must be noted that merely connecting a single constant flow pump to the two bearings, as shown in figure 12.5b, does not produce an effective arrangement because the pressures p_1 and p_2 would always be equal to each other regardless of the variation of film thickness in the bearings, provided of course that the resistance to flow in the conduits which supply the recesses is small.

Although the principle of using a separate constant flow pump to supply each bearing in a multibearing system is technically sound, it suffers from the obvious disadvantage that a high-pressure pump must be provided for each bearing.* This solution to the problem of carrying offset loads may be justifiable in installations which include bearings of large dimensions, but in machinery of more modest size it is both practically and economically out of the question for there to be as many pumps as bearings. A method has been devised which makes it possible to operate a number of bearings from a single high-pressure pump by introducing compensating elements into a constant pressure, rather than a constant flow, system. This important modification is discussed at length in the next section.

12.6 CHARACTERISTICS OF COMPENSATED BEARINGS

The fundamental features of a constant pressure system used in conjunction with a single hydrostatic bearing have already been illustrated in figure 12.1b. Following the discussion in the previous section it will be appreciated that a more typical arrangement would involve a number of bearings as shown, for example, in figure 12.6a. In this figure three separate thrust bearings, which need not necessarily be of the same size as shown, are supplied with fluid from a single high-pressure pump through individual compensating elements. It is important to note that the pressures in the three recesses may assume different values if the upper member of the bearing becomes inclined; this contrasts with the case shown in figure 12.5b and will be explained in due course.

In this section the principle of operation and an elementary analysis of a single compensated bearing of the long rectangular shape treated in section 12.4 will be considered as part of a constant pressure system. First, it is clear from the discussion in section 12.4 that the pressure in the recess of the bearing shown in figure 12.1b must increase as the film thickness decreases, even though fluid is delivered to the compensating element at a constant supply pressure. Now, the flow resistance provided by the compensating element must be overcome by the pressure difference which is established across it. From considerations of continuity the rate of flow of fluid through

* It is in fact possible to construct control valves and mechanical flow dividers which ensure that fluid is supplied at a rate which is more or less constant. Any number of these devices may therefore be connected between an equal number of bearings and a *single* high pressure pump. These then act effectively as constant flow sources.

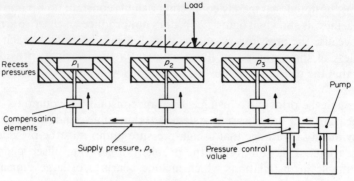

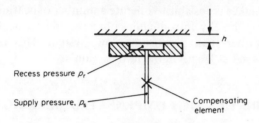

(b) Schematic diagram of a compensated bearing

Figure 12.6   Compensated bearings

the compensating element must equal the rate at which fluid flows out of the recess. It will be shown below that the way in which the recess pressure, film thickness and characteristics of the compensating element are related accounts for the satisfactory operation of a constant pressure hydrostatic bearing system.

Compensating elements, which are also known as control elements or restrictors, are usually, but not exclusively, of two types

    (a) capillary tubes
    (b) orifices.

The characteristics of the former element have already been described in section 12.3 and therefore in the first instance we will consider the analysis of the long rectangular bearing of figure 12.3 in a capillary-compensated constant pressure system.

12.6.1   Capillary Compensation

This system is shown schematically in figure 12.6b where the pressure drop across the capillary tube is clearly $p_s - p_r$, and it therefore follows from

318

equation 12.3 that the rate of flow of fluid through it is

$$\frac{\pi d^4}{128\eta}\left(\frac{p_s - p_r}{l_t}\right)$$

Similarly regardless of the way in which fluid is supplied to the recess, it follows from section 12.4 that fluid flows out of it at a rate

$$\frac{h^3 L p_r}{6\eta l}$$

By continuity these two rates of flow must be the same and equal to the rate of flow which is necessary to operate the bearing, hence

$$\frac{\pi d^4}{128\eta l_t}(p_s - p_r) = \frac{h^3 L}{6\eta l} p_r$$

or

$$p_r = \frac{p_s}{1 + h^3/K} \tag{12.7}$$

where

$$K = \frac{3\pi d^4 l}{64 l_t L}$$

Having thus determined the recess pressure in terms of the bearing's dimensions, its load capacity may now be found from the expression developed in section 12.4, that is

$$W = p_r L(b + l)$$

substituting for p_r from equation 12.7 gives

$$W = \frac{p_s L(b + l)}{(1 + h^3/K)} \tag{12.8}$$

This expression indicates that the load capacity of the bearing does in fact increase as the film thickness decreases. The physical explanation of this behaviour is that the pressure drop across the capillary decreases to maintain continuity of flow as the recess pressure increases in response to a reduction in the bearing's film thickness. This same principle is used very effectively in fluid gauging equipment and in the construction of pneumatically-operated process controllers.

Following the procedure of section 12.4 we differentiate expression 12.8 with respect to h to evaluate the stiffness of the bearing

$$\frac{\partial W}{\partial h} = -\frac{3}{K}\left(\frac{h}{1 + h^3/K}\right)^2 p_s L(b + l) \tag{12.9}$$

319

Now, both of the expressions for W and $\partial W/\partial h$ assume a more complicated form than the corresponding ones which were derived for a constant flow system in section 12.4, but certain common features may be discerned, for example, the presence of the factor $L(b + l)$. This observation will be pursued in section 12.8.

Before considering the characteristics of compensated bearings further, it is necessary to introduce the method of dimensionless presentation, which is conveniently and conventionally used in the discussion of hydrostatic bearings. Although this refinement could have been avoided in view of the simplicity of the bearings under discussion, it is nevertheless introduced here since a knowledge of it is desirable when more complicated systems are considered.

We may define the ratio of the recess pressure, p_r, to the supply pressure, p_s, as a dimensionless recess pressure, $\bar{p}$, and the ratio of the film thickness, h, to some datum or design thickness, h_o, as $\bar{h}$ so that expression 12.7 becomes, in dimensionless form

$$\bar{p} = \frac{1}{1 + (h_o^3/K)\bar{h}^3} \tag{12.10}$$

If the bearing normally operates at the design film thickness h_o then the value of the dimensionless recess pressure $\bar{p}$ is an important parameter which is known as the pressure ratio r of the bearing, that is, the ratio of the recess pressure to the supply pressure at the design film thickness h_o is the bearing's pressure ratio r. Thus putting $\bar{p} = r$ and $\bar{h} = 1$ for these conditions in expression 12.10 gives

$$r = \frac{1}{1 + (h_o^3/K)}$$

from which

$$(h_o^3/K) = \frac{1 - r}{r} \tag{12.11}$$

Substituting for (h_o^3/K) from equation 12.11 in equation 12.10 gives

$$\bar{p} = \frac{1}{1 + \left(\dfrac{1 - r}{r}\right)\bar{h}^3} \tag{12.12}$$

This is an important general expression which relates the dimensionless recess pressure of any capillary compensated bearing, not just one of the shape shown in figure 12.3, to its dimensionless film thickness for given values of pressure ratio.

It may be seen from figure 12.3b that the load capacity of the bearing will attain its greatest value when the pressure in its recess equals the supply pressure, consequently it is convenient to define a dimensionless load, $\overline{W}$, as the ratio of the load actually carried by the bearing under a given set of conditions to the load it would carry if the recess pressure rose to the supply pressure, that is, the maximum load. We know that the load capacity, W, is given by

$$W = p_r L(b + l)$$

and remembering that the recess pressure may be derived from expression 12.12 it follows that

$$\overline{W} = \frac{W}{p_s L(b + l)} = \frac{1}{1 + \left(\dfrac{1 - r}{r}\right)\bar{h}^3} \qquad (12.13)$$

It is of interest that the dimensionless recess pressure given by expression 12.12 is identical with this value of dimensionless load.

Similarly, it is possible to express the stiffness of the bearing in dimensionless form by dividing both sides of expression 12.9 by the factor

$$\left(\frac{p_s L(b + l)}{h_o}\right)$$

which has the dimensions of stiffness; the dimensionless stiffness $\partial \overline{W}/\partial \bar{h}$ then follows as

$$\frac{\partial \overline{W}}{\partial \bar{h}} = \frac{\dfrac{\partial W}{\partial h}}{\left(\dfrac{p_s L(b + l)}{h_o}\right)}$$

$$= -\frac{3h^2 h_o}{K} \times \left| \frac{1}{1 + \left(\dfrac{1 - r}{r}\right)\bar{h}^3} \right|^2$$

putting $h^2 = h_o^2 \bar{h}^2$ and substituting for h_o^3/K from equation 12.11 yields the dimensionless stiffness as

$$\frac{\partial \overline{W}}{\partial \bar{h}} = -3\left(\frac{1 - r}{r}\right)\left| \frac{\bar{h}}{1 + \left(\dfrac{1 - r}{r}\right)\bar{h}^3} \right|^2 \qquad (12.14a)$$

Since it is understood that the load capacity of a bearing increases with decreasing film thickness the negative sign in the above expression is often discarded by defining the absolute value of the dimensionless stiffness, $\overline{S}$, as $\overline{S} = -\partial \overline{W}/\partial \bar{h}$. Similarly the dimensional form follows from equation 12.9 as $S = -\partial W/\partial h$.

It is of interest to use expression 12.12 to eliminate the pressure ratio of

321

the bearing r from expression 12.14a to yield the following alternative expression for the stiffness of the bearing

$$\frac{\partial \overline{W}}{\partial \overline{h}} = -\frac{3}{\overline{h}} \, \overline{p}(1 - \overline{p}) \tag{12.14b}$$

It must be noted that in this expression the dimensionless pressure $\overline{p}$ is the pressure which exists in the recess at the particular value of dimensionless film thickness $\overline{h}$ for which the stiffness is being evaluated. This dimensionless pressure is of course found from expression 12.12 in each case.

Finally, the total flow rate necessary to operate the bearing at any value of film thickness may be expressed non-dimensionally as a proportion of the rate of flow which would occur if the recess pressure became equal to the supply pressure at the design film thickness. As the flow rate Q in the general case is given by

$$Q = \frac{h^3 L p_r}{6\eta l} = \left(\frac{h_o^3 L p_s}{6\eta l}\right) \overline{p}\overline{h}^3$$

and $\overline{Q}$ has been defined as

$$Q \bigg/ \left(\frac{h_o^3 L p_s}{6\eta l}\right)$$

substitution of the above value of Q gives $\overline{Q} = \overline{p}\overline{h}^3$ but $\overline{p}$ is itself a function of $\overline{h}$ as given by equation 12.12, thus

$$\overline{Q} = \frac{\overline{h}^3}{1 + \left(\dfrac{1 - r}{r}\right)\overline{h}^3} \tag{12.15}$$

It is interesting to note that this expression reveals the dimensionless flow rate at the design film thickness to be equal to the pressure ratio of the bearing, that is, $\overline{Q} = r$ when $h = h_o$. This agrees of course with the definition of $\overline{Q}$.

It has been shown that the important parameters of a capillary compensated bearing of simple shape may be expressed in dimensionless form. It is worth recalling at this stage that the dimensional values of these parameters are easily determined in a given case as follows

$$\text{recess pressure} \quad p_r = (p_s)\overline{p}$$

$$\text{load capacity} \quad W = (p_s L(b + l))\overline{W}$$

$$\text{stiffness} \quad S = -\frac{\partial W}{\partial h} = \left(\frac{p_s L(b + l)}{h_o}\right)\overline{S}$$

$$\text{flow rate} \quad Q = \left(\frac{h_o^3 L p_s}{6\eta l}\right)\overline{Q} \tag{12.16}$$

It is clear from the foregoing that the compensating element of a constant pressure system simply provides a resistance to flow between the pressure control valve and the bearing's recess. Because the flow through a capillary compensator is proportional to the pressure drop across it (expression 12.3), it is recognised as a linear device. Although the rate of flow through an orifice is proportional to the square root of the pressure difference across it, hence its classification as a non-linear element, it too finds application as a compensating element. The characteristics of an orifice-compensated bearing are considered in the next section.

12.6.2 Orifice Compensation

It is shown in textbooks on fluid mechanics[2] that the rate of flow of an incompressible fluid through a short, sharp-edged orifice of diameter d and discharge coefficient C_d is

$$\frac{\pi}{4} C_d d^2 \left(\frac{2(p_2 - p_1)}{\rho}\right)^{1/2}$$

where ρ is the density of the fluid and $(p_2 - p_1)$ the relevant pressure difference. If the capillary compensator which has been considered as the compensating element in the bearing of figure 12.6b is now replaced by an orifice then the condition of continuity of flow becomes

$$\frac{\pi}{4} C_d d^2 \left(\frac{2(p_s - p_r)}{\rho}\right)^{1/2} = \frac{h^3 L p_r}{6\eta l}$$

The investigation of this bearing may be facilitated by analysing it from the outset in terms of dimensionless quantities, thus in the nomenclature of the previous section we put $p_r = \bar{p} p_s$ and $h = \bar{h} h_o$ in the above expression, which then reduces to

$$K(1 - \bar{p}) = \bar{p}^2 \bar{h}^6 \tag{12.17}$$

where, in this case

$$K = \frac{9\pi^2}{2\rho p_s} \left(\frac{C_d d^2 \eta l}{h_o^3 L}\right)^2$$

The dimensionless recess pressure is obtained by evaluating the positive root of the equation for $\bar{p}$ which is explicit in (equation 12.17)

$$\bar{p} = \frac{K}{2\bar{h}^6} \left[\left(1 + 4\left(\frac{\bar{h}^6}{K}\right)\right)^{1/2} - 1\right]$$

This expression is comparable to equation 12.10 and may be further generalised by using the fact that $\bar{p}$ must be equal to the pressure ratio of the bearing

323

r when the dimensionless film thickness $\bar{h}$ is unity. Substituting these conditions in equation 12.17 gives

$$K = \frac{r^2}{1 - r} \qquad (12.18)$$

and the expression for the dimensionless pressure ratio becomes

$$\bar{p} = \frac{1}{2\bar{h}^6} \left(\frac{r^2}{1 - r} \right) \left[\left(1 + 4 \left(\frac{1 - r}{r^2} \right) \bar{h}^6 \right)^{1/2} - 1 \right] \qquad (12.19)$$

It is again useful to note that this expression is true for all orifice controlled bearings and not merely for the kind shown in figure 12.3b.

Consideration of the manner in which the dimensionless stiffness of a capillary-compensated bearing, expression 12.14a, was derived reveals that $\partial \bar{W}/\partial \bar{h} = \partial \bar{p}/\partial \bar{h}$ and it follows that the dimensionless stiffness of an orifice-compensated bearing is obtained by differentiating equation 12.19 with respect to $\bar{h}$

$$\frac{\partial \bar{W}}{\partial \bar{h}} = -\frac{3}{\bar{h}^7} \left(\frac{r^2}{1 - r} \right) \left| \left(1 + 4 \left(\frac{1 - r}{r^2} \right) \bar{h}^6 \right)^{1/2} - 1 - \frac{2 \left(\frac{1 - r}{r^2} \right) \bar{h}^6}{\left(1 + 4 \left(\frac{1 - r}{r^2} \right) \bar{h}^6 \right)^{1/2}} \right)$$

$$(12.20\text{a})$$

This is a much more complicated expression than the equivalent one for a capillary-compensated bearing given in expression 12.14a but it too may be conveniently simplified by first removing the expressions under the square root signs by using equation 12.19. After some further simplification the above expression reduces to

$$\frac{\partial \bar{W}}{\partial \bar{h}} = -\frac{6}{\bar{h}} \bar{p} \left(\frac{1 - \bar{p}}{2 - \bar{p}} \right) \qquad (12.20\text{b})$$

As in expression 12.14b the pressure $\bar{p}$ in this expression must be determined from expression 12.19 for the particular film thickness $\bar{h}$ under consideration.

The dimensionless flow variable $\bar{Q}$ for an orifice-compensated bearing is defined in exactly the same way as for a capillary-restricted bearing. From equation 12.19

$$\bar{Q} = \frac{1}{2\bar{h}^3} \left(\frac{r^2}{1 - r} \right) \left[\left(1 + 4 \left(\frac{1 - r}{r^2} \right) \bar{h}^6 \right)^{1/2} - 1 \right] \qquad (12.21)$$

Since the dimensionless parameters $\bar{p}$, $\partial \bar{W}/\partial \bar{h}$ and $\bar{Q}$ have been determined, the dimensional values of p_r, W, $\partial W/\partial h$ and Q for an orifice-controlled bearing may be evaluated immediately from expression 12.16. This is true even though the expressions of 12.16 were derived initially for a capillary-controlled bearing.

12.6.3 Constant Flow Compensation

Although it may seem that a constant flow system of the type discussed in section 12.4 is not describable in terms of a supply pressure, it is usual for the delivery pressure of the pump which operates it to be limited to some value, say p_s. Thus, if the recess pressure in the bearing's recess is p_r at some operating film thickness h then, as before, a dimensionless recess pressure may be defined as p_r/p_s and a dimensionless film thickness as h/h_o, where h_o is the film thickness under design conditions. Now from expression 12.4 we may write

$$\bar{p} = \left(6\,\frac{Q\eta l}{Lp_s h_o^3}\right)\frac{1}{\bar{h}^3}$$

The pressure ratio of the bearing may again be defined as the ratio r of the recess pressure to the supply pressure at the design film thickness, that is, when $\bar{h} = 1$. Thus from the above

$$r = \left(\frac{6Q\eta l}{Lp_s h_o^3}\right) \tag{12.22}$$

and

$$\bar{p} = \frac{r}{\bar{h}^3} \tag{12.23}$$

The dimensionless stiffness of the bearing then follows immediately from equation 12.23 as

$$\frac{\partial \overline{W}}{\partial \bar{h}} = -\frac{3r}{\bar{h}^4} \tag{12.24a}$$

or in terms of the recess pressure

$$\frac{\partial \overline{W}}{\partial \bar{h}} = -\frac{3\bar{p}}{\bar{h}} \tag{12.24b}$$

It is, of course, a trivial matter to show that the dimensionless flow rate $\overline{Q}$ is a constant and is given by

$$\overline{Q} = r = \bar{p}\bar{h}^3 \tag{12.25}$$

The dimensional values of this bearing's parameters may be found by use of expressions 12.16.

12.7 COMPARISON OF CHARACTERISTICS

It is significant, and of course convenient, that the expressions for the recess pressure, load capacity, stiffness and flow rate which were derived in dimensionless form in the previous section for capillary, orifice and constant flow compensated bearings are functions of only two variables, the dimensionless film thickness $\bar{h}$ and the pressure ratio r. In this section these dimensionless parameters will be discussed at some length to ensure that a clear

325

appreciation of the characteristics of the various types of compensation is gained. Because the dimensional values of the variables of interest may be obtained for any particular bearing of the long rectangular shape which has so far been considered, by use of expressions 12.16, this discussion will be presented entirely in terms of the dimensionless variables and, to avoid repetition, the adjective *dimensionless* will be dispensed with in this section.

12.7.1 Choice of Pressure Ratio

Because the stiffness of a bearing determines the deflection which occurs in a machine following the imposition of a load, it is recognised as one of the most important parameters. It is therefore reasonable and convenient to assume that it is desirable to maximise the stiffness of a bearing at some particular combination of operating conditions. Now as the stiffness of a capillary compensated bearing is given by equation 12.14b it follows that for a given film thickness the stiffness is a function of only the recess pressure. By differentiating this expression with respect to $\bar{p}$ and equating to zero it is revealed that the stiffness will be a maximum when $\bar{p} = 0.500$. Substitution of this value of $\bar{p}$ in expression 12.14b gives the maximum stiffness of a capillary compensated bearing *at the film thickness $\bar{h}$* as

$$\bar{S}_{max} = \frac{0.75}{\bar{h}} \tag{12.26}$$

Similarly, differentiating expression 12.20b with respect to $\bar{p}$ shows that the stiffness of an orifice-compensated bearing is a maximum when

$$\bar{p}^2 - 4\bar{p} + 2 = 0$$

The positive root of this expression is $2 - (2)^{1/2} = 0.586$ and substitution of this value of $\bar{p}$ in expression 12.20b gives the following expression for the maximum stiffness of an orifice compensated bearing *at the film thickness $\bar{h}$*

$$\bar{S}_{max} = \frac{1.0296}{\bar{h}} \tag{12.27}$$

It is clear from expression 12.24b that the stiffness of a bearing with constant flow compensation does not exhibit a maximum, but increases monotonically with recess pressure.

Without sacrificing too much generality, we may now fix the film thickness at unity so that the preferred values of recess pressure derived above then become the pressure ratios r of the bearing. By evaluating the stiffness of a capillary, orifice and constant flow controlled bearing for various values of pressure ratio from expressions 12.14b, 12.20b and 12.24b respectively, the curves of figure 12.7 may be drawn. The maximum values of the stiffness of capillary and orifice compensated bearings are shown at the preferred values of pressure ratio and it is noticeable that the stiffness of both types of

326

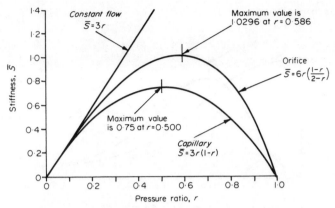

Figure 12.7   Effect of pressure ratio on the stiffness of a simple hydrostatic thrust bearing with constant flow, capillary and orifice compensation

bearing is relatively insensitive to variations in pressure ratio of, say, ± 0.05 about these values. The stiffness of a bearing with constant flow compensation changes rapidly with pressure ratio as mentioned earlier.

An important practical observation is that under the most favourable conditions the stiffness of an orifice controlled bearing is about 33 per cent greater than that of the best equivalent capillary-controlled bearing, but of course the higher recess pressure necessary increases the flow rate of fluid required to operate the former bearing.

12.7.2   Effects of Changes in Film Thickness

In the last section the performance of the various types of bearing was considered and the effects of varying the pressure ratio while the film thickness was maintained constant at the design value of unity resulted in the curves of figure 12.7 being drawn and discussed. Assuming that the preferred values of pressure ratio of 0.5 and 0.586, for capillary and orifice-controlled bearings respectively, would be used at the design film thickness of unity, it is of interest to study the behaviour of the bearings when the film thickness departs from the design value in response to changes in loading.

First considering a capillary-controlled bearing which operates at a pressure of 0.5 under design conditions, the variation of recess pressure and hence load capacity may be obtained by putting $r = 0.5$ in expressions 12.12 or 12.13. The curve for $\bar{p}$ and $\overline{W}$ drawn in figure 12.8a then results, and it is clear that the recess pressure rapidly approaches the supply pressure once the film thickness decreases below, say, 0.6. Alternatively this observation may be interpreted as indicating that a high proportion of the maximum load capacity of the bearing is developed once the film thickness becomes less than about 0.6.

327

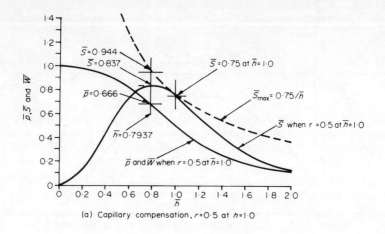

(a) Capillary compensation, $r=0.5$ at $h=1.0$

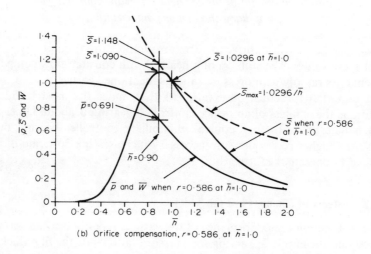

(b) Orifice compensation, $r=0.586$ at $\bar{h}=1.0$

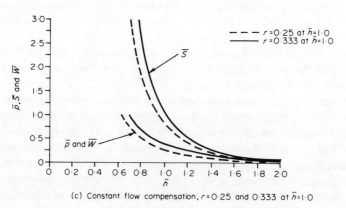

(c) Constant flow compensation, $r=0.25$ and 0.333 at $\bar{h}=1.0$

Figure 12.8   Characteristics of compensated bearings

328

The way in which the stiffness changes with film thickness may be studied by putting $r = 0.5$ in expression 12.14a which then becomes

$$\bar{S} = -\frac{\partial \bar{W}}{\partial \bar{h}} = 3\left(\frac{\bar{h}}{1 + \bar{h}^3}\right)^2$$

By substituting various values of $\bar{h}$ in this expression the curve of $\bar{S}$ drawn in figure 12.8a is obtained. Naturally, as given by equation 12.26, the stiffness has a value of 0.75 at $\bar{h} = 1$, but it is of importance that as the film thickness decreases from this value the stiffness reaches a maximum before decreasing rapidly as the film thickness approaches zero. By differentiating the above expression with respect to $\bar{h}$ and equating to zero, the value of $\bar{h}$ at which this maximum occurs is found to be 0.7937 (actually $(\frac{1}{2})^{1/3}$). Substitution of this value of $\bar{h}$ into the above expression gives the value of the maximum stiffness as 0.837, but this behaviour will only arise in a bearing which has a *pressure ratio of* 0.5 *at a film thickness of unity*.

It is of interest to note that the value of the recess pressure under these conditions as found from equation 12.12 is 0.666 (actually 2/3). This figure is often incorrectly quoted in the literature as the value of pressure ratio which yields the maximum value of stiffness for a capillary-compensated bearing; this anomaly is explained below.

Consideration of expression 12.26 shows that a hyperbolic relationship exists between the maximum value of stiffness $(-\partial \bar{W}/\partial h)_{\max}$, and the film thickness $\bar{h}$. This relationship is also plotted in figure 12.8a, but it must be realised that a bearing only exhibits this characteristic if its compensating element is continually altered to ensure that the *recess pressure assumes a value of* 0.5 *at all values of film thickness*. Thus the true maximum value of the stiffness attainable at the film thickness of 0.7937, or for that matter at any other film thickness, is given by expression 12.26. On substituting $\bar{h} = 0.7937$ in this expression the maximum stiffness at this film thickness is shown to be 0.944. This is of course greater than the value of 0.837 which would be obtained at this particular value of film thickness when the bearing's compensating element is chosen merely to give a recess pressure of 0.5 at a film thickness of unity only.

Since there is considerable confusion in the published treatments and explanations of the characteristics which have been plotted in figure 12.8a, it is worth summarising the above discussion as follows: if the recess pressure of a capillary-compensated bearing is adjusted to be 0.5 at all values of film thickness, then its stiffness is given by the hyperbolic relationship of expression 12.26, that is, the stiffness increases rapidly as the film thickness is reduced. However, if the recess pressure is chosen to be 0.5 at a film thickness of unity, say, the stiffness of the bearing is described by expression 12.14a and a reduction in film thickness will at first cause an increase in stiffness until a maximum is reached at a film thickness of 0.7937, but further decreases in film thickness result in a rapidly decreasing stiffness. In all cases the greatest

329

possible stiffness which can be developed by a capillary-compensated bearing is that given by the hyperbolic relationship of expression 12.26.

In the case of an orifice-compensated bearing a similar approach may be made and the variation of recess pressure and load capacity with film thickness may be investigated, for example, by putting $r = 0.586$ in expression 12.19. The curve for $\bar{p}$ and $\bar{W}$ drawn in figure 12.8b then results, and it is noticeable that as the film thickness decreases a large proportion of the maximum load capacity of the bearing is developed even more rapidly than in the case of a capillary-controlled bearing.

The variation of stiffness with film thickness may be determined by putting $r = 0.586$ in expression 12.20a, or by using equation 12.20b in conjunction with equation 12.19, and the curve of $\bar{S}$ drawn in figure 12.8b results. Again a maximum in this curve is reached before the stiffness decreases rapidly with film thickness for values of $\bar{h}$ less than, say, about 0.8. Differentiation of expression 12.20a with respect to $\bar{h}$, with r put equal to 0.586, yields an intractable expression for the value of $\bar{h}$ at which the maximum of the stiffness curve occurs. However, a study of figure 12.8b reveals that this occurs at a film thickness of approximately 0.9 and that the maximum in stiffness is about 1.09.

The recess pressure under these conditions may be found from expression 12.19 as 0.691, a figure which similarly appears incorrectly in the literature as the value of pressure ratio which produces maximum stiffness in an orifice-compensated bearing.

Consideration of the hyperbolic relationship of expression 12.27 shows that the maximum stiffness $\bar{S}_{max}$ attainable at this film thickness is slightly higher at 1.148.

In parallel with the discussion of the behaviour of a capillary compensated bearing it is now clear that the maximum stiffness $\bar{S}_{max}$ of an orifice compensated bearing will only be attained if its recess pressure is continually adjusted, as the film thickness varies, to the value of 0.586. If the compensating element is merely chosen to yield a pressure ratio of 0.586 at a film thickness of unity, then the stiffness will increase to the maximum of 1.090 as the film thickness decreases to about 0.900, but if the recess pressure were then adjusted to 0.586 at this film thickness, the stiffness would increase to the value of 1.148 given by the hyperbolic relationship of expression 12.27.

Following the considerations of section 12.7.1 it is clear that the choice of pressure ratio to be used in the discussion of the characteristics of a bearing with constant flow compensation is not as obvious as it was for the other two types of bearing which have been dealt with. For purposes of comparison, however, it follows from expression 12.24b that at a film thickness of unity the stiffness of a bearing with constant flow compensation may be made equal to that of a capillary controlled bearing, that is 0.75, by taking a pressure ratio of 0.25. The variation of recess pressure and load capacity then follows from expression 12.23 and is plotted in figure 12.8c along with

330

the corresponding stiffness for the same pressure ratio as given by expression 12.24b. It must be noted that these curves are terminated at the film thickness where the recess pressure reaches a value of unity.

Similarly it is also useful to consider the behaviour of the bearing at a pressure ratio of 0.333 which gives approximately the same stiffness as an orifice compensated bearing, about 1.0, at a film thickness of unity. The variation of recess pressure and stiffness with film thickness at this pressure ratio is also shown in figure 12.8c.

Although these choices of pressure ratio may seem a little arbitrary, they yield characteristics which may be compared with those of the other two types of bearing; this is done in the next section.

12.7.3 General Comparison of Types of Compensation

In comparing the three types of bearing, by studying their behaviour over a range of film thickness, some values must clearly be assigned to their respective pressure ratios. Following the developments of the previous section the pressure ratio of capillary- and orifice-controlled bearings have been taken as 0.500 and 0.586 respectively, because these values give the corresponding values of $\bar{S}_{max}$ at a film thickness of unity. The behaviour of a constant flow compensated bearing will be studied at the two pressure ratios of 0.250 and 0.333 which, as has been shown in the previous section, produce values of stiffness equal to the maximum obtainable stiffness of capillary and orifice-controlled bearings respectively at a film thickness of unity.

Since the curves which describe the behaviour of the three types of bearing under these conditions have already, with the exception of the relationships for flow rate, been presented they will now be drawn together and discussed as a whole.

In figure 12.9a the variation of recess pressure and load capacity is shown over a range of film thickness for the four cases mentioned above. As has been noticed previously an orifice-controlled bearing develops a larger proportion of its maximum load capacity than a capillary-controlled bearing at all values of film thickness, but otherwise there is little difference between these two curves. However, the load capacity of the bearings which operate with constant flow control at pressure ratios of 0.250 and 0.333, increase much more rapidly with decreasing film thickness than either of the bearings which work in a constant pressure system, but since the recess pressure should not by definition exceed unity, the decrease in film thickness is severely limited. Thus if the film thickness of a bearing with a pressure ratio of 0.25 decreases below about 0.63 then the operating pressure of the pump which supplies the fluid would exceed the practical limit imposed on it.

This consideration must be applied to the other characteristics of interest and the variation of flow rate with film thickness shown in figure 12.9b for bearings with constant flow control are similarly terminated when the film thickness reaches the value at which the recess pressure attains a value of one.

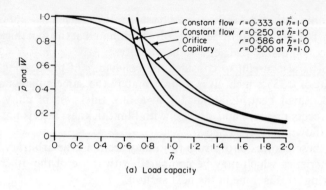

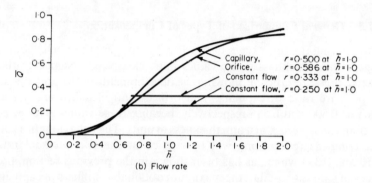

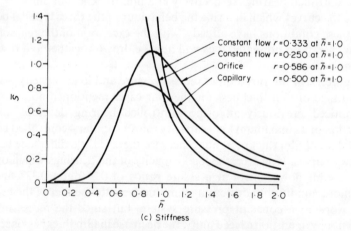

Figure 12.9   Comparison of bearing characteristics

Above these film thicknesses the flow rates are, of course, constant as would be expected with this form of control. It is also noticeable that there is comparatively little difference between the flow rate of fluid necessary to operate capillary- and orifice-controlled bearings, but it is again of practical importance that as the film thickness decreases, more fluid is returned directly from the constant pressure control valve, figure 12.1b, to the reservoir of the system.

Several important features are apparent in figure 12.9c, although greater stiffness is provided by orifice rather than capillary compensation at film thicknesses of about unity, there is a more rapid decrease with decreasing film thickness in the former case. It may therefore be useful in a practical case to utilise the more uniform stiffness of a capillary controlled bearing. This possibility must be balanced against the previously observed consideration that a bearing with orifice control exhibits a higher load capacity than a capillary-compensated bearing over a wide range of film thickness.

There are two striking features revealed by the curves for the bearings with constant flow control drawn in figure 12.9c. It is first clear that the stiffness becomes progressively larger at an increasing rate as the film thickness decreases, and second that the magnitude of their stiffness is considerably greater than that of either an orifice- or a capillary-controlled bearing at film thicknesses of just less than unity. It must be recalled that since the recess pressure cannot exceed unity, the range of operation of a bearing with constant flow control is limited and therefore the very rapid increase in stiffness shown in this figure must terminate at film thicknesses of about 0.63 and 0.70 for pressure ratios of 0.25 and 0.333 respectively as indicated in figure 12.9b. Nevertheless, it may be useful in some applications to make use of the very high stiffness which may be obtained with bearings of this type, provided that the often severe limitation on the variation of film thickness is tolerable.

It should be remembered that the preceding discussion of the various types of externally pressurised bearings has been based on particular values of pressure ratios which, although chosen for a sound reason, clearly determine the characteristics of the bearings. Since changes in the pressure ratios affect the characteristics of the bearings, the observations which have been made in this section could perhaps be questioned by considering the behaviour of bearings at other values of pressure ratio. However, the general features and important tendencies of the characteristics of externally pressurised bearings, at values of pressure ratio which would be likely to be used in practice, have been introduced and discussed.

12.8 FLOW, LOAD AND POWER FACTORS FOR OTHER SHAPES OF BEARING

The foregoing discussions have been based on the treatment of the long rectangular thrust pad of figure 12.3, and it is clear that bearings of other shapes may be similarly analysed by solving Laplace's equation 12.1 for the appropriate plan shape and boundary conditions. However, it will be shown in this section that considerable rationalisation of the presentation of such

333

solutions may be made by the introduction of load, flow and power factors which depend only on the shape of a bearing and not its actual physical dimensions.

12.8.1 Flow, Load and Power Factors for a Long Rectangular Pad

Regardless of the means of compensation it has been shown that the load capacity of a long rectangular pad of the type drawn in figure 12.3 is given by

$$W = L(l + b)p_r$$

Noting that the *overall* area of the pad A is $L(b + 2l)$ we may rewrite the above as

$$W = (Ap_r)\bar{a} \tag{12.28}$$

where $\bar{a} = (l + b)/(2l + b)$, the load factor for a long rectangular pad, or

$$\bar{a} = \frac{1 + (b/l)}{2 + (b/l)} \tag{12.29}$$

Clearly (b/l) may vary from nearly zero for wide lands, to large values for narrow lands and the variation of $\bar{a}$ over a useful range of the ratio (b/l) is shown in figure 12.10.

Similarly the rate of flow of fluid necessary to operate the pad is given by expression 12.4 as

$$Q = \left(\frac{h^3 p_r}{\eta}\right)\left(\frac{L}{6l}\right)$$

Again, the factor $(L/6l)$ may be recognised as the dimensionless coefficient which introduces the shape of the pad into the expression for flow rate and thus we may write

$$Q = \left(\frac{h^3 p_r}{\eta}\right)\bar{q} \tag{12.30}$$

where

$$\bar{q} = \frac{L}{6l} \tag{12.31}$$

$\bar{q}$ is known as the flow factor for a long rectangular thrust bearing and its variation for a particular range of the ratio (L/l) is also shown in figure 12.10.

It is interesting to note that the load capacity of any bearing of this form is given by expression 12.28 and that this involves only the overall area of the bearing A, the recess pressure p_r, and the particular value of load factor $\bar{a}$ corresponding to the shape of the bearing as defined by the ratio (b/l). Similarly the flow rate through the bearing is given by expression 12.30. Thus the behaviour of a large range of bearings of different sizes and shapes may be deduced from these expressions in conjunction with the curves of figure 12.10.

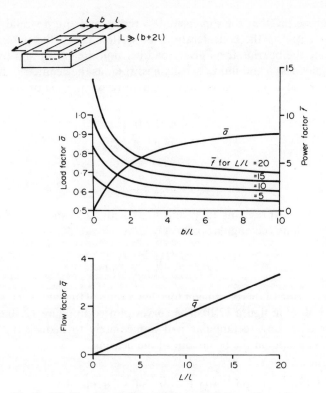

*Figure 12.10   Load, flow and power factors for a long
rectangular thrust bearing*

It is often necessary to consider if a thrust bearing has the most suitable proportions or shape for the particular application in mind. What is considered to be suitable varies of course from application to application, but it is often appropriate to consider how the pumping power necessary to operate a bearing for a given load per unit overall area, varies with changes in its shape. Now the required power P is simply the product of the flow rate and supply pressure, hence from expression 12.30

$$P = \left(\frac{h^3 p_s^2}{\eta}\right)\bar{p}\bar{q}$$

and noting from expression 12.28 that

$$p_s = \left(\frac{W}{A}\right) \times \frac{1}{\bar{p}\bar{a}}$$

we may write

$$P = \frac{h^3}{\eta}\frac{1}{\bar{p}}\left(\frac{W}{A}\right)^2\left(\frac{\bar{q}}{\bar{a}^2}\right)$$

335

Remembering that the dimensionless recess pressure $\bar{p}$ would be chosen independently on the considerations of section 12.7.1, the power required to operate the bearing for a given load per unit area W/A and fixed values of film thickness h and fluid viscosity η is, as far as the geometric shape of the bearing alone is concerned, dependent only on the factor $\bar{q}/\bar{a}^2$ which is termed the power factor $\bar{f}$ thus

$$P = \frac{h^3}{\eta} \frac{1}{\bar{p}} \left(\frac{W}{A}\right)^2 \bar{f} \tag{12.32}$$

where

$$\bar{f} = \frac{\bar{q}}{\bar{a}^2} \tag{12.33}$$

Combining expressions 12.29 and 12.31 with 12.33 shows that the power factor for a long rectangular pad is given by

$$\bar{f} = \left(\frac{L}{6l}\right)\left(\frac{2 + b/l}{1 + b/l}\right)^2 \tag{12.34}$$

The variation of this parameter for a few values of the ratio L/l is shown as a function of b/l in figure 12.10. The curves plotted in figure 12.10 enable the designer of a long rectangular pad immediately to deduce the effects of changes in shape on the behaviour of the bearing.

12.8.2 Flow, Load and Power Factors for a Circular Pad

A very useful form of hydrostatic thrust bearing in the form of an annulus is shown in figure 12.11a. Fluid is supplied to the central recess of radius R_1, and it flows radially outward to the periphery of the bearing at radius R_2.

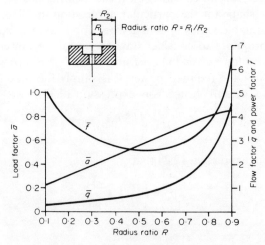

Figure 12.11   Load, flow and power factors
for a circular thrust bearing

336

If the uniform clearance between the pad and its plane mating member is h, the fluid viscosity η and the recess pressure p_r it may be shown[3] that the closed form solution of Laplace's equation in the case of this particular shape yields the following expressions for load capacity W and flow rate Q

$$W = \pi R_2^2 p_r \left(\frac{1 - R^2}{2 \log_e(1/R)} \right)$$

$$Q = \frac{h^3}{\eta} p_r \left(\frac{\pi}{6} \times \frac{1}{\log_e(1/R)} \right)$$

where $R = R_1/R_2$, the radius ratio of the bearing which is sufficient to describe its shape completely.

Recognising πR_2^2 as the overall area of the bearing and putting

$$\bar{a} = \frac{1 - R^2}{2 \log_e(1/R)} \tag{12.35a}$$

and

$$\bar{q} = \frac{\pi}{6} \times \frac{1}{\log_e(1/R)} \tag{12.35b}$$

the above expressions for W and Q may be written as

$$W = (Ap_r)\bar{a}$$

and

$$Q = \left(\frac{h^3 p_r}{\eta} \right) \bar{q}$$

These latter two expressions are identical to expressions 12.28 and 12.30 respectively, which were derived for a long rectangular pad but of course the values of $\bar{a}$ and $\bar{q}$ must be correctly chosen. It follows that the power consumption of a circular thrust bearing is given by expression 12.32, provided of course that the power factor $\bar{f}$ is given the value

$$\bar{f} = \frac{\bar{q}}{\bar{a}^2} = \frac{2\pi}{3} \frac{\log_e(1/R)}{(1 - R^2)^2} \tag{12.35c}$$

The variation of the factors $\bar{a}$, $\bar{q}$ and $\bar{f}$ for a circular pad are given over a wide range of radius ratio R in figure 12.11b, and it is interesting to observe that a minimum exists in the power factor at a radius ratio of about 0.54. This type of deduction is often useful during the design stage when the proportions of a bearing are being chosen.

12.8.3 Load, Flow and Power Factors for a Typical Bearing of more Complicated Shape

It is a relatively simple matter to obtain closed form expressions for the load, flow and power factors for the long rectangular and circular types of thrust bearing which have been dealt with so far. However, there are many practically useful plan forms which do not yield to simple analysis and the three factors must then be evaluated by more advanced methods, for example, with the aid of a conducting sheet analogue or by digital computation.

An example of this class of bearing is the square thrust pad of side $4l$ shown in figure 12.12a. This pad contains four equi-spaced recesses of side $2s$ and its overall area A is $16l^2$. Assuming that the pressures in the four recesses all

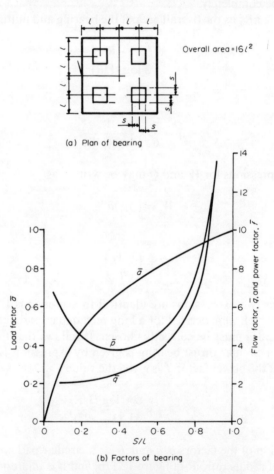

(a) Plan of bearing

(b) Factors of bearing

Figure 12.12   Load, flow and power factors for a square thrust bearing with four equispaced square recesses

338

attain the same value p_r, it follows that its load capacity will be given by expression 12.28, its flow rate by expression 12.30 and its power consumption by expression 12.32, provided that the appropriate values of $\bar{a}$, $\bar{q}$ and $\bar{f}$ respectively are used in these expressions. Now it is clear that the flow patterns and pressure distribution in this bearing will be of a complicated form and no accurate closed form expressions for these three factors may be derived. Nevertheless since these parameters have been evaluated by an advanced method[4] they may be presented in graphical form as functions of the ratio (s/l) which alone is capable of defining the shape of the bearing. These curves are drawn in figure 12.12b and the performance of any size of bearing of this shape may be deduced immediately by entering the appropriate values of $\bar{a}$, $\bar{q}$ and $\bar{f}$ from these curves into expressions 12.28, 12.30 and 12.32 respectively. It is again of practical importance that the power consumption, for a given load per unit overall area, will be a minimum when the ratio (s/l) has a value of approximately 0.4.

The load, flow and power factors for many other useful shapes of thrust bearing are available in the literature but some care is necessary in using them since certain constants are not always included in the definitions of these parameters, for example the constant $\pi/6$ of expression 12.35b is sometimes included in expression 12.30 rather than in $\bar{q}$. The use of load, flow and power factor offers a considerable amount of simplification and, more importantly, generalisation in even these complicated cases; an example of their application to the case of a single tilted thrust bearing is presented in the following section.

Finally since the general expressions for the load capacity, stiffness and flow rate given in expression 12.16 have been used in the discussion of compensation in section 12.7 it is of interest to relate these parameters to the load and flow factors which have been introduced here.

It will be recalled from expressions 12.16 that the load capacity of a long rectangular bearing is given by

$$W = p_s L(b + l)\overline{W}$$

Introducing the load factor $\bar{a}$ and overall area A

$$W = p_s A\bar{a}\overline{W}$$

or

$$\left(\frac{W}{p_s A}\right) = \bar{a}\overline{W}$$

The parameter $(W/p_s A)$ is an alternative form of dimensionless load and is often used in work on hydrostatic bearings. It may be interpreted as the effectiveness of a bearing because it is equal to the ratio of the load actually carried by a bearing to the load which would be carried if its overall area A

were subjected to the supply pressure p_s. Thus putting $W' = (W/p_s A)$ the relationship between the two dimensionless load parameters is found to be

$$W' = \bar{a}\overline{W} \qquad (12.36a)$$

and remembering from section 12.6.2 that $\overline{W} = \bar{p}$

$$W' = \bar{a}\bar{p} \qquad (12.36b)$$

Similar reasoning shows that the dimensionless stiffness defined in expression 12.16 is related to W' by

$$-\frac{\partial W'}{\partial \bar{h}} = \frac{-\partial \overline{W}/\partial \bar{h}}{(p_s A/h_o)} = \bar{a}\bar{S} \qquad (12.37a)$$

or

$$\frac{\partial W'}{\partial \bar{h}} = \bar{a}\frac{\partial \bar{p}}{\partial \bar{h}} \qquad (12.37b)$$

The dimensionless flow rate is related to the flow factor by

$$Q' = \frac{Q}{\left(\dfrac{h_o^3 p_s}{\eta}\right)} = \bar{q}\overline{Q} \qquad (12.38)$$

12.9 SLIDING EFFECTS IN THRUST BEARINGS

So far our attention has been limited to the operation of hydrostatic thrust bearings under static conditions, that is, where there is no relative motion of the two bearing surfaces. However, in practice, bearings are required to operate under dynamic conditions and we must now consider the effects introduced by relative motion of the bearing surfaces.

Any general relative motion of the bearing surfaces can be divided into two components, a normal velocity component and a sliding component and the effects of these may be considered separately. Normal velocities occur when a bearing is subjected to a time-dependent load and have the effect of modifying the pressure distribution in the bearing—this is termed squeeze action. The resulting changes in pressure give rise to a force which is proportional to the squeeze velocity and opposite in direction to the motion of the bearing surfaces. Hydrostatic bearings therefore possess inherent viscous damping. A thorough treatment of this feature of hydrostatic bearings is beyond the scope of this book, but in the majority of applications dynamic loads are of secondary importance and are often ignored. However, sliding effects cannot be so readily disposed of.

340

12.9.1 Friction in Hydrostatic Thrust Bearings

The simple circular thrust pad shown in figure 12.13a is very suitable for supporting a rotating member such as a shaft. When the shaft is stationary the lubricant flows radially outwards from the recess but rotation introduces a circumferential velocity. The circumferential velocity varies linearly across the lubricant film, as shown in figure 12.13a and this shearing action

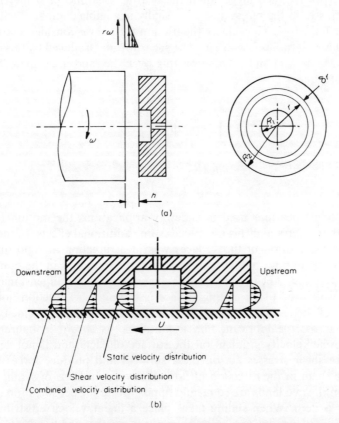

Figure 12.13   Effects of sliding velocity in thrust bearings

gives rise to shear stresses which oppose the rotation. For a newtonian lubricant the shear stress τ is given by[2]

$$\tau = \eta \frac{\partial u}{\partial y}$$

where η is the dynamic viscosity of the lubricant and $\partial u/\partial y$ is the velocity gradient. In this case, at any general radius r the velocity gradient does not vary in the circumferential direction and is equal to $\omega r/h$; the shear stress at radius r is given by

$$\tau = \eta \frac{\omega r}{h}$$

This shear stress acts on the surface of the rotating member and produces a torque which opposes the motion, that is, a friction torque. Normally the recess depth is much bigger than the bearing clearance and therefore the shear stresses in the recess area are usually negligible compared with those over the land area. To evaluate the friction torque we consider a circumferential element, radius r, width δr. The shear torque produced by this element is $(\tau \times 2\pi r \, \delta r \times r)$ and integrating this over the land area gives the total friction torque T

$$T = \int_{R_1}^{R_2} 2\pi \tau r^2 \, \mathrm{d}r = \int_{R_1}^{R_2} \frac{2\pi\eta\omega r^3 \, \mathrm{d}r}{h}$$

$$= \frac{\pi\eta\omega(R_2^4 - R_1^4)}{2h} \qquad (12.39)$$

For thrust bearings used in slideway applications the motion is linear rather than rotary and this introduces some additional effects. In the rotary example the motion of the surfaces was in a direction at right angles to the direction of flow, but with linear sliding the motion is in the same direction as the flow. Let us consider the example of a one-dimensional thrust pad in which the plane mating surface is sliding in a direction normal to the axis of the recess. In the static case with no sliding the velocity distribution across the lubricant film is parabolic as shown in figure 12.13b. Clearly, the velocity gradient on the surface of each land is not zero, and therefore shear stresses act on these surfaces and produce forces tending to drive them in the direction of flow. However, because of symmetry the forces on the two lands are equal and opposite and the net shear force on the bearing is zero. When sliding takes place a linear velocity distribution is superimposed on the parabolic distribution so that the resulting velocity distribution is as shown in figure 12.13b. It is important to note that the velocity distributions for the two lands are now different and a net friction force is now generated which opposes the motion of the sliding surface. Since the system is linear we can claculate the friction force by considering only that factor which brings it about, namely the linear velocity distribution resulting from the sliding action. The shear stress due to the sliding action is given by

$$\tau = \eta \frac{U}{h}$$

and since this is constant over the land area, the total friction force F is given by

$$F = \frac{A_l \eta U}{h} \tag{12.40}$$

where A_l is the total land area. The above equation again assumes that the recess depth is large compared with the film thickness so that shear forces over the recess area can be ignored.

Although the arguments have been developed for a one-dimensional bearing, equation 12.40 can be applied to bearings of any shape.

12.9.2 Effect of Sliding on Flow Rate

In addition to the friction effects, sliding also modifies the flow in the bearing clearance. Under stationary conditions the flows from the two sides of the one-dimensional bearing (figure 12.13b) are equal. When sliding takes place a shear flow is superimposed on the static flow. If the total static flow rate into the bearing per unit length is Q then the static flow rate over each land is $Q/2$. The shear flow rate is equal to the mean shear velocity multiplied by the flow area, and for the linear velocity distribution the shear flow per unit length of bearing is $Uh/2$. With sliding the shear flow on the upstream land of the bearing reduces the static flow so that the net outflow over this land is $Q/2 - Uh/2$. On the downstream side the shear flow increases the existing flow giving $Q/2 + Uh/2$. While this situation exists, the total flow into the bearing is

$$\left(\frac{Q}{2} + \frac{Uh}{2}\right) + \left(\frac{Q}{2} - \frac{Uh}{2}\right) = Q$$

that is, the inflow is the same as it was under static conditions. Since the shear flow is dependent on the sliding velocity it is possible to have the situation where $Uh/2 = Q/2$ and the flow from the upstream side of the bearing is then zero. If the sliding velocity is increased beyond this point, the outflow from the upstream side becomes negative that is, the flow is inwards towards the recess. This can only happen if there is sufficient lubricant present on the sliding surface approaching the bearing to provide the necessary inflow. If the approaching surface is dry or carries insufficient lubricant the above arguments no longer apply, so that the conditions on the upstream land are modified with the result that the lubricant film no longer extends to the edge of the bearing. This phenomenon is known as film recession, and the extent of the recession increases with the sliding velocity. In the presence of film recession the load capacity of the bearing is reduced and, more importantly, there is the danger of air being entrained into the recess with the risk that the bearing could collapse. This must of course be prevented and a simple method is to introduce a low pressure source of lubricant at

343

the upstream edge of the bearing to supply the required inflow and thereby prevent air entrainment.

Film recession is not peculiar to one-dimensional bearings, it can occur with any shape of bearing, but the criterion for the onset of film recession in other cases is more complicated than it is with the one-dimensional bearing.

12.9.3 Non-parallel Operation of Thrust Bearings

It is convenient and common practice to design hydrostatic thrust bearings to operate with a uniform film thickness, that is, with the two bearing surfaces parallel, since this simplifies both the manufacture and analysis of performance predictions. However, it is highly unlikely that parallel operation is ever achieved in practice. It is reasonably straightforward to produce thrust bearing surfaces which are flat, but with manufacturing tolerances and assembly errors it is very difficult to ensure that the bearing surfaces are parallel when assembled, and even if they are, structural deflections due to loading and thermal effects will occur in operation. Deflections of the bearing surfaces can lead to complex variations in bearing clearance and these in turn can produce significant changes in the bearing properties under both static and sliding conditions.

A simple but common example of non-uniform clearance is that which occurs due to tilt, that is, where both surfaces are plane but inclined relative to each other as shown in figure 12.14. In order to assess the possible effects of tilt on the performance of a bearing we shall consider a simple circular thrust pad and, initially, limit our attention to the static operation. When the bearing surfaces are parallel the pressure distribution is symmetrical about the bearing's centre but this is modified in the presence of tilt. Over the recess area, tilt has negligible effect and the pressure remains essentially

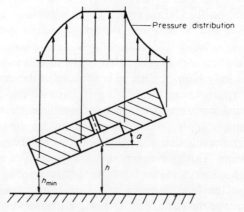

Figure 12.14 Effect of tilt on pressure distribution

344

uniform. Over the land where the flow takes place through a converging clearance the pressures are increased and where the clearance is diverging the pressures are decreased. The unsymmetrical pressure distribution produces a moment which tends to force the bearing into a parallel state, but the magnitude of this moment is so small that its effect can usually be neglected, however, the accompanying changes in load capacity and flow rate may be very significant.

It has been shown that the load capacity and flow rate of a parallel thrust bearing can be expressed in the form[5]

$$W = (p_r A)\bar{a}$$

$$Q = \left(\frac{p_r h^3}{\eta}\right)\bar{q}$$

and these equations can be extended to tilted operation. For parallel conditions $\bar{a}$ and $\bar{q}$ are functions of the bearing geometry only, which is described entirely by the radius ratio, in this case, expression 12.35

$$\bar{a}_o \doteq \frac{(1 - R^2)}{2 \log_e\left(\dfrac{1}{R}\right)}$$

$$\bar{q}_o = \frac{\pi}{6 \log_e\left(\dfrac{1}{R}\right)}$$

where the subscript o has been introduced to refer to parallel conditions. When tilt occurs $\bar{a}_o$ and $\bar{q}_o$ are modified and the effect can be described in terms of a tilt parameter

$$\bar{\alpha} = \frac{\alpha R_2}{h}$$

where α is the angle of tilt (in radius) and h is the effective clearance at the centre of the bearing as shown in figure 12.14. In the presence of tilt, $\bar{a}$ and $\bar{q}$ are both greater than the corresponding values for parallel surfaces, as shown in figures 12.15a and b, the changes being dependent on the radius ratio. The increase in $\bar{a}$ is greatest for small radius ratios and can be up to 15 per cent while changes in $\bar{q}$ can be as large as 100 per cent and increase with radius ratio.

The data of figures 12.15a and b allow a bearing to be designed specifically for tilted operation, but this is not normal practice at present and it is more useful to consider how the operation of a bearing designed for parallel conditions may be modified by the inevitable occurrence of tilt: these effects are demonstrated in figures 12.15c–f. It should be noted that we are now considering the effects of tilt on the bearing system as a whole, and these are dependent on the pressure ratio whereas the load and flow coefficients for

345

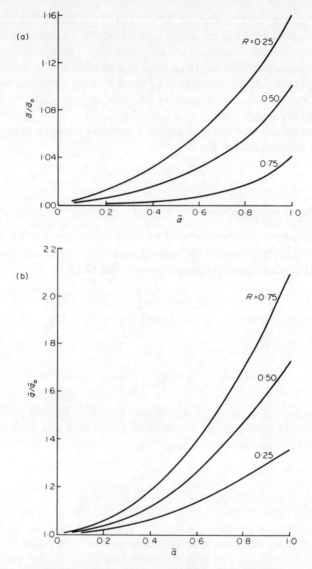

Figure 12.15   Characteristics of tilted pads

the pad are not. The curves given in figures 12.15c–f are for a capillary-compensated bearing supplied from a constant pressure source, but similar curves would be obtained with orifice compensation. Except for small radius ratios, the central clearance is always less than the parallel design clearance (figure 12.15c). These changes in clearance are generally small, but when they are combined with the effects of tilt they can produce dangerously small minimum clearances even for modest values of the tilt parameter as shown in figure 12.15d. Changes in the design flow rate are generally

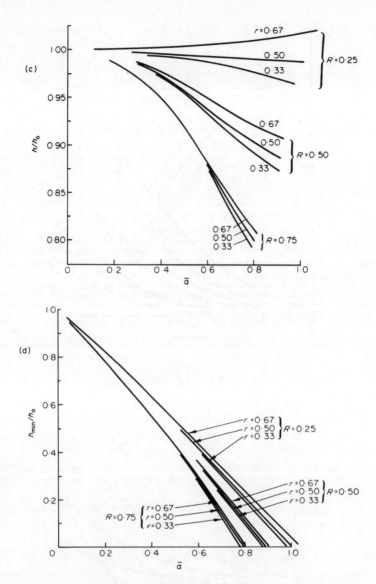

Figure 12.15

smaller than might be expected from the variation in flow parameter but the
increase in flow can still be as much as 20–30 per cent (figure 12.15e). The
static stiffness of a tilted bearing depends upon the combination of radius
ratio, pressure ratio and tilt parameter and may be greater or smaller than
the design stiffness.

The data of figure 12.15 are derived for a tilted circular thrust pad but they
are indicative of the effects which tilt will have on the static behaviour of
other shapes of bearing.

347

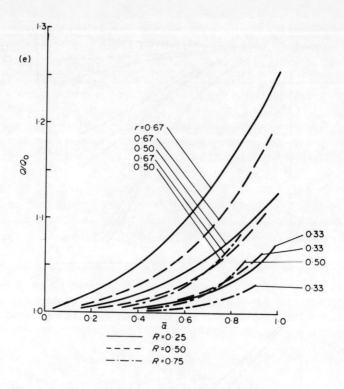

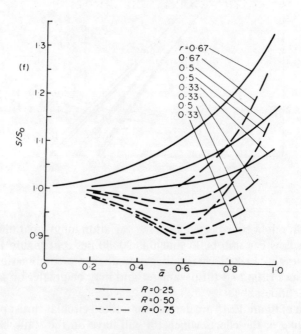

Figure 12.15

348

When sliding is introduced, the effects of tilt are more pronounced. The non-uniform clearance leads to the generation of hydrodynamic pressures which are superimposed on the hydrostatic pressures[6] and these can make a significant difference to the bearing. When the tilt and sliding produce a converging lubricant film, positive hydrodynamic pressures are generated which enhance the load carrying capacity. Therefore the same load can be supported with a smaller recess pressure, but this in turn leads to a greater pressure drop across the compensating restrictor with a consequent increase in flow, which is accompanied by an increased film thickness. When the tilt and sliding lead to a diverging lubricant film, the hydrodynamic action reduces the pressure over the lands and decreases the load capacity. In order to support the same load the recess pressure must be increased, the flow rate consequently decreases as does the film thickness.

It is possible for the pressure on the lands to fall locally to the vapour pressure of the lubricant (approximately absolute zero for oil) as a result of the hydrodynamic action, and this produces cavitation which further modifies the bearing behaviour. The characteristics of a tilted sliding bearing may also be further modified by boundary recession unless precautions are taken as already mentioned. A more detailed treatment of the behaviour of tilted sliding bearings is given in references 6 and 7.

12.10 HYDROSTATIC JOURNAL BEARINGS

An obvious and practically useful development of the types of hydrostatic thrust pads which have been described so far is their application to the support of rotating circular shafts. The principles of operation of hydrostatic journal bearings are the same as those of thrust bearings but their behaviour, analysis and design are complicated by two main factors

(i) Except for the unloaded, concentric position the fluid film thickness in the bearing is not uniform.
(ii) Shaft rotation results in comparatively high relative velocities being established between the shaft and bearing. This considerably alters the pressure distribution within the bearing.

Hydrostatic journal bearings offer two advantages over all other types of fluid film journal bearings: their load capacity and radial stiffness are high even at zero speed, and very small frictional torque exists during running. The fact that the frictional torque is zero at zero shaft speed is unique, and hydrostatic journal bearings are often chosen for particular applications by virtue of this fact alone.

The evolution of journal bearing configurations and the behaviour of some important types will be described in this section.

349

12.10.1   Evolution of Journal Bearings

It has been recognised that two main classes of hydrostatic journal bearings exist

(i) *multipad*[8] bearings which consist of one or more thrust pads with curved surfaces:

(ii) *multirecess*[9] bearings in which individual pads are almost indistinguishable from each other.

A single-pad journal bearing is shown in figure 12.16a. This consists of a simple thrust bearing shaped to accommodate the curved surface of the shaft, but obviously severe limitations exist on the direction in which radial loads may be applied. A typical multipad bearing is shown in figure 12.16b and, although a useful load capacity is provided for all radial directions, two

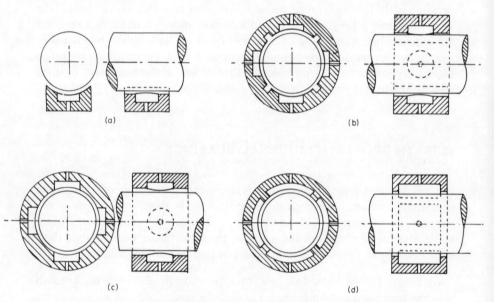

(a)   (b)

(c)   (d)

Figure 12.16   Evolution of hydrostatic journal bearings

undesirable features of this design are apparent. First, the drainage grooves between adjacent pads are regions of low pressure and therefore contribute nothing to the bearing's load capacity. Second, proper analysis, and therefore good design, is difficult since the flow patterns and pressure distribution in the lands of the bearing are difficult to predict even when the shaft is stationary and concentric.

An improvement may be made in the first of these features by developing the configuration shown in figure 12.16c. The absence of drainage grooves in this bearing results in a continuous land being formed between the recesses, so that a larger effective area of high pressure is formed and the flow

350

rate required to operate the bearing is reduced. When this type of bearing is taken to an extreme the shape of figure 12.16d is obtained with the following accrued advantages

(i) The relatively large pockets form regions of high pressure which provide good load capacity and stiffness.
(ii) Because the lands are narrow compared with their length simple one dimensional viscous flow theory, section 12.4, may be used to describe the flow of fluid within the bearing.
(iii) Following the assumption of (ii) above and the reasoning of section 12.9.2, it becomes possible to deal with the effects of shaft rotation on the flow of fluid within the bearing in a simple manner.

In the following sections a brief treatment of the types of journal bearings introduced above will be given.

12.10.2 Journal Bearings with Isolated Pads—Multipad Bearings

The method of analysing bearings of this general type, figure 12.16b, is introduced in this section by considering the bearing of figure 12.17 which consists of N equi-spaced, isolated pads of identical shape and overall area A. Assume that a load W is applied to the non-rotating shaft at some angle ϕ to the vertical line which has been drawn through the centreline of pad 1, and that the pads are numbered consecutively in the clockwise direction, the angular pitch of the pads thus being $2\pi/N$. It may further be assumed that the centre line of the shaft will move to some position within the bearing

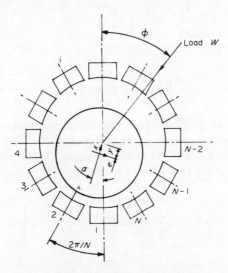

Figure 12.17 Hydrostatic journal bearing with N isolated pads

351

specified by the eccentricity ratio e and the attitude angle α. If the radial clearance between the shaft and bearing in the unloaded concentric position is c then the eccentricity ratio e in the loaded condition may be defined as the ratio of the distance between the centrelines of the shaft and journal to the clearance c.

Study of the geometry of the bearing shows that the decrease in the original film thickness c along the centreline of the ith pad is given by

$$\frac{\delta h_i}{c} = e \cos\left[\frac{2\pi}{N}(i - 1) - \alpha\right]$$

Assuming for the moment that the stiffness of each pad is given by expression 12.37a as

$$\frac{\partial W}{\partial h} = \left(\frac{p_s A}{c}\right)\left(\frac{\partial W'}{\partial h}\right)$$

$$= \left(\frac{p_s A}{c}\right)\bar{a}\left(\frac{\partial \bar{p}}{\partial \bar{h}}\right)$$

where $\bar{a}$ is the load factor for the individual pads and $\partial\bar{p}/\partial\bar{h}$ depends on the type of compensation adopted. It will be recalled that $\partial\bar{p}/\partial\bar{h}$ may be obtained in the general case from expressions 12.14b, 12.20b and 12.24b for capillary, orifice and constant flow compensation respectively.

Thus the additional radial load W_i produced by the ith pad because of the movement of the shaft is simply given by

$$W_i = \left(\frac{\partial W}{\partial h}\right)_i \times \delta h_i$$

or

$$W_i = \left(\frac{p_s A}{c}\right)\bar{a}\left(\frac{\partial \bar{p}}{\partial \bar{h}}\right) \times ce \cos\left[\frac{2\pi}{N}(i - 1) - \alpha\right] \qquad (12.41)$$

Referring to figure 12.17 it is seen that the components of this force in the direction of and at right angles to the line of centres are $W_i \cos\left[(2\pi/N)(i - 1) - \alpha\right]$ and $W_i \sin\left[(2\pi/N)(i - 1) - \alpha\right]$ respectively. The total loads produced by all N pads in these two directions, W_x and W_y respectively, follow immediately as

$$W_x = \sum_{i=1}^{N} W_i \cos\left[\frac{2\pi}{N}(i - 1) - \alpha\right]$$

and

$$W_y = \sum_{i=1}^{N} W_i \sin\left[\frac{2\pi}{N}(i - 1) - \alpha\right]$$

352

Substituting for W_i from equation 12.41 yields

$$W_x = \left(\frac{p_s A}{c}\right)\bar{a}ce\sum_{i=1}^{N}\left(\frac{\partial\bar{p}}{\partial\bar{h}}\right)\cos^2\left[\frac{2\pi}{N}(i-1)-\alpha\right] \qquad (12.42)$$

and

$$W_y = \left(\frac{p_s A}{c}\right)\bar{a}ce\sum_{i=1}^{N}\left(\frac{\partial\bar{p}}{\partial\bar{h}}\right)\cos\left[\frac{2\pi}{N}(i-1)-\alpha\right]\sin\left[\frac{2\pi}{N}(i-1)-\alpha\right]$$

It will be appreciated that as the value of $(\partial\bar{p}/\partial\bar{h})$ will vary from pad to pad, the quantities within the summation signs of expression 12.42 must be evaluated at length in the general case. However, it may be shown that if the eccentricity ratio e is less than about 0.2 or 0.3, then there is comparatively little variation in $(\partial\bar{p}/\partial\bar{h})$ which may therefore be given the value it assumes at zero eccentricity. Noting that

$$\sum_{i=1}^{N}\cos^2\left\{\frac{2\pi}{N}(i-1)-\alpha\right\} = \frac{N}{2}$$

and

$$\sum_{i=1}^{N}\cos\left\{\frac{2\pi}{N}(i-1)-\alpha\right\}\sin\left\{\frac{2\pi}{N}(i-1)-\alpha\right\} = 0$$

for all values of N and α, expressions 12.42 reduce to

$$W_x = \left(\frac{p_s A}{c}\right)\frac{\bar{a}N}{2}\left(\frac{\partial\bar{p}}{\partial\bar{h}}\right)ce$$

and $W_y = 0$.

Thus the shaft deflects along the direction of the applied load and it follows that the load angle ϕ must equal the attitude angle α. It follows also that the zero eccentricity stiffness of the bearing, S_o, may be deduced from the above expression as

$$S_o = \frac{W_x}{ce} = \left(\frac{p_s A}{c}\right)\frac{\bar{a}N}{2}\left(\frac{\partial\bar{p}}{\partial\bar{h}}\right)$$

and a nondimensional zero eccentricity stiffness, $\bar{S}_o$, may be defined as

$$\bar{S}_o = \frac{S_o}{(p_s A/c)} = \frac{\bar{a}N}{2}\left(\frac{\partial\bar{p}}{\partial\bar{h}}\right) \qquad (12.43)$$

The flow rate of fluid Q_t necessary to operate a multipad bearing is simply found by applying expression 12.38 to each of the N pads and forming a sum such as

$$Q_t = \left(\frac{c^3 p_s}{\eta}\right)\bar{q}\sum_{i=1}^{N}\bar{Q}_i$$

where $\bar{Q}_i$ must be evaluated for each pad from expressions 12.15, 12.21 and 12.25 for capillary, orifice and constant flow compensation respectively.

The benefit of the generalised description of hydrostatic bearings which has been developed in this chapter is now apparent since expression 12.43 enables the zero eccentricity stiffness of a hydrostatic journal bearing with any number of individual pads of any shape to be determined for any type of compensation. It is useful to note from expression 12.43 that the zero eccentricity stiffness of a journal bearing which consists of N equi-spaced and identical pads is simply $N/2$ times the radial stiffness of one of the pads.

The above treatment of multipad journal bearings is necessarily brief and many important features have been discounted. For example it is more difficult to evaluate the load factor $\bar{a}$ for a pad with curved surfaces than it is for a plane pad of the same projected shape, because the pressure in the fluid film acts radially and not perpendicularly to the plane shape of the pad. At high values of eccentricity ratio e the non-uniformity of the film thickness cannot be ignored and it is recognised that the load factor of each pad will be dependent on the eccentricity ratio and the attitude angle of the shaft. Finally, the effects of shaft rotation can also be important, so that in principle a solution of the Reynolds equation, rather than Laplace's equation, must be obtained for each pad before the behaviour of the complete journal bearing is determined.

In spite of these and other complications the simple analysis which has been presented in this section is adequate for many purposes.

12.10.3 Journal Bearings with Interacting Pads—Multirecess Bearings

The analysis and design of multirecess hydrostatic journal bearings, in which no drainage grooves exist between adjacent recesses, have been considered by many workers[10]. Although this classification includes bearings of the type shown in figure 12.16c only the more highly developed and practically useful shape with narrow lands shown in figure 12.16d will be considered here. A rigorous analysis involves the simultaneous determination, for every given shaft position, of as many recess pressures as there are recesses, and consequently the results of this form of inverse analysis are generally presented graphically. It is not possible to deal with this approach here, and therefore the results of a more restricted analysis of a bearing with four recesses will be presented and discussed[11].

In figure 12.18 the general form of the type of bearing under discussion is shown along with the dimensions required to specify its geometry. The directions of shaft rotation and applied load are indicated, and of importance are the definitions of the load angle ϕ and attitude angle α implied in this diagram. The position of the shaft is thus given by the eccentricity ratio e and the attitude angle α. An important geometric variable for this type of bearing is the aspect ratio m which is defined as $m = l_r l/Dl_c$. This parameter may be taken as a measure of the ratio of the circumferential flow rate be-

354

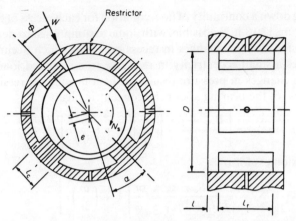

Figure 12.18   Hydrostatic journal bearing with four recesses

tween adjacent recesses to the axial flow rate from each recess, and clearly determines the degree of interaction between the recesses. Two other parameters are necessary to describe the bearing, that is, the pressure ratio r and the speed variable ω.

The pressure ratio r is defined as the ratio of the pressures in the recesses of the bearing to the supply pressure at zero eccentricity and is related to the bearing's dimensions and fluid properties through the following expressions

For orifice compensation $\dfrac{r}{(1-r)^{1/2}} = \dfrac{2\lambda}{\pi}$

For capillary compensation $\dfrac{r}{1-r} = \dfrac{2\lambda}{\pi}$

where

$$\lambda = \frac{3(2\pi)^{1/2}C_d d^2 \eta l}{(\rho p_s)^{1/2} c^3 D}$$

for orifices and

$$\lambda = \frac{3\pi d^4 l}{32 l_t c^4 D}$$

for capillaries. Here η and ρ are the dynamic viscosity and density respectively of the working fluid, c the radial clearance, d the diameter of the orifice or capillary, C_d the discharge coefficient of the orifice, l_t the length of each capillary restrictor, and p_s the supply pressure.

Similarly the speed variable ω is given in terms of the above dimensions, and parameters, for all types of compensation as

$$\omega = \frac{6\pi \eta N_s l l_r}{c^2 p_s}$$

where N_s is the speed of shaft rotation in revolutions per unit time.

355

By writing down a continuity of flow equation for each recess of the bearing shown in figure 12.18 it is possible, with some assumptions, to derive closed form expressions for the changes in recess pressures which occur when the shaft assumes a small eccentricity in response to an applied load. Integration of these changes in pressure results in the following expression for the radial stiffness of the bearing about the concentric position.

$$S_o = \left[\frac{D(l_r + l)p_s}{c}\right]\left[\frac{6r}{a}\right]\left[1 + \left(\frac{\omega}{3r}\right)^2\right]^{1/2}$$

or, in dimensionless form

$$\bar{S}_o = \frac{S_o}{\left\{\frac{D(l_r + l)p_s}{c}\right\}} = \frac{6r}{a}\left[1 + \left(\frac{\omega}{3r}\right)^2\right]^{1/2} \tag{12.44}$$

where

$$a = \frac{\pi}{4}\left(\frac{r}{1 - r}\right) + \frac{\pi}{2} + 2m$$

for orifice compensation, or

$$a = \frac{\pi}{2}\left(\frac{r}{1 - r}\right) + \frac{\pi}{2} + 2m$$

for capillary compensation, and

$$a = \frac{\pi}{2} + 2m$$

for constant flow compensation.

It will be seen that the dimensionless stiffness $\bar{S}_o$ is a function of aspect ratio m and pressure ratio r, and for a non-rotating shaft its variation takes the form shown in figure 12.19 for the common cases of orifice and capillary compensation.

It is clear from these curves and expression 12.44 that the stiffness of a hydrostatic journal bearing with four recesses increases with shaft speed, decreases with increases in aspect ratio, is a function of pressure ratio and is in general greater for orifice compensation than capillary compensation. It may also be shown that in the presence of shaft rotation the attitude angle α is not equal to the load angle ϕ but is related to it as follows

$$\alpha = \tan^{-1}\left(\frac{3r \tan \phi + \omega}{3r - \omega \tan \phi}\right) \tag{12.45}$$

It is also clear from figure 12.19 that the nearer the aspect ratio approaches the practically unobtainable value of zero the greater will be the stiffness of the bearing. For each aspect ratio a pressure ratio may be chosen which yields the greatest value of stiffness and the locus of this optimum value of

356

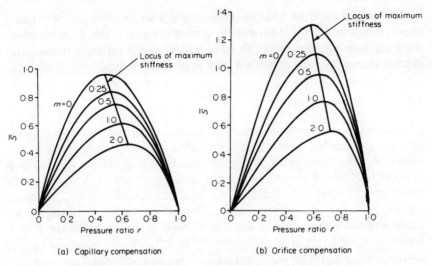

(a) Capillary compensation   (b) Orifice compensation

Figure 12.19   Variation of stiffness of a four-recess journal bearing at zero speed

stiffness and pressure ratio is shown for both cases of compensation in figure 12.19.

The flow rate necessary to operate a four-recess bearing, Q_t, depends only on the flow of fluid in the axial direction and for small values of eccentricity this may be expressed as

$$Q_t = \left(\frac{\pi D c^3}{6\eta l}\, p_s\right) r$$

The results which have been discussed above were obtained on the assumption that the eccentricity ratio is small, and it is clearly necessary to know the limitations of this restriction[12]. In figure 12.20 the behaviour of a bearing

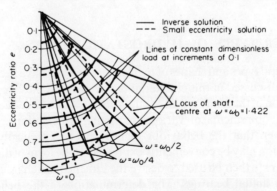

Figure 12.20   Behaviour of a hydrostatic journal bearing at various values of shaft speed

357

with an aspect ratio of 1.345 is shown for a load angle of $\phi = 45°$ and a pressure ratio of $r = 0.5$. Two sets of curves are drawn in this diagram which show the displacement of the shaft at several speeds of shaft rotation for different values of a dimensionless load $\overline{W}$, defined here as

$$\overline{W} = \frac{W}{D(l_r + l)p_s}$$

as given first by the small displacement analysis described above and second by the more rigorous inverse analysis which is in principle applicable at high values of eccentricity ratio. The shaft speeds considered are described in terms of fractions of an optimum speed variable, ω_o, which is related to the rate at which energy is dissipated in the bearing, but details of its significance are not of importance in this limited discussion. For convenience the attitude angle is measured from the line of loading and not from the centre-line of a circumferential land as before. The full lines represent the inverse solution and the broken lines the small eccentricity solution.

Up to an eccentricity ratio of about 0.3 very little discrepancy exists between the two solutions in either the prediction of eccentricity ratio or attitude angle. However, the small eccentricity solution predicts a constant stiffness and attitude angle for each shaft speed while gradual variations are revealed by the inverse solution. It is significant that the small eccentricity solution gives either an overestimate or an underestimate of the load capacity, depending on the shaft speed. Thus, neither the increasing stiffness at zero speed or the decreasing stiffness at the optimum speed variable are predicted. It is emphasised that the manner in which the load capacity changes with eccentricity depends on the pressure ratio, speed variable and load angle. However, the comparison between the two methods of solution presented here for one set of circumstances is generally true.

Many other features of the detailed operation of this type of bearing have been studied in the extensive literature on this topic which may be consulted if required.

12.11 OTHER TYPES OF HYDROSTATIC BEARINGS

Although several types and shapes of hydrostatic bearing have been described, analysed and discussed in this chapter, it is obvious that these represent only a small selection of the many useful configurations which have been developed to meet the requirements of various particular applications. It has been shown, however, that the behaviour of bearings of practically any shape and complication may be conveniently expressed in the form of dimensionless factors, which may then be used to derive the dimensional characteristics of a wide range of actual bearings. The determination of the behaviour of a hydrostatic bearing system must of course involve the characteristics of the control or compensating devices used, but again a formal presentation of the

358

cardinal features of three important methods of compensation have been presented in dimensionless form.

It must be remembered, however, that many features which are sometimes of importance in hydrostatic bearings have not been considered at all; these include dynamic response, cavitation, energy dissipation, active compensation, inherent compensation, and the effects of non-uniform film thickness in bearings of complicated shape.

REFERENCES

1. H. C. Rippel. Design of hydrostatic bearings—Pt I. *Machine Design*, (Aug. 1963), 108.
2. R. H. F. Pao. *Fluid Mechanics.* Wiley, New York, (1965).
3. D. D. Fuller. *Theory and Practice of Lubrication for Engineers.* (Ch. 3.) Wiley, New York, (1956).
4. A. M. Loeb and H. C. Rippel. Determination of optimum proportions for hydrostatic bearings. *Trans. Am. Soc. Lubric. Engrs*, **1** (2), (1958), 241.
5. R. B. Howarth. Effects of tilt on the performance of hydrostatic thrust pads. *Proc. Instn mech. Engrs*, **185**, (1970–71), 51/71.
6. R. B. Howarth and M. J. Newton. Investigation of the effects of tilt and sliding on the performance of hydrostatic thrust bearings. Instn mech. Engrs/Instn Prod. Engrs Conf. Externally Pressurized Bearings, London, (1971).
7. M. J. Newton and R. B. Howarth. Film recession in hydrostatic thrust bearings. *Proc. Instn mech. Engrs*, **187**, (1973) 57/73.
8. H. C. Rippel. Design of hydrostatic bearings—Pt 9. *Machine Design*, (Nov. 1963), 199.
9. H. C. Rippel. Design of hydrostatic bearings—Pt 10. *Machine Design*, (Dec. 1963), 158.
10. P. B. Davies. A general analysis of multirecess hydrostatic journal bearings. *Proc. Instn mech. Engrs*, **184**, Pt. 1, (1970).
11. P. B. Davies and R. Leonard. The dynamic behaviour of multirecess hydrostatic journal bearings. *Proc. Instn mech. Engrs*, **181**, Pt. 3L, (1969–70), 139.
12. P. B. Davies. Modes of failure in multirecess hydrostatic journal bearings. Proc. 10th Int. M.T.D.R. Conf. Pergamon Press, (1969), 425.

13

Selection of Tribological Solutions

13.1   INTRODUCTION

Before considering tribological solutions a designer must first consider the possibility of eliminating the tribological difficulty by using an alternative design. Designs with fewer moving parts not only eliminate some of the tribological headaches, but invariably provide a cheaper and more elegant solution to the overall problem. In chapter 1 this principle was illustrated by a rather graphic example, but many other less glamorous examples can be quoted. The use of elastometers and flexible linkages to provide a range of oscillatory motions without rubbing interfaces is an established practice, while the recently developed Wankel engine clearly illustrates the same basic principle.

Despite the designer's best endeavours, however, there are still many residual tribological contacts and this chapter indicates some of the guide lines in choosing the most appropriate tribological solution to any particular problem. In a sense this chapter demonstrates the relevance of much in the foregoing chapters, so that the reader may obtain a perspective view of the authors' intentions.

From the designer's standpoint he will be interested in achieving a tribological solution to the particular problem which satisfies the basic operational parameters of load capacity and speed within a specified environment. It is therefore necessary to consider the effects of load and speed within any

360

specified environment and to identify those physical conditions which place limits on the performance.

In practice the majority of tribological problems in industry will arise within an atmospheric environment. None the less special problems are created in systems operating in high vacuums, nuclear reactors and some chemical plants, because of the absence of self-generating oxide films. The tribology in these so called 'hostile environments' presents special problems which are not dealt with in this text, although the earlier chapters will have indicated why they arise and, more usefully, how they might be solved. In what follows we shall consider the effects of load and speed in defining the limits of performance of the various tribological solutions assuming a normal atmospheric environment.

13.2 LOAD AND SPEED

The full ranges of load and speed are too complex to be considered in detail since they have both magnitude and direction, each of which may be constant or vary in some prescribed pattern. There are four broad classes of loading, while the velocity has eight combinations due to the additional effect of combined rolling and sliding as occurs between mating gear teeth. The full pattern of load and speed options may be classified as shown in figure 13.1, which also shows some typical examples from the *thirty-two* possible combinations of load and speed patterns. When it is recalled that no attempt

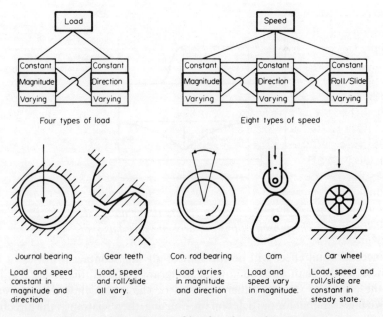

Figure 13.1   Types of load and speed variation

361

has been made to define the wide range of patterns of load and speed variations, we can see how these two variables alone can lead to an enormous
range of tribological situations.

In the following it will generally be assumed that the load and speed are
constant in every sense, that is, we shall concern ourselves with steady-state
arguments. None the less the effects of load and speed variations can be
inferred by extrapolation of the ideas presented, particularly when one
recalls the importance of the contact history of discrete particles. Consider
the journal bearing shown in figure 13.2 where the load is constant in magnitude but oscillates between the limits $\pm\alpha$ and produces the pressure distribution shown. The stress history of particle 1 on the shaft and particle 2 on the
housing will be as shown, and will clearly depend on the relative value of α.

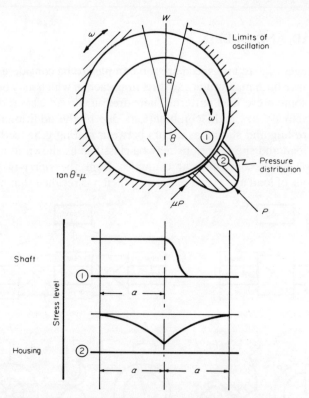

*Figure 13.2   Oscillating constant normal load; case
where oscillations are at same speed as shaft*

Since the total effect will be the sum of all such contact histories of the
particles of the shaft and the housing, diagrams such as this give a real feel
for the effects of varying either the load or the speed. Indeed in the case
shown it is seen that for particles on the housing, the variation in the direction
of the load reduces the contact stress and therefore the frictional and thermal

362

effects which would occur, as compared with the effects when the load direction does not vary. It is from this type of argument that we can see that the service conditions for a connecting rod bearing are somewhat less severe than they would be for a bearing in which the same load was applied in a constant direction.

Using our accumulated knowledge we shall now consider the types of limits on load and speed which are likely to arise in tribological contacts. In the following the emphasis is on the shapes of the load–speed limitations; the actual position of each boundary would be defined by such factors as the maximum allowable stress, temperature and wear rate. In all these diagrams the load and speed are plotted logarithmically to indicate the shapes of the actual limits and also to cover several orders of magnitude of the variables.

Strength Considerations (figure 13.3a)

Considering the geometry of the contact and the materials to be used we can define the maximum load which may be safely applied to the contact. Thus for a cylindrical or spherical contact the results from chapter 3 define our safe working load, while with other geometries even simple calculations will yield the required answer.

Inertia Characteristics (figure 13.3b)

Since the components of any tribological contact are in motion they are subjected to stress fields arising from their inertia. This defines a maximum speed above which failure could ensue. A simple example of this limitation arises with rotating shafts which can be made to burst due to the induced centrifugal stress at high rotational speeds. With more complex rolling bearings the effects of centrifugal and gyroscopic forces can also impose maximum speeds above which operation becomes unsatisfactory and failures may occur due to bursting of the cage and similar phenomena.

Frictional Instabilities (figure 13.3c)

At very low operating speeds some tribological devices will suffer from the type of instability discussed in chapter 7. For operation with such systems the designer must either provide a satisfactory low speed combination or restrict the system from such a low speed operation.

Heat Sources (figure 13.3d)

The heat release due to friction may be distributed between both the moving and stationary contacting bodies. This has been discussed in chapter 3 so that the thermal limit, as identified by a maximum allowable temperature for

363

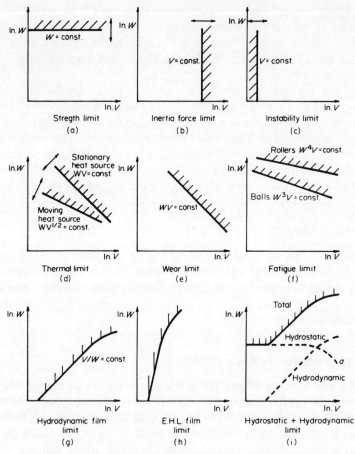

Figure 13.3 Load and speed limits on performance

the two components, would be as shown in figure 13.3d, where the load–speed relationships for the two cases are based on equations 3.46 and 3.47 in chapter 3.

Wear Limit (*figure 13.3e*)

Many wear mechanisms such as adhesive or abrasive wear have been seen in chapter 5 to be reasonably represented by the product of load and velocity. The limit to define a specified wear value is therefore as shown in figure 13.3e. This limit is often defined by the product of pressure and velocity, the P–V limit, since the load per unit area, that is, pressure, is a more useful design parameter than the load, which is only meaningful when associated with the area over which it operates.

364

Fatigue Limit (figure 13.3f)

Fatigue limits are particularly relevant where such mechanisms lead to failure, as in rolling contact bearings. In such bearings the fatigue will depend on both the stress level and the frequency of loading. Practical tests show that the relationship between load W and life L in revolutions is given by W^3L for ball bearings and W^4L for roller bearings. Since L is clearly related to speed V one can define the fatigue limit by $W^3V = $ Constant for ball-bearings as shown in figure 13.3f.

Hydrodynamic Films (figure 13.3g)

When load is carried by virtue of the pressure generated by hydrodynamic action, it has been seen in chapter 10 that the film thickness is some function of $\eta V/W$. In such cases the operational limit is clearly defined by the need for a continuous film to carry the applied load at the given speed using the appropriate lubricant. This immediately defines the limit of operation of such systems by the line V/W is a constant. At the higher sliding velocities the heat released due to viscous shearing will result in a reduction in the viscosity of the fluid. This will result in a departure from linearity of the bounding limit as shown in figure 13.3g.

Elastohydrodynamic Films (figure 13.3h)

From the results of chapter 11 it will be noted that this type of hydrodynamic film is less dependent on the load than the pure hydrodynamic film and this leads to a limit for film formation of the type shown in figure 13.3h.

Hydrostatic Films (figure 13.3i)

In chapter 12 it was shown that by using an external pumping system a load carrying fluid film could be created, the maximum load carrying capacity being entirely dependent on the external pressure available. This leads to a limiting characteristic of the type shown in figure 13.3i by the line *a*, the slight reduction at high speeds being again due to reductions in viscosity due to frictional heating. In such devices it is also clear that at higher speeds some additional hydrodynamic load carrying capacity may be generated so that the total load capacity will be the sum of both the hydrostatic and hydrodynamic effects as shown in figure 13.3i.

The foregoing has not attempted to provide numerical values for the various limits, since these clearly depend on such factors as the geometry, the materials and the properties of the lubricant. None the less, for any particular system the information in the preceding chapters of this book would allow such calculations to be carried out. One interesting application of these ideas is their use in determining the characteristics of various types of journal bearing.

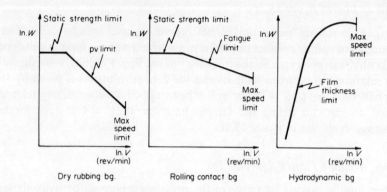

Dry rubbing bg.　　Rolling contact bg　　Hydrodynamic bg

Figure 13.4　Limits on three types of journal bearing

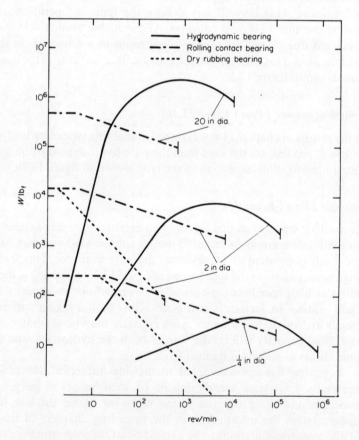

Figure 13.5　Comparison of three bearing types

13.3 THE SELECTION OF JOURNAL BEARINGS

We shall consider three widely used types of journal bearings, the hydrodynamic bearing, the ball bearing, and the dry rubbing bearing based on
PTFE compounds. Using the limits described in figure 13.3 it is readily seen
that the basic characteristics for each of these bearing types will have the
form shown in figure 13.4. If we now calculate each of these limits for a given
diameter of shaft we may plot the ensuing characteristic on a load–speed
graph as shown in figure 13.5, the load and speed ranges being chosen to
cover almost the complete range of engineering applications. From such a
plot we can clearly see the advantages in load carrying capacity of ball bearings over other types at speed of 1–2000 rev/min, which explains why they
are so often employed in this type of situation, for example, small electric
motors and the like. With large diameter shafts at the same order of speed,
the enhanced load capacity of hydrodynamic bearings is clearly demonstrated. This explains why such bearings are so often used in large steam
turbines, etc. In such cases it is also common practice to use a form of auxiliary
hydrostatic jacking to preclude metallic contact at the lower speeds during
start up and stopping of such equipment. Bearing selection charts of this
kind have been constructed for a wide range of shaft sizes for both journal
and thrust bearings and are published by the Institution of Mechanical
Engineers. (Design Data, Item 65007, 66023).

13.4 MATCHING OF TRIBOLOGICAL SOLUTIONS

Finally one should always try to match the characteristics of the chosen
tribological solution to the particular problem in hand. Thus with large
gearbox bearings the load–speed characteristics might be of the form shown
in figure 13.6, and it is clear that the characteristics of the hydrodynamic

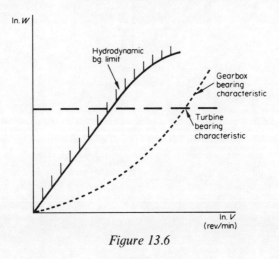

Figure 13.6

367

bearing are a very good match for this requirement. With steam turbines the load is substantially constant at all speeds, so that it would appear that a form of hydrostatic bearing might be the most appropriate choice in such situations.

13.5 CONCLUSIONS

The foregoing might be considered as the author's attempt to justify the content of the preceding chapters, and to show the essential unity of the subject of tribology. The most important result from the introduction of this word is that we are made to concentrate on the essential problem of carrying load, and to do this we should select the most appropriate tribological solution. These solutions arise from the studies of engineering and science in many disciplines as has been seen from the contents of this book, but this should not detract from the essential unity of the subject which is concerned with *the function to be performed*.

REFERENCES

1. M. J. Neale. *Tribology Handbook*. Butterworth, London, (1973).
2. M. J. Neale and A. B. Crease. Rubbing bearings for aircraft, a survey of applications, materials and needs. M.O.D. Report P.E.S. and T.M., July, 1972.
3. *Engineering Science Data*. Items 68018, 66023, 65007, 67033, Instn mech. Engrs.

Appendix A

Reynolds Equation

The equations of motion for a viscous fluid are basic to the study of lubricant films. These are normally referred to as the Navier–Stokes equations[1] which, for a newtonian fluid, take the form

$$\rho \frac{Du}{Dt} = X - \frac{\partial p}{\partial x} + \frac{\partial}{\partial x}\left[\eta\left(2\frac{\partial u}{\partial x} - \frac{2}{3}\Delta\right)\right]$$

$$+ \frac{\partial}{\partial y}\left[\eta\left(\frac{\partial u}{\partial y} + \frac{\partial v}{\partial x}\right)\right] + \frac{\partial}{\partial z}\left[\eta\left(\frac{\partial u}{\partial z} + \frac{\partial w}{\partial x}\right)\right]$$

$$\rho \frac{Dv}{Dt} = Y - \frac{\partial p}{\partial y} + \frac{\partial}{\partial y}\left[\eta\left(2\frac{\partial v}{\partial y} - \frac{2}{3}\Delta\right)\right]$$

$$+ \frac{\partial}{\partial x}\left[\eta\left(\frac{\partial u}{\partial y} + \frac{\partial v}{\partial x}\right)\right] + \frac{\partial}{\partial z}\left[\eta\left(\frac{\partial v}{\partial z} + \frac{\partial w}{\partial y}\right)\right]$$

$$\rho \frac{Dw}{Dt} = Z - \frac{\partial p}{\partial z} + \frac{\partial}{\partial z}\left[\eta\left(2\frac{\partial w}{\partial z} - \frac{2}{3}\Delta\right)\right]$$

$$+ \frac{\partial}{\partial x}\left[\eta\left(\frac{\partial u}{\partial z} + \frac{\partial w}{\partial x}\right)\right] + \frac{\partial}{\partial y}\left[\eta\left(\frac{\partial v}{\partial z} + \frac{\partial w}{\partial y}\right)\right]$$

(A.1)

where

$$\frac{D}{Dt} = \frac{\partial}{\partial t} + u\frac{\partial}{\partial x} + v\frac{\partial}{\partial y} + w\frac{\partial}{\partial z}$$

$$\Delta = \frac{\partial u}{\partial x} + \frac{\partial v}{\partial y} + \frac{\partial w}{\partial z}$$

X, Y, Z = components of body forces in coordinate directions
u, v, w = components of velocity in coordinate directions

369

There is no general solution to the Navier–Stokes equations but in most lubricated situations the conditions are such that the equations can be greatly simplified. The usual situation involves two surfaces moving relative to each other and separated by a film of lubricant as shown in figures A.1 and A.2.

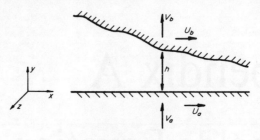

Figure A.1

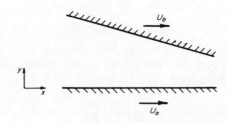

Figure A.2

In order to simplify the Navier–Stokes equations a number of justifiable assumptions must be made as listed below

(a) The flow is laminar; no vortices and no turbulence anywhere in the lubricant film.

(b) The body forces are zero or at least negligible compared with the viscous forces, that is, $X = Y = Z = 0$.

(c) Inertia forces are negligible compared with the viscous forces, that is,

$$\frac{Du}{Dt} = \frac{Dv}{Dt} = \frac{Dw}{Dt} = 0$$

(d) The clearance is small compared with the dimensions of the lubricant film in the x and z directions. This allows the curvature of the fluid film to be ignored and also rotational velocities may be replaced by translational velocities.

(e) At any location (x, z) the pressure, density and viscosity are constant across the lubricant film, that is

$$\frac{\partial p}{\partial y} = \frac{\partial \rho}{\partial y} = \frac{\partial \eta}{\partial y} = 0$$

370

(f) No slip at the bearing surfaces, that is, at the bearing surfaces the velocity of the lubricant is identical with the surface velocity.

(g) Compared with the two velocity gradients $\partial u/\partial y$ and $\partial w/\partial y$ all other velocity gradients are negligible. This is a justifiable assumption since u and w are usually much greater than v, and y is a much smaller dimension than x and z.

With these assumptions the Navier–Stokes equations reduce to

$$\frac{\partial^2 u}{\partial y^2} = \frac{1}{\eta}\frac{\partial p}{\partial x} \qquad (A.2a)$$

$$\frac{\partial^2 w}{\partial y^2} = \frac{1}{\eta}\frac{\partial p}{\partial z} \qquad (A.2b)$$

Integrating equation A.2a with respect to y and using the boundary conditions $u = U_a$ when $y = 0$ and $u = U_b$ when $y = h$ gives

$$u = \frac{1}{2\eta}\frac{\partial p}{\partial x}[y(y - h)] + \frac{(h - y)}{h}U_a + \frac{y}{h}U_b \qquad (A.3a)$$

Integrating equation A.2b with respect to y and using the boundary conditions $w = 0$ when $y = 0$ and $w = 0$ when $y = h$ gives

$$w = \frac{1}{2\eta}\frac{\partial p}{\partial z}[y(y - h)] \qquad (A.3b)$$

Equation A.3b is the velocity distribution for a fluid flowing under laminar conditions through a small clearance h as a result of the pressure gradient $\partial p/\partial z$. In fluid mechanics this type of pressure induced flow is referred to as Poiseuille flow. In equation A.3a the first term on the right-hand side represents the Poiseuille flow while the two remaining terms describe the velocity distribution which results from the motion of the bearing surfaces alone, that is with constant pressure. In fluid mechanics this type of flow is described as Couette flow.

The flow of lubricant must satisfy the continuity requirements expressed by the continuity equation which, for steady flow, takes the form

$$\frac{\partial(\rho u)}{\partial x} + \frac{\partial(\rho v)}{\partial y} + \frac{\partial(\rho w)}{\partial z} = 0$$

Integrating the continuity equation across the lubricant film

$$\int_0^h \frac{\partial(\rho u)}{\partial x}\,dy + \int_0^h \frac{\partial(\rho v)}{\partial y}\,dy + \int_0^h \frac{\partial(\rho w)}{\partial z}\,dy = 0 \qquad (A.4)$$

371

Substituting u and w from equation A.3 into equation A.4 and completing the integration yields

$$\frac{\partial}{\partial x}\left(\frac{\rho h^3}{\eta}\frac{\partial p}{\partial x}\right) + \frac{\partial}{\partial z}\left(\frac{\rho h^3}{\eta}\frac{\partial p}{\partial z}\right) = 6(U_a - U_b)\frac{\partial(\rho h)}{\partial x}$$

$$+ 6\rho h\frac{\partial(U_a + U_b)}{\partial x} + 12\rho(V_b - V_a)$$

If the bearing surfaces are inelastic in the direction of sliding U_a and U_b will be independent of x, that is

$$\frac{\partial(U_a + U_b)}{\partial x} = 0$$

and the above equation reduces to

$$\frac{\partial}{\partial x}\left(\frac{\rho h^3}{\eta}\frac{\partial p}{\partial x}\right) + \frac{\partial}{\partial z}\left(\frac{\rho h^3}{\eta}\frac{\partial p}{\partial z}\right) = 6(U_a - U_b)\frac{\partial(\rho h)}{\partial x} + 12\rho(V_b - V_a) \quad \text{(A.5)}$$

Equation A.5 is the general form of Reynolds equation which is fundamental to analysis of all lubrication problems.

In some situations it is possible to simplify Reynolds equation. If for example it is reasonable to assume that the lubricant is incompressible, that is, the density is constant, then equation A.5 reduces to

$$\frac{\partial}{\partial x}\left(\frac{h^3}{\eta}\frac{\partial p}{\partial x}\right) + \frac{\partial}{\partial z}\left(\frac{h^3}{\eta}\frac{\partial p}{\partial z}\right) = 6(U_a - U_b)\frac{\partial h}{\partial x} + 12(V_b - V_a) \quad \text{(A.6)}$$

If, in addition, it may be assumed that the viscosity of the lubricant does not change then equation A.6 can be further simplified to

$$\frac{\partial}{\partial x}\left(h^3\frac{\partial p}{\partial x}\right) + \frac{\partial}{\partial z}\left(h^3\frac{\partial p}{\partial z}\right) = 6\eta(U_a - U_b)\frac{\partial h}{\partial x} + 12\eta(V_b - V_a) \quad \text{(A.7)}$$

Some complex situations exist where it is difficult to ascribe the correct values to the components of velocity. The difficulty arises because there may exist velocity components in the y direction even though there is no normal motion of the bearing elements. (Consider, for example, the case of two rolling cylinders where, at all positions except the point of closest approach, there is a component of the surface velocity in the y direction.) We overcome this problem by rewriting the Reynolds equation so that these components are automatically included. If we consider a case such as that shown in figure A.3, we see that there is no normal motion of the two solids and the components of velocity are

$$\text{in the } x \text{ direction } U_a, \ U_b\cos\left(\text{arc tan}\frac{dh}{dx}\right) \simeq U_b$$

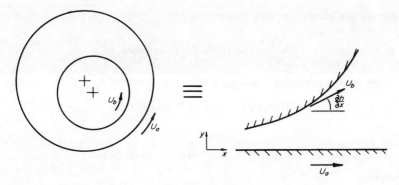

Figure A.3

in the y direction 0, $U_b \sin\left(\arctan \dfrac{dh}{dx}\right) \simeq U_b \dfrac{dh}{dx}$

The right-hand side of equation A.7 becomes

$$6\eta(U_a - U_b)\frac{\partial h}{\partial x} + 12\eta\left(U_b\frac{\partial h}{\partial x} - 0\right) = 6\eta(U_a + U_b)\frac{\partial h}{\partial x}$$

We may therefore rewrite the Reynolds equation for the case where there is no normal motion of the bodies and *the film thickness is invariant with time* (see chapter 10) as

$$\frac{\partial}{\partial x}\left(\frac{h^3}{\eta}\frac{\partial p}{\partial x}\right) + \frac{\partial}{\partial z}\left(\frac{h^3}{\eta}\frac{\partial p}{\partial z}\right) = 6(U_1 + U_2)\frac{\partial h}{\partial x}$$

where U_1 and U_2 are the actual surface velocities, not the components.

If the two solids are moving in the y direction, the effect of this can be added into the equation, thus

$$\frac{\partial}{\partial x}\left(\frac{h^3}{\eta}\frac{\partial p}{\partial x}\right) + \frac{\partial}{\partial z}\left(\frac{h^3}{\eta}\frac{\partial p}{\partial z}\right) = 6(U_1 + U_2)\frac{\partial h}{\partial x} + 12(V_2 - V_1) \qquad \text{(A.8)}$$

where V_2 and V_1 are the respective velocities in the y direction of the two solids.

Sliding Motion

If we neglect pressure variation in the z direction, assume constant viscosity and have pure sliding ($V_1 = V_2 = 0$), the Reynolds equation becomes

$$\frac{d}{dx}\left(h^3 \frac{dp}{dx}\right) = 6\eta(U_1 + U_2)\frac{\partial h}{\partial x}$$

373

Integrating with respect to x gives

$$h^3 \frac{\mathrm{d}p}{\mathrm{d}x} = 6\eta(U_1 + U_2)h + \text{constant}$$

If $\mathrm{d}p/\mathrm{d}x = 0$ at some film thickness h^*

$$0 = 6\eta(U_1 + U_2)h^* + \text{constant}$$

or

$$\text{constant} = -6\eta(U_1 + U_2)h^*$$

Therefore

$$\frac{\mathrm{d}p}{\mathrm{d}x} = 6\eta(U_1 + U_2)\frac{(h - h^*)}{h^3} \qquad (A.9)$$

Normal Motion

$$U_1 = U_2 = 0$$

$$\frac{\mathrm{d}}{\mathrm{d}x}\left(h^3 \frac{\mathrm{d}p}{\mathrm{d}x}\right) = 12\eta(V_2 - V_1)$$

and integrating with respect to x

$$h^3 \frac{\mathrm{d}p}{\mathrm{d}x} = 12\eta(V_2 - V_1)x + \text{constant}$$

If $x = x^*$ where the pressure is a maximum that is, where $\mathrm{d}p/\mathrm{d}x = 0$, then

$$\text{constant} = -12\eta(V_2 - V_1)x^*$$

and

$$\frac{\mathrm{d}p}{\mathrm{d}x} = 12\eta(V_2 - V_1)\frac{(x - x^*)}{h^3} \qquad (A.10)$$

REFERENCES

1. O. Pinkus and B. Sternlicht. *Theory of Hydrodynamic Lubrication*, McGraw-Hill, New York, (1961).

Appendix B
Some Typical Problems

Example 1

A sample of a surface profile consists of 100 triangular asperities having a constant flank angle α and maximum peak valley heights of 1, 2, 3, ..., etc. microns. Assuming that the valleys have a point distribution, that is, all valleys lie at the same level, find the position of the centre-line and calculate the C.L.A. value of the profile. Plot the all-ordinate distribution of this profile and hence determine the R.M.S. and C.L.A. values. Does the order in which the peaks occur affect your answer?

Solution

Let the centre-line be at distance h above the level of the valleys. By definition of the centre-line h will be given by

$$h = \frac{\text{total area enclosed by profile}}{\text{profile length}}$$

or

$$h = \frac{\sum\limits_{1}^{100} z^2 \tan \alpha}{2 \sum\limits_{1}^{100} z \tan \alpha}$$

$$\simeq \frac{\int\limits_{0}^{100} z^2 \tan \alpha \, dz}{2 \int\limits_{0}^{100} z \tan \alpha \, dz} \simeq \frac{100}{3} \, \mu m$$

375

$$\text{C.L.A.} = \frac{2\sum\limits_{h}^{100}(z-h)^2\tan\alpha}{2\sum\limits_{1}^{100}z\tan\alpha},$$

$$\simeq \frac{2\int\limits_{h}^{100}(z-h)^2\,dz}{2\int\limits_{0}^{100}z\,dz} = \frac{16(100)^3}{81(100)^2} = 19.75\ \mu m$$

For this surface the all-ordinate distribution curve is basically a right-angled triangle. The centre-line of the distribution is at height $100/3$ μm. This is converted to a probability curve by making the area unity, that is, by using a scaling factor $1/100$. The first moment m_1 of one-half of the probability distribution curve about the centreline is

$$2\int\limits_{0}^{2/3}z\psi(z)\,dz$$

that is,

$$\text{C.L.A.} = \tfrac{16}{81} \times 100 = 19.75\ \mu m$$

The second moment m_2 about the centre-line is

$$m_2 = \int\limits_{-1/3}^{2/3}[z^2\psi(z)\,dz]^{1/2}$$

Therefore

$$\text{R.M.S.} = 23.6\ \mu m$$

Example 2

Consider the profiles of surfaces with (i) rectangular, and (ii) triangular height distributions. Derive the standardised distribution functions for both cases, and show that C.L.A. $= (\sqrt{3}/2)$ R.M.S. for the rectangular distribution and C.L.A. $= (\sqrt{6}/3)$ R.M.S. for the triangular distribution. Also show that for a gaussian distribution C.L.A. $= (2/\pi)^{1/2}$ R.M.S.

Solution

The distribution functions are given by

(i) $\psi(z) = \dfrac{1}{2(3)^{1/2}\sigma}$

(ii) $\psi(z) = \dfrac{1}{(6)^{1/2}\sigma} - \dfrac{z}{6\sigma^2} \qquad 0 < z < (6)^{1/2}\sigma$

$ = \dfrac{1}{(6)^{1/2}\sigma} + \dfrac{z}{6\sigma^2} \qquad -(6)^{1/2}\sigma < z < 0$

376

For the rectangular distribution

$$\text{C.L.A.} = 2 \int_{0}^{(3)^{1/2}\sigma} z\, \frac{1}{2(3)^{1/2}\sigma} = \frac{(3)^{1/2}}{2}\sigma$$

$$\text{R.M.S.} = \int_{-(3)^{1/2}\sigma}^{(3)^{1/2}\sigma} \left[z^2 \left(\frac{1}{2(3)^{1/2}\sigma} \right) \right]^{1/2} = \sigma$$

For the triangular distribution

$$\text{C.L.A.} = 2 \int_{0}^{(6)^{1/2}\sigma} z \left(\frac{1}{(6)^{1/2}\sigma} - \frac{z}{6\sigma^2} \right) dz = \frac{\sqrt{6}}{3}\sigma$$

$$\text{R.M.S.} = \int_{0}^{(6)^{1/2}\sigma} \left[z^2 \left(\frac{1}{(6)^{1/2}\sigma} - \frac{z}{6\sigma^2} \right) dz + \int_{-(6)^{1/2}\sigma}^{0} z^2 \left(\frac{1}{(6)^{1/2}\sigma} + \frac{z}{6\sigma^2} \, dz \right) \right]^{1/2} = \sigma$$

For the gaussian distribution

$$\psi(z) = \frac{1}{\sigma(2\pi)^{1/2}} e^{-z^2/2\sigma^2}$$

$$\text{C.L.A.} = 2 \int_{0}^{\infty} z \times \frac{1}{\sigma(2\pi)^{1/2}} e^{-z^2/2\sigma^2} \, dz = \left(\frac{2}{\pi} \right)^{1/2} \sigma$$

$$\text{R.M.S.} = \left[\frac{1}{\sigma(2\pi)^{1/2}} \int_{-\infty}^{\infty} z^2 e^{-z^2/2\sigma^2} \, dz \right]^{1/2} = \sigma$$

Therefore

$$\text{C.L.A.} = \left(\frac{2}{\pi} \right)^{1/2} \text{R.M.S.}$$

Example 3

Discuss briefly the advantages and disadvantages of profilometry. The bearing area of a surface represents the ratio of the real area of contact to the nominal area; show that for surfaces with randomly distributed asperities this is given by the line density l/L, where L is the total length of the surface profile and l is that part of L which lies in regions of real contact.

Solution

Consider the surface to consist of closely placed parallel profiles dividing the whole area into n differential strips in both the x and y directions.

377

If the line density of the ith strip in the x direction is denoted by X_i, then for infinitesimally narrow strips of areas A_i the mean line density of the whole surface in the x direction will be given by

$$\bar{X} = \frac{\sum\limits_{i=0}^{n} X_i A_i}{\sum\limits_{i=0}^{n} A_i}$$

This is clearly equal to the ratio of the real contact area a to the nominal area A, that is, $\bar{X} = a/A$. Similarly, the mean line density in the y direction will be given by

$$\bar{Y} = \frac{\sum\limits_{i=0}^{n} Y_i A_i}{\sum\limits_{i=0}^{n} A_i} = \frac{a}{A}$$

For surfaces with randomly distributed asperities, the line density of contact of any profile l/L will be equivalent to the mean line density of contact, hence

$$\bar{X} = \bar{Y} = \frac{l}{L} = \frac{a}{A}$$

Example 4

The contact of a rough surface and a smooth plane produces circular contact spots of radius a. If a particular asperity carries a load P uniformly distributed over the contact spot, find the vertical displacement at any point inside the contact area and show that the maximum displacement is given by

$$w = \frac{2(1 - v^2)P}{\pi a E}$$

Solution

Referring to equation 3.13, the displacement of a point D due to the elemental load acting on area dA is given by (figure B.1)

$$w = \frac{(1 - v^2)P}{\pi^2 a^2 E} \iint ds \, d\phi$$

The length of the chord AB is $2a \cos \theta$ and ϕ varies from 0 to $\pi/2$ so that

$$w = \frac{4(1 - v^2)P}{\pi^2 a^2 E} \int\limits_0^{\pi/2} a \cos \theta \, d\phi$$

378

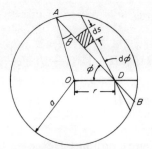

Figure B.1

since $a \sin \theta = r \sin \phi$ we have

$$w = \frac{4(1 - v^2)P}{\pi^2 aE} \int_0^{\pi/2} \left(1 - \frac{r^2}{a^2} \sin \phi \right)^{1/2} d\phi$$

The maximum displacement occurs at the centre of the contact area, that is at $r = 0$, thus

$$w_{\max} = \frac{2(1 - v^2)P}{\pi aE}$$

Example 5

A surface contact model consists of a smooth plane in contact with a rough plane covered with spherical asperities of the same radius β and having a uniform distribution of peak heights. If the asperity area density is η derive expressions for the real area of contact A and the total load P as functions of the normal approach δ for cases of both plastic and elastic modes of asperity deformation. Show that $A \propto P$ for the plastic case and $A \propto P^{4/5}$ for the elastic case.

Solution

The uniform distribution of asperity peaks is given by

$$\psi(z) = \frac{1}{2(3)^{1/2}\sigma}$$

For a plastic deformation mode

$$A = 2\pi\beta\eta a \int_d^{(3)^{1/2}\sigma} (z - d) \frac{1}{2(3)^{1/2}\sigma} dz$$

$$= \frac{\pi\beta\eta a}{2(3)^{1/2}\sigma} \delta^2$$

379

where a is the nominal area.

$$P = 2\pi\beta\eta aH \int\limits_{d}^{(3)^{1/2}\sigma} (z - d)\frac{1}{2(3)^{1/2}\sigma}\,dz$$

$$= \frac{\pi\beta\eta aH}{2(3)^{1/2}\sigma}\delta^2$$

where H is the flow pressure (constant). Hence

$$A = \frac{1}{H}P \quad \text{or} \quad A \propto P$$

For an elastic deformation mode

$$A = \pi\eta\beta a \int\limits_{d}^{(3)^{1/2}\sigma} (z - d)\frac{1}{2(3)^{1/2}\sigma}\,dz$$

$$= \frac{\pi\eta\beta a}{4(3)^{1/2}\sigma}((3)^{1/2}\sigma - d)^2 = \frac{\pi\eta\beta a}{4(3)^{1/2}\sigma}\delta^2$$

and

$$P = \frac{4}{3}\eta\beta^{1/2}aE' \int\limits_{d}^{(3)^{1/2}\sigma} (z - d)^{3/2}\frac{1}{2(3)^{1/2}\sigma}\,dz$$

$$= \frac{4\eta\beta^{1/2}aE'}{15(3)^{1/2}\sigma}((3)^{1/2}\sigma - d)^{5/2} = \frac{4\eta\beta^{1/2}aE'}{15(3)^{1/2}\sigma}\delta^{5/2}$$

or

$$A = \text{constant} \times P^{4/5}$$
$$A \propto P^{4/5}$$

Example 6

Two contacting discs of the same radius R roll under a normal load with angular velocities ω_1 and ω_2 where $\omega_2 > \omega_1$. One of the discs is assumed to have a smooth surface while the surface of the other disc consists of spherical asperities having equal heights and the same radius β and assumed to deform plastically under a constant flow pressure H. If μ is the coefficient of friction and J is the mechanical equivalent of heat, find the surface temperatures produced at a single contact spot when the load produces a normal approach δ between the two discs.

380

Solution

The contact spot may be considered as a stationary heat source with respect to disc 1 and a moving heat source, having a sliding velocity $(\omega_2 - \omega_1)R$, with respect to disc 2.

The total rate of heat Q generated at the contact spot is given by

$$Q = \frac{\mu P g(\omega_2 - \omega_1)R}{J}$$

Now, if λ is that part of Q supplied to disc 1, then the heat supplied to disc 2 will be $(1 - \lambda)Q$.

Using equations 3.40 and 3.42 we have

$$\theta_1 = \frac{\lambda Q}{4a\alpha_1}$$

and

$$\theta_2 = \frac{0.318(1 - \lambda)Q}{a^{3/2}(\alpha \rho c V)^{1/2}}$$

For a plastically deforming asperity the area of the contact spot A is given by

$$A = 2\pi\beta\delta = \pi a^2$$

thus, the radius of the contact spot is

$$a = (2\beta\delta)^{1/2}$$

and the load supported by the contact spot is

$$P = 2\pi\beta\delta H$$

The surface temperatures can now be obtained as

$$\theta_1 = \frac{2\lambda\mu\pi\beta\delta H \, g(\omega_2 - \omega_1)R}{4(2\beta\delta)^{1/2}J\alpha_1} = \frac{\pi\lambda\mu H \, g(2\beta\delta)^{1/2}}{4\alpha_1 J}(\omega_2 - \omega_1)R$$

and

$$\theta_2 = \frac{0.318(1 - \lambda)\mu(2\pi\beta\delta H)g(\omega_2 - \omega_1)R}{(2\beta\delta)^{3/4}J[\alpha_2\rho_2 c_2(\omega_2 - \omega_1)R]^{1/2}}$$

$$= \frac{0.318\pi(1 - \lambda)(2\beta\delta)^{1/4}\mu H \, g}{(\alpha_2\rho_2 c_2)^{1/2}J}[(\omega_2 - \omega_1)R]^{1/2}$$

Example 7

A solid brass cylinder of 10 mm diameter with a conical end supports a load of 1 kg and rubs against a smooth steel disc of 100 mm diameter, rotating in air at 1 revolution per second. Describe qualitatively the expected variation of friction and wear with time of rubbing.

Solution Hint

Initially the pressure over the contact is greater than $H/3$. See sections 4.1.3, 4.4.1, and 5.2.1a.

Example 8

Discuss the relative merits of the Edwards and Halling, and the Bowden and Tabor friction theories.

Use the Bowden and Tabor theory to plot the friction coefficient μ against c (where c is the ratio of the critical shear stresses for surface film and substrate) for different junction angles. Compare the results with those plotted in figure 4.12.

Solution Hint

The Bowden and Tabor theory states that the friction force is the sum of an adhesion term and a ploughing term. The adhesion term is given in section 4.5.1 and the ploughing term in section 4.7.

Example 9

Show from first principles that the adhesive wear rate is proportional to load and inversely proportional to the hardness of the softer of the two rubbing materials.

Discuss the effect of surface films on the wear rate of metals.

Solution Hint

See section 5.2.1a and 5.2.1b.

Example 10

Explain why 'wear rate' is not a meaningful parameter in describing the useful life of rolling element bearings.

Solution Hint

See section 5.2.3a.

Example 11

Derive a relationship between the coefficient of friction and the wear rate for two body abrasion.

Solution Hint

See sections 4.7 and 5.2.2a.

Example 12

In a given rubbing situation it is found that the maximum allowable shear stress for zero wear in 2000 passes is τ. Using the IBM model for zero wear obtain a graphical relationship between the number of passes and the maximum allowable shear stress.

Solution Hint

See section 5.2.8a.

Example 13

A plastic journal bearing 20 mm in diameter and 100 mm long carries a radial load of 150 N. The wear rate of the bearing reaches the maximum permissible level at a rotational speed of 500 rev/min.

(a) What is the maximum radial load which can be carried at 750 rev/min.?
(b) How should the dimensions of the bearing be changed to enable it to carry the original load at 800 rev/min.?

Solution Hint

See section 6.5.3.
(a) 100 N.
(b) The length should be increased to 160 mm.

Example 14

A machine drive system is represented by figure 7.4, with values $M = 300$ kg and $k = 15 \times 10^4$ Nm^{-1}. The slideways coefficient of friction μ is measured at various constant speeds; and the results show that, approximately, μ varies linearly from 0.13 at 0.002 ms^{-1} to 0.12 at 0.004 ms^{-1}. Dynamic data for the friction variation is not available and the design is to be based on the constant speed data.

Derive the equation of motion for the driven mass, at a drive speed $v = 0.003$ ms^{-1}, and calculate the minimum value of the dashpot coefficient in Nm^{-1}s if the overall damping ratio for the system, relevant to small transient oscillations, is to be at least 0.2.

Solution

The gradient of friction–velocity curve $= -14\,700$ Nm^{-1}s. The equation of motion, measuring X from the initial position of the mass with the spring

383

undeflected, is

$$300 \frac{d^2 X}{dt^2} = -f\left(\frac{dX}{dt} - 0.003\right) - 15 \times 10^4 (X - 0.003t) - \left(412 - 14\,700 \frac{dX}{dt}\right)$$

$$\zeta = \frac{f - 14\,700}{2(300 \times 15 \times 10^4)^{1/2}}$$

$$f = 17\,400 \text{ Nm}^{-1}\text{s}$$

Example 15

The system of example 14 is redesigned and the effective stiffness is increased to a value $k = 12 \times 10^6$ Nm^{-1}. It is decided to obtain more realistic data for the friction resistance offered by the slideways and dynamic tests are carried out at various speeds and frequencies of oscillation. During oscillation it is observed that, approximately, the friction resistance varies linearly with speed and a sample of the results for the negative gradient, λ, of the friction velocity curve is as follows

(i) $v = 0.002$ ms^{-1}, 20 Hz, $\lambda = 3000$ Nm^{-1}s
(ii) $v = 0.002$ ms^{-1}, 40 Hz, $\lambda = 2000$ Nm^{-1}s
(iii) $v = 0.004$ ms^{-1}, 20 Hz, $\lambda = 2000$ Nm^{-1}s
(iv) $v = 0.004$ ms^{-1}, 40 Hz, $\lambda = 1000$ Nm^{-1}s

Using linear interpolation between the values of the sample, calculate the approximate value of the dashpot coefficient f to give the overall damping ratio 0.2, in regard to small amplitude transient oscillations occurring at a speed of 0.003 ms^{-1}.

Solution

The undamped natural frequency for the redesigned system is 32 Hz. Assuming that oscillations occur at this frequency

(i) at $v = 0.002$ ms^{-1} $\lambda = 2400$ Nm^{-1}s
(ii) at $v = 0.004$ ms^{-1} $\lambda = 1400$ Nm^{-1}s

Interpolating for

$$v = 0.003 \text{ ms}^{-1} \qquad \lambda = 1900 \text{ Nm}^{-1}\text{s}$$

$$\zeta = \frac{f - 1900}{2(300 \times 12 \times 10^6)^{1/2}}$$

$$f = 25\,900 \text{ Nm}^{-1}\text{s}$$

384

Example 16

A drive system similar to that shown in figure 7.4 with friction characteristics as in figure 7.5a has the following parameter values

$$F_s = 50 \text{ N}$$
$$F_c = 30 \text{ N}$$
$$M = 20 \text{ kg}$$
$$k = 5 \times 10^4 \text{ Nm}^{-1}$$
$$\lambda = 0$$
$$f = 400 \text{ Nm}^{-1}\text{s}$$

The static friction, F_s, is assumed to be independent of the time of sticking.

(a) Determine whether or not stick–slip occurs at a speed $v = 4 \times 10^{-3}$ ms^{-1}.

(b) Estimate the minimum speed above which stick–slip does not occur.

Solution

(a) For the parameter values and at a speed of 4×10^{-3} ms^{-1}, the values of ζ, P, γ (see section 7.5) are

$$\zeta = \frac{f - \lambda}{2(Mk)^{1/2}} = 0.2 \qquad \omega_n = \frac{k}{M} = 50 \text{ rad s}^{-1}$$

$$\gamma = \sqrt{1 - \zeta^2} = 0.98 \qquad P = \frac{F_s - F_c}{Mv\omega_n} = 5$$

From equation 7.35, $\omega_n \gamma T_1 = 3\pi/2$. Inserting these values in equation 7.33 shows that the slip velocity is negative at time T_1, proving that sticking will occur, and hence stick–slip oscillations will ensue.

(b) Trial values for v, using the method of part a show that the approximate critical velocity is 7.3×10^{-3} ms^{-1}.

Example 17

Two cylinders of equal weight W and axial length l are placed at the top of a plane inclined to the horizontal at an angle θ. It may be assumed that the plane is rigid and that the two cylinders are defined by diameters D_1 and D_2, elastic properties $E_1 v_1$ and $E_2 v_2$ and hysteresis loss coefficients ε_1 and ε_2. Determine the ratio of the distances rolled by the two cylinders after any time interval if both start from rest.

If the two cylinders in the above case have solid rubber tyres, one with $\varepsilon = 0.10$, the other with $\varepsilon = 0.05$, but E and v are the same for both rubbers, what would be the ratio of their radii if they travelled the same distance in the same time?

385

Solution

Accelerating torques on cylinders are

$$(W \sin \theta - F_1)R_1 = I_1 \ddot{\phi}_1$$

where $I_1 = (3/2)WR_1^2$

$$(W \sin \theta - F_2)R_2 = I_2 \ddot{\phi}_2$$

where $I_2 = (3/2)WR_2^2$ also

$$a_1 = \frac{4WR_1}{\pi l} \times \frac{(1 - v_1^2)^{1/2}}{E_1}$$

$$a_2 = \frac{4WR_2}{\pi l} \times \frac{(1 - v_2^2)^{1/2}}{E_1}$$

From equation 8.1

$$F_1 = \frac{2}{3} \frac{Wa_1 \varepsilon_1}{\pi R_1}$$

$$F_2 = \frac{2}{3} \frac{Wa_2 \varepsilon_2}{\pi R_2}$$

Distance rolled from rest is $\phi R = \ddot{\phi} Rt$ after t secs. Thus ratio of distance rolled by two cylinders is

$$\frac{S_1}{S_2} = \frac{(W \sin \theta - F_1)\dfrac{R_1 R_1 t}{I_1}}{(W \sin \theta - F_2)\dfrac{R_2 R_2 t}{I_2}}$$

$$= \frac{\sin \theta - \dfrac{2}{3}\left(\dfrac{\varepsilon_1}{\pi R_1}\right)a_1}{\sin \theta - \dfrac{2}{3}\left(\dfrac{\varepsilon_2}{\pi R_2}\right)a_2}$$

where a_1 and a_2 are defined above.
If $S_1 = S_2$ and $\varepsilon_1 a_1/R_1 = \varepsilon_2 a_2/R_2$, substituting gives

$$\frac{R_1}{R_2} = 4$$

Example 18

A sphere of diameter D and density ρ is made from a material of modulus E, Poisson's ratio v and hysteresis loss coefficient ε. The sphere rolls along a horizontal plane of the same material with a linear velocity v. If it is assumed

386

that the heat released due to rolling friction is equally distributed between the sphere and the plane what would be the temperature rise of the plane? Assume that the thermal properties of the material are conductivity α, mechanical heat equivalent J, and specific heat c. What can you say about the underlying assumption above as time elapses?

Solution

In equation 8.2

$$F = \frac{3}{16} W \frac{\varepsilon a}{R}$$

where

$$a = \frac{3}{2} WR \left(\frac{1 - v^2}{E} \right)^{1/3}$$

Rate of working is Fv, so for the plane

$$Q = \frac{1}{2} \frac{Fv}{J}$$

From equation 3.42 for moving heat source

$$\theta = \frac{0.318Q}{a(a\alpha\rho cv)^{1/2}}$$

Since

$$W = \tfrac{4}{3}\pi R^3 \rho$$

these equations yield

$$\theta = \frac{0.125R^2\varepsilon}{J} \left(\frac{\rho v}{a\alpha c} \right)^{1/2}$$

Substituting for a gives

$$\theta = 0.092 \times \frac{\varepsilon}{J} \times \left(\frac{\rho^2 v^3 R^8 E}{\alpha^3 c^3 (1 - v^2)} \right)^{1/16}$$

After one revolution the temperature of the sphere is higher than the plane so that the assumption is less valid as further distance is travelled.

Example 19

A 10 cm diameter steel cylinder of axial length 1 cm is subjected to a normal load of 200 N and a resisting torque of 0.5 Nm. It rolls along a steel plane at a uniform velocity of 0.1 m/s. Calculate the discrepancy in distance travelled in 100 seconds over what would have been expected from purely geometric

387

considerations. State the percentage of the contaet area subjected to micro-slip under these conditions. The coefficient of friction is 0.15, $E = 200 \times 10^9$ Nm^{-2} and $v = 0.3$ for steel.

Solution

$$\text{Normal load} \qquad P = 200 \text{ N}$$
$$\text{Tangential traction } T = 10 \text{ N}$$

$$\frac{T}{\mu P} = \frac{10}{30}$$

From equation 8.12

$$\left(\frac{\alpha}{a}\right)^2 = 1 - \frac{T}{\mu P}$$

$$\frac{\alpha}{a} = \left(\frac{2}{3}\right)^{1/2} = 0.815$$

that is, 81.5 per cent of contact width is stick area so that 18.5 per cent of contact width is slip area. Now

$$a = \left(\frac{8PR(1 - v^2)}{\pi l E}\right)^{1/2}$$

$$= \left(\frac{8 \times 200 \times 0.05 \times 0.91}{\pi \times 0.01 \times 200 \times 10^9}\right)^{1/2} = 1.08 \times 10^{-4} \text{ m}$$

In equation 8.14

$$\frac{\Delta U}{U} = \frac{0.15 \times 1.08 \times 10^{-4}}{0.05} [1 - (1 - \tfrac{1}{3})^{1/2}]$$

$$= 6.16 \times 10^{-3}$$

Due microslip

$$\Delta U = 6.16 \times 10^{-3} \times 0.1$$

therefore

$$\Delta s = \Delta U t = 0.616 \times 10^{-3} \times 100 = 0.0616 \text{ m}$$

Cylinder travels 0.0616 m less than 10 m in 100 sec.

Example 20

A 1 cm diameter steel ball rolls around the inner race of a ball-bearing having a groove radius of 0.6 cm and an overall diameter to the bottom of the groove of 6 cm. The normal load on the ball is 100 N and the torque applied to the ball gives rise to an angular velocity ratio between the ball and the race of value ϕ. Determine the size of the contact zone and the value of ϕ when $c = 0$, that is, no slip occurs along the lines where y is one-quarter the value of the major diameter of the contact ellipse. For this condition sketch the shape of the stick and slip zones in the contact ellipse. Use $E = 200 \times 10^9$ Nm^{-2}, $v = 0.3$ and $\mu = 0.2$.

Solution

$$R_{11} = R_{12} = 0.5 \times 10^{-2} \text{ m}$$
$$R_{21} = 3.0 \times 10^{-2} \text{ m}$$
$$R_{22} = -0.6 \times 10^{-2} \text{ m}$$

From equation 3.23

$$B - A = 1.0 \times 10^2$$
$$B + A = 1.333 \times 10^2$$

Hence

$$\cos \gamma = \frac{1.0}{1.33} = 0.75$$

$$\gamma = 41°24'$$

From figure 3.15

$$k_a = 2.08 \qquad k_b = 0.6$$

Using equation 3.22 gives

$$a = 7.73 \times 10^{-4} \text{ m} \qquad b = 2.24 \times 10^{-4} \text{ m}$$

From equation 3.24

$$R_c = \frac{2 \times 0.5 \times 0.6 \times 10^{-4}}{(0.5 + 0.6)10^{-2}} = 0.546 \times 10^{-2} \text{ m}$$

From equation 8.21

$$\pm Kc = \pm \frac{\mu}{R_{11}} c = \frac{(\phi - 6) - \dfrac{y^2}{2R_c R}(1 + \phi)}{(\phi + 6)}$$

Putting $c = 0$ and $y = b/2$ and $R_c = 0.546 \times 10^{-2}$ gives a value of ϕ as 6.00275/0.99725. Inserting this value of ϕ, $\mu = 0.2$ and $R_{11} = 0.5 \times 10^{-2}$ m shows variation of c with y so that the stick–slip pattern may then be constructed.

Example 21

A circular hydrostatic thrust pad, 15 cm diameter has a central circular recess 7.5 cm diameter. The pad is designed to support a mass of 2500 kg with a uniform clearance of 0.05 mm. The lubricant is incompressible with dynamic viscosity 0.1 poise and is supplied to the pad from a constant pressure source

389

at 5×10^6 N/m^2 via an external capillary restrictor. Calculate the volumetric flow rate of lubricant and the static stiffness of the pad under these conditions. If, in operation, the bearing surfaces are not parallel, but inclined at an angle of 5×10^{-4} radians, calculate the new flow rate and the minimum clearance between the bearing surfaces.

Solution

For Parallel Operation First, calculate $\bar{a}$ and $\bar{q}$ from equations 12.35 and then p_r from equation 12.28. The flow rate and stiffness can then be calculated from equations 12.30 and 12.37b.

The relationship between flow rate and pressure drop in a capillary takes the form $Q = K \Delta P$ where K is a constant which can be calculated from the parallel conditions.

For Tilted Operation From equation 12.35

$$p_r = \frac{W}{A\bar{a}} \tag{a}$$

Also

$$Q = K(p_s - p_r) = \frac{P_1 h^3}{\eta} \bar{q}$$

hence

$$p_r = \frac{p_s}{1 + \dfrac{(\alpha R_2)^3}{K\eta} \dfrac{\bar{q}}{\bar{\alpha}^3}} \tag{b}$$

After substituting the known values equations a and b give two expressions for p_r as functions of $\bar{\alpha}$ ($\bar{q}$ and $\bar{a}$ are functions of $\bar{\alpha}$). Using the data of figures 12.15a and 12.15b the two graphs of p_r against $\bar{\alpha}$ can be plotted and the intersection defines the values of p_r and $\bar{\alpha}$ under tilted condition. The new flow rate may now be calculated from the pressure drop across the restrictor. The solution for $\bar{\alpha}$ gives the central clearance and allows the minimum clearance to be calculated.

Example 22

An orifice-controlled hydrostatic journal bearing of the type shown in figure 12.18 supports a shaft of 8 cm diameter, has an overall length of 6 cm, a recess length of 4 cm and the width of the lands which separate the recesses is 2 cm. Fluid is supplied to the bearing at a pressure of 10 N/mm^2 and in the unloaded condition the recess pressures are 6 N/mm^2. If the radial clearance between the shaft and bearing is 0.04 mm find the zero eccentricity stiffness of the bearing. Estimate the rate of flow required to operate the bearing if the viscosity of the fluid is 300 cP.

Solution

From section 12.10.3, the aspect ratio of bearing m is 0.25 and pressure ratio r is 0.6.

From expression 12.44, or directly from figure 12.19, the dimensionless stiffness of the bearing $\bar{S}_o$, is 1.10 and the dimensional stiffness follows as 1.1×10^6 N/mm.

The flow rate required is 5.4 cm^3/s.

Example 23

A step bearing 20 cm long, 50 cm wide, moving at 50 cm/s is lubricated with an oil of viscosity 10×10^{-3} N s/m^2. Design the bearing such that it has optimum proportions to carry a load of 200 N. For this condition calculate the coefficient of friction and the power consumption. If the bearing as designed above were lubricated with a gas of viscosity 10^{-5} N s/m^2, calculate the bearing number $\varLambda$.

Solution

From section 10.3.2 we have the ratio L_1/L_2. The smaller film thickness may be calculated from

$$W = \frac{6\eta U L}{h_o^2} 0.0342$$

and hence the step height from the optimum value of a

$$L_1 = 14.35 \text{ cm} \qquad L_2 = 5.63 \text{ cm}$$
$$\text{step height} = 0.28 \text{ mm}$$

Surface tractions may be calculated from equations 10.24 and 10.25 since we know dp/dx for each section. The traction on the lower surface is equal to the traction on the upper plus the pressure at the step multiplied by the step area. Why?

$$\text{Coefficient of friction} = 0.067$$
$$\text{Power consumption} = 0.67 \text{ W}$$

For gas bearing $\varLambda \simeq 0.0018$ assuming an atmospheric pressure of 1 bar. This is so low that we could use incompressible theory to describe the action of this bearing.

391

Example 24

A rotor weighing 2000 N, rotating at 3000 rev/min is supported symmetrically by two hydrodynamic journal bearings. The journal shaft is 6 cm diameter and each bearing bush is 6 cm long. Oil is supplied having a viscosity of 2×10^{-3} N sec/m². Can the bearing be designed to run such that the minimum film thickness does not fall below 10 μm? What if the minimum film thickness is permitted to fall to 7 μm? If either of these conditions is possible, select a suitable clearance.

Solution

The only parameter we can change is the radial clearance C. The Sommerfeld number $S = (\eta N/W)LD(R/C)^2$, the maximum permissible eccentricity ε depends on the clearance, but ε also depends on the Sommerfeld number. Since this precludes an analytical solution, we must guess a value of clearance and see how the calculated eccentricity compares with the permissible value.

We may write $S = \text{constant}/WC^2$ and with our guessed value of C we calculate the maximum permissible eccentricity. The value W in the expression is the load on a length L of an infinitely long bearing, while our bearing has $L/D = 1$. We must therefore use chart 10.24a to ascertain the side leakage factor. To achieve a load/bearing of 100 N we will have to have conditions which would permit an infinitely long bearing to carry a larger load. Therefore, $W > 1000$ N. Calculate S and hence, from table 10.2, find ε. This will probably not be the same as the initially assumed value. What can be concluded from comparison of the values? Select another clearance and repeat.

You will find that it is not possible to achieve the first condition, but that the second may be obtained with a maximum radial clearance of about 26 μm.

Example 25

A steel strip 60 cm wide is being transported on rollers 10 cm diameter rotating at 600 rev/min. At any given instant the strip is supported by 10 rollers which may be assumed to carry 100 N each. If each contact is fully lubricated by oil of viscosity 5×10^{-3} N s/m² and the strip moves at 50 cm/s, calculate the minimum thickness of the oil film at each roller, the total sideways force on the strip and the total rate of energy input to the rollers. Would an increase in viscosity decrease the energy consumption?

Solution

Find U_1 and U_2 and substitute in equation 10.12 to find h_o. Note that we have to assume a value of h_o to determine α from table 10.1. Do not forget that the width is 60 cm.

$$h_o = 3.8 \ \mu m$$

Find P_x from equation 10.32 and the frictional forces from 10.31. Once again do not forget the width.

$$P_x = 3.17\,\text{N}$$
$$\eta(U_1 - U_2)A = 0.12\,\text{N}$$
$$F_1 = -3.18\,\text{N}$$
$$F_2 = -3.16\,\text{N}$$

Total force on steel = 31.6 N
Energy = Number of rollers × torque × angular velocity $\simeq$ 100 J/s.

Example 26

Two rollers each of length 50 cm, diameter 40 cm and rotational speed $30/\pi$ rev/min are loaded together. Two lubricants are available

lubricant A $\eta = 10 \times 10^{-3}\,\text{Ns/m}^2$ $\alpha = 0.2 \times 10^{-7}\,\text{m}^2/\text{N}$
B $\eta = 12 \times 10^{-3}\,\text{Ns/m}^2$ $\alpha = 8 \times 10^{-7}\,\text{m}^2/\text{N}$

If the two rollers are made from steel, calculate the elastohydrodynamic film thickness for the following combinations of load and lubricant

(1) 2.5×10^5 N, A
(2) 2.5×10^3 N, A
(3) 10^4 N, B,
(4) when one of the rollers is replaced by a rubber roller of the same dimensions, the load is 10^4 N and the lubricant is A. For steel $E = 208 \times 10^9\,\text{N/m}^2$, $\mu = 0.3$; for rubber $E = 10 \times 10^7\,\text{N/m}^2$; $\mu = 0.5$.

Solution

Before calculating the film thickness, it is necessary to decide in which regime the situation lies. For each case calculate q_m, q_α and p_o (section 11.22). Evaluate q_m/q_α and q_m/p_o. It will be found that each case lies in a different regime. The dimensionless film thickness may then either be calculated by means of the appropriate equation from 11.4–11.7 or more simply may be read off the chart (figure 11.4). Film thicknesses are

(1) 0.072 μm
(2) 0.20 μm
(3) 0.059 μm
(4) 0.049 μm

Author Index

Subject Index

399